400 DAYS – A Call To Duty

A documentary of a citizen-soldier's experience during the Iraq War 2008/2009

Volume 2

LTC Mitchell R. Waite, Ph.D.

Copyright © 2010 LTC Mitchell R. Waite, Ph.D.

Hardcover ISBN: 978-1-60910-236-4
Paperback ISBN: 978-1-60910-235-7

All rights reserved. No part of this publication may be reproduced, stored in a retrieval system, or transmitted in any form or by any means, electronic, mechanical, recording or otherwise, without the prior written permission of the author.

Printed in the United States of America.

BookLocker.com, Inc.
2010

TABLE OF CONTENTS

CHAPTER 10: *400 DAYS – Days 161-180*
April 15 – May 4, 2009 1

CHAPTER 11: *400 DAYS – Days 181-200*
May 5 – May 24, 2009 46

CHAPTER 12: *400 DAYS – Days 201-220*
May 25 – June 13, 2009 94

CHAPTER 13: *400 DAYS – Days 221-254*
June 14 – July 17, 2009 140

CHAPTER 14: *400 DAYS – Days 255-280*
July 18 – August 12, 2009 191

CHAPTER 15: *400 DAYS – Days 281-300*
August 13 – September 1, 2009 235

CHAPTER 16: *400 DAYS – Days 301-320*
September 2 – September 21, 2009 272

CHAPTER 17: *400 DAYS – Days 321-340*
September 22 – October 21, 2009 310

CHAPTER 18: *400 DAYS – Days 341-360*
October 12 – October 31, 2009 348

CHAPTER 19: *400 DAYS* - EPILOGUE 376

POST WORD 397

INDEX 401

REFERENCES 405

MILITARY ACRONYMS 424

SYMBOLS 429

USMC Photo taken by LCPL Jonathan P. Sotelo

Note: Part of the proceeds from this book will go to the *Wounded Warrior Project* to assist our brave men and women who have given so much to our country.

Note: The pictures within this text are used only to highlight key educational commentary for the enrichment of the reader. All pictures were taken by the author or obtained through public domain sources with permission to use.

Note: All information contained within this book is unclassified and obtained from open source media highlights.

CHAPTER 10
400 DAYS – Days 161-180
April 15 – May 4, 2009

While this is the most resilient, professionally seasoned combat force that I've been associated with in the 38 years of my own service, we're stretched, and we're deploying at a rate we can't continue and still sustain the all-volunteer force, and we don't have enough time at home to prepare for other things.

--Chief of Staff of the Army General George Casey
2007-present
also served as MNF-I Commander
2004-2007

Day 161: Wednesday-April 15, 2009/Day 130-Iraq was a no BUA day and so the day should be relatively normal. It looks like rain this morning as we have had more rain in the last couple of weeks than we did my first three months here! Well, despite the fact that it creates quite a mess, they really do need the rain because they are expecting another dry summer.

We have a big engineer summit coming up this weekend so preparations are in full swing. Yesterday I assisted some of the guys from the Gulf Region Central (GRC) office, which is right here on Victory Base Complex and where my room is located, to continue arrangements for this big conference. Engineers from around Iraq will be attending and many General Officers as well. General Odierno was originally scheduled to speak, but has since backed out and will be replaced by his second in command, Lieutenant General Brown from the British Army. I am sure I will be assisting with final preparations up until this Sunday when the summit begins right here at Al Faw palace.

Today I will also continue to work on the final chapter of my EMS leadership book and my first assignment for Naval War College. I hope to have the assignment done within the next week and then ship it off so I can move on to the next module. My buddy LTC Tony Jocius from the 416th decided to join Naval War College right after I did while we were at Ft. McCoy. Tony works at GRC as the Operations Officer (S3) and stays very busy. I will say that Naval War College is interesting thus far, although I have facilitated some EMS courses and taken others in a similar venue and I would do more with the course than our instructor has, but it is very flexible and you can move as quickly as you so desire with a few caveats. There are assignments that are due and a loose schedule to adhere to. From my perspective, the program workload is catered toward people with busy schedules, but there is a lot of reading. Writing obviously does not bother me in the least. Der! Unlike most people I talk to, my problem is not writing a few five to six page papers, but rather, keeping the paper that short! I normally have to cut out chunks of text to get the assignment down to size! But despite some of these shortcomings, I still enjoy the course so far, but Tony is struggling with it as he is trying to fit it in around his other duties so he may have to defer the program until he gets back to the states. I do not blame him at all, and I am fortunate to find the time to do the course in the scope of my position, but then again, I did not select this job!

I hope to submit my EMS leadership book to my editor, which I have not yet selected, sometime this summer. As this book is not related to my work here and is an entirely different topic, the same rules and policies do not apply as it does with this book, whereby I must wait until I am off of active duty orders in order to sign any publishing contracts.

The situation concerning the Somalia pirates becomes more interesting with each passing day. The Navy and their Seals did an incredible job with the Captain Phillips situation. Some people questioned whether this rescue may lead to an increase in violence from the pirates. Not sure if these same people believe the Navy should have let Captain Phillips just die so that this perceived increase in violence would not exist! Stupidity must be contagious! But there are always those morons out there that want to pour rain on everyone's parade as opposed to being happy for Captain Phillips, his family, and the incredible mission just completed by the Navy. There is no guarantee that these pirates would not begin to kill hostages down the road anyway. But there has been talk of hitting these pirates on land at their home bases **(CENTCOM, 14 APR 09: *US military considers attacks on pirates' land bases*)**. While I would say let's go for it, we must keep in mind that we are already engaged in two wars. I would also state that some other countries need to step it up a bit. These pirates are nothing more than thugs and criminals. This is how they need to be dealt with. Not sure why these shipping companies keep paying ransom or do not put armed security personnel on the vessels, but that is their choice. The people that navigate in this area of the world know of the risk and do so willingly. So I am all for going onto land and taking care of these criminals once and for all, but there should be many other countries involved in the operation. If the countries that have had ships seized and hostages taken do not want to be involved, then I say the US protects its own ships and everyone else can do the same. If they choose not to do so, then the ships can move at their own peril. But I have this feeling that this pirate adventure is about to come crashing down all around them and unlike the Johnny Depp movies, there will be no happy ending to this pirate tale.

Ol' Captain Jack – arr matie

▌▌ ▌▌ **Day 162: Thursday-April 16, 2009/Day 131-Iraq** was an early BUA morning. While we had no rain yesterday, it threatened several times and a few sprinkles were noticed. We have the threat of isolated storms over the course of the next few days as well.

Secretary Gates is moon walking away from his claim of a few days ago that al-Qaeda was making one last gasp here in Iraq. Much like the sporting world, it is entirely possible that bold statements such as the Secretary made a few days ago became al-Qaeda locker room material and provided them with some incentive to prove that they are not yet dead and gone. The result since the Secretary's statement was an uptick in violence, which resulted in more casualties. Now the Secretary is stating that "what I should have said is that I hope it's al-Qaeda in Iraq's last gasp. I don't know if it is" **(MNFI, 16 APR 09:** *McClatchy;* **Stars & Stripes, 16 APR 09,** *Gates corrects al-Qaida remark,* **written by Nancy A. Youssef)**. It takes a big man to admit this and temper previous comments. I doubt that Donald Rumsfeld would have made such a retraction!

The situation in Iraq continues to evolve, but how the story will end is still anybody's guess. The Turkish President, Abdullah Gul, who just paid Iraq a visit recently commented on Iraqi national unity:

> Iraq is a friend and a neighbor state and it is a colorful country that is made of Arabs, Turkmen, Kurds, Sunni, and Shi'a and in addition there are different religions in Iraq such as Islam and Christianity. The unity of Iraq is

very, very important and it is necessary that all factions of Iraq live under the umbrella of one state.
-MNFI, 16 APR 09: *Kurdish Aspect*

These are important words from a fellow Middle Eastern state and Iraq is beginning to stretch its area of influence and make an important statement that it is getting in position to become a serious player on the world stage. But as they do so, they most assuredly do not want to look like the 51st state of the United States, so it is wise to distance themselves slightly from the US to look better in the eyes of the other Middle Eastern countries, yet maintain close enough ties in a more discreet manner. Mustapha Alani, Defense Studies Director for the Gulf Research Center stated that:

Iraqi government leaders are trying to prove they are independent of the United States and they are trying to draw a clear foreign policy, and we can see other countries are beginning to take Iraq seriously.
-MNFI, 16 APR 09: *NY Times*

So the situation is taking shape here, but no one really knows if the situation is reversible at this point or not. The situation is most assuredly still fragile, but again, the real test of the strength of the Iraqi government will come when the US leaves for good. Then the rest of the story (God Bless Paul Harvey) will begin to unfold.

Shari and I are back on track after a couple of tense days over family matters. I believe we are over this hurdle, and as much as I do not like the situation I find myself in over here, I also realize it is very difficult for Shari as well. This kind of stress is the sort that people can never truly understand unless they have gone through it themselves. This has been very tough on Shari, as I am certain it is for all spouses with loved ones deployed. I realize I have mentioned this on several occasions over the course of this large book, but I believe it to be important enough to mention often. This is a year taken away from Shari and I that we will never get back, and while Skye is great, it is not like being home. So imagine yourself being away from someone you love for an entire year! This is something that most politicians that vote on sending young men and women to war really do not grasp unless they have made such a sacrifice. But such is life and the path I have chosen as a member of the military, and Shari and I will get through this as I am almost halfway home now. Thank goodness!

Day 163: Friday-April 17, 2009/Day 132-Iraq was our late BUA day, but because of a change in schedule over the next few days I decided to go in a bit early to get some things done before I met our G3, LTC Chuck Samaris, who was coming over to VBC for the engineer summit which begins this weekend. The summit actually begins on Sunday and lasts for two days. I am to pick Chuck up at GRC and then we will go visit the proposed new GRD HQ location, as they must move from the IZ to VBC before we depart here in October. Then we will go over to the boathouse, which is close to the palace and visit with some of the new personnel from 1st Corps that have arrived and who I worked with at the MRX at Ft. Lewis, WA last October. So this will most assuredly mess up my normal Friday battle rhythm so I want to get my entry into this book, get my run in, and work on a few other projects before I pick Chuck up. Tomorrow I will assist the GRC personnel, and those arriving from GRD HQ, to set up the main briefing room for the conference. I may also need to play peacemaker between our General's secretary and a young Lieutenant from GRC who have not seen eye to eye on some details concerning the conference. As I used to be a basketball ref and a baseball ump many years ago, I know how to intervene when necessary!

On Sunday morning I am also scheduled to be the escort for the acting US Ambassador Wall who is speaking at the conference. Not certain why I am selected to escort the ambassador into the palace, as you would think he would have an entourage of people to assist with this, but it should be a unique experience to say the least. I may also have to escort the Iraqi Deputy Prime Minister into the palace later in the day. This too should be quite an interesting experience. So even though my battle rhythm will be out-of-whack the next few days (hopefully back to normal on Tuesday), I look forward to this engineer summit and the unique experiences that will accompany it.

I also Skyped Shari this AM and she is getting ready to head to Chicago with all of her pals from Schmidt's Jewelry in Wisconsin Rapids. They are going to a jewelry conference and I am sure they will have a good time. Shari was already preparing for the event as her sister Tina was over to the house when I Skyped and they were having a few Miller Lites (a good old Wisconsin beer). Shari and Tina were feeling pretty good, which put Shari is a very squirrelly mood. This always makes our conversations more interesting and fun. Sure do miss her.

As I have reported to you before, according to the provisions within the new security agreement signed by the US and Iraq last year, all US troops are to be out of all major cities in Iraq by the end of June. But there are still a few cities that are problem areas and it is uncertain whether or not the Iraqi Security Forces are ready to handle this task on their own. But it will ultimately be up to the Iraqi government to request US troops to stay in some of these more dangerous areas. An Iraqi civilian spokesman, Dr. Tahseen Sheikhly, stated that the government will consider asking US troops to remain in some cities after June 30th **(MNFI, 17 APR 09: TIME)**. I believe the US is prepared to do this if requested, so we shall see very soon how far along Iraq thinks it has come, especially in these more difficult areas of the country.

In relation to the political development within Iraq, Nehme Al-Bekari, a Dawa Party lawmaker stated, "We have to be honest with the people. The Iraqi political experiment is not mature enough. We are still students in the school of democracy" **(MNFI, 17 APR 09: *NY Times*)**. General Odierno commented that this is very true and that many people think the government here should be a reflection of the US government and that is a very unrealistic expectation. I would agree that Americans often have unrealistic expectations of where Iraq should be at this point in time. Iraq will never be an America East, or a mini-US in the Middle East, nor should it be. Yet this entire war has been based upon such unrealistic expectations and quite frankly, the more infighting I see within our own government I hope the Iraqis do go in a different direction and have a more effective and functional government than the one we believe we have!

One other thing I feel worth mentioning today is how many contractors we have over here. I will just give you round figures to provide you with an idea on the scope of the situation, but there are a little over 140,000 contract personnel in Iraq right now! These people are involved in such functions as base support, security, construction, transportation, and the like. This equates to about a 1:1 ratio in relation to the number of military personnel in Iraq! This seems a little odd to me in addition to the fact that typically contractors cost a lot more than military troops. The other odd thing is that while we are trying to put Iraqis to work to keep them from joining the insurgency, only about 12% of this total workforce is comprised of Iraqi Nationals! Does something not seem wrong to you about all of this? General Odierno is looking into this to not only reduce the number of contractors here, but also the exorbitant cost to the taxpayer that goes with this support. And it is highly unlikely that Iraq is footing the bill for all of this as most of these contracts were awarded by the US government! Talk about no accountability!

Day 164: Saturday-April 18, 2009/Day 133-Iraq was back to a bit more of a normal battle rhythm day. There was a transformation meeting over at GRC which most of our brass would be at for most of today. This transformation relates to how GRD will begin to drawdown in conjunction with the rest of the military forces in Iraq. So Chuck Samaris would be over there all day which leaves me to my typical Saturday routine for the most part. I would be going to lunch with power row again today as Sean is within three days of leaving us, so we must have one last celebration. Then I will just need to check in at the palace to ensure all preparations are moving along for the engineer summit which begins tomorrow. I will try and pull the plug early, although there is a cigar club meeting tonight that I will try and attend. Because I will not get to enjoy my Sunday morning off as usual due to the summit, I will sneak in a few moments of personal time as I can over the course of the next few days to make up for this inequity. But this conference will bring some of the guys from my unit here to VBC so it will be nice to see some of them, and the conference is a nice break from the normal monotony around here. So it is all good!

Shari made it to Chicago alright and she and girls were heading down to the hotel lounge for a night cap when I called. I am sure this will be quite a weekend for her at this jewelry conference. Hopefully, she does a lot of looking and not much buying!

Iraqi President Jalal Talabani stated that he would endorse US troops staying in Iraq past the end of 2011 if conditions warrant. But he went on to state that he felt Iraqi Security Forces would be able to maintain security around Iraq without assistance by this time **(MNFI, 18 APR 09: *al-Baghdadiya*)**.

Another interesting story worth telling concerns a suicide bombing against an Iraqi Army base in Anbar Province a few days ago. Initial media reports from the scene, which was reported on CNN and Fox News to name but a few, was that this suicide bomber killed 15 Iraqi soldiers. In the BUA yesterday, it was reported by the Corps (the coalition warfighters), that no one but the bomber was killed and 38 Iraqi soldiers were injured. General Odierno was a bit perturbed to say the least at such a disparity in reporting. In the BUA this morning, a spokesman for the Ministry of Defense verified the Corps report of yesterday. So what should this tell you? Media reports are often incorrect and unsubstantiated. In their haste to get 'the scoop' on everyone else, inaccuracies are bound to occur, although there was quite a disparity in reporting on this particular story. It also tells you that US sources seem to be more accurate than those from other sources in Iraq. And as you see all too often in the American media, each major media outlet has their own spin on the same story. Fox News spins it differently than CNN. CNN spins it differently than MSNBC. So you, the receiver of the message, are tainted by whatever station you watch the most. I am not sure that there is such an animal as a completely objective media outlet that tells both sides of the story! Most, if not all of these outlets have a hidden agenda. This is called information operations and it is intended to persuade your way of thinking. This happens all over the world. However, in some countries the populace may only be able to get one media outlet which is controlled by the ruling government. The populace in these areas then only gets one side of the story. At least in the US, because of the freedoms we enjoy, you can turn the channel and get several variations of the same story! I would suggest that you do turn the channel occasionally if you are not prone to do so already, in order to get all sides of a story. Then you can formulate your own thoughts based upon a more multi-dimensional view. But most people are creatures of habit and will watch their favorite channel and never turn to another one. They may prefer the good looking female reporters on CNN, or they may like the spunky Bill O'Reilly on Fox News. I enjoy Keith Olbermann of MSNBC because of his days at ESPN. He is smart, witty, and funny to listen to. But his point of view is certainly slanted as are all such shows and I do not agree with everything that he says about a given subject. But one thing I will state about the Armed Forces Network is that they broadcast all of these points of view on the same channel. At certain times of the day they broadcast CNN, at other times it is Fox News. This allows people to explore all sides of a story, which is the wisest thing to do before formulating an opinion. Certainly, I have tried to give you perspectives on this war from many different sources and angles, but there can be no doubt that much of the information is also my opinion. I like to think my opinion is pretty impartial, but you can never take all bias out of any writer's thoughts. So I just urge you to look at other sources, research a topic for yourself, and do not simply take my word on any given topic. Hopefully this book will stimulate thought and meaningful discussion besides entertain you. If this occurs, then I feel I have done my job as an author. But not all of which I write should be considered gospel or the eleventh commandment! Just as I defend our freedoms willingly, I also have my own opinions on issues, but it most assuredly is not the only opinion and it may be one that you do not necessarily agree with. But that is alright. This is what makes our country the 'land of the free' and the greatest on earth today. So judge the value of this information on your own and challenge yourself to find the truth. As Agent Mulder (X-Files) would tell you, the truth is out there.' It is just a matter of how much research, time, and energy you want to spend to find it!

The truth is out there

Day 165: Sunday-April 19, 2009/Day 134-Iraq was not my typical Sunday. Because of the big engineer summit here at Al Faw the next few days there would be no rest and relaxation on my favorite day of the week. I was hoping to get the acting US Ambassador into the palace without issue, and then the Iraqi Deputy Prime Minister, and then sit in on the more interesting topics of the summit. I would move like the wind and go back and forth from the meeting to the Strategic Operations Center so I could learn a few things and get some of my work done so I would not fall behind over the next few days.

In news today, General Odierno was just on CBS News, which I missed, but he did inform Katie Couric that the Iraqi government is trying to absorb ex-insurgents into the Iraqi Security Force (ISF) **(MNFI, 19 APR 09: *CBS*)**. This program, known as the Sons of Iraq, has been a very interesting story here and one that has played a huge role in neutralizing al-Qaeda. Basically, the reader's digest version of this story is that many tribes banded together to fight against al-Qaeda. These tribesmen felt al-Qaeda's brutality against innocent civilians was far worse than what coalition forces were doing, so they stopped fighting us and took up arms against the insurgents. Part of the deal to keep them from reverting back to the insurgency was the Iraqi government assimilating them into the Iraqi Security Forces. This would provide them with a job and money to take care of their families. But getting this many men jobs within the ISF has been a real challenge, but one that is exceedingly important. Money speaks loudly over here, and if the Government of Iraq cannot pay them, the insurgents may be able to and we will be right back to square one!

It looks like US diplomatic efforts may not be working all that well with Iran. There was the recent jailing of two reporters in Iran, including an Iranian-American journalist who was just convicted of espionage and sentenced to eight years in prison. Secretary of State Clinton has been working this issue, but thus far has had no success. It was also reported that Iraqi Security Forces in Basrah seized a boatload of weapons, including rockets that originated in Iranian territorial waters **(MNFI, 19 APR 09: *al-Sharqiya*)**. So whether it is a case of diplomacy working very slowly, or simply not working at all, or perhaps the fact that the Iranian government has no control over their borders, is a matter of great debate. But the bottom-line is that weapons from Iran are still steadily flowing into Iraq and that does not make for good relations between Iraq and Iran, or the US and Iran. Iran may still be trying to leverage this weapons flow against their nuclear program, but then it becomes which form of poison is worse!

My man Thomas Ricks, author of **Fiasco** and **The Gamble,** wrote an article in the **Washington Post** recently concerning how the military could save money **(Washington Post, *Why we should get rid of West Point*, 19 APR 09, written by Thomas E. Ricks)**. Mr. Ricks believes that not only is it more costly to send people to the military academies ($300,000 versus $130,000) in comparison to an ROTC program, but the service academies tend to be a bit more incestuous in nature. Everyone at a service academy is in the military, and therefore they often share the same bias. This can lead to something termed 'group think', which does not lead to a diverse set of opinion or discussion. In an ROTC program at a four-year university, these military students are integrated into the student population with aspiring attorneys, doctors, political science majors, etc. Most certainly, one would conclude that this would lead to a more diverse experience and perhaps a more holistic view of the global situation. While an interesting argument to be sure, I have already outlined my dissatisfaction with the way the military selects people for War College, which Mr. Ricks also suggests we eliminate for the same reasons. However, in the tradition-laden military establishment I cannot see his argument ever becoming reality. But as someone who has been on both active duty and in the reserve component of the military, and worked with service academy graduates, I will state that there are excellent officers in both components, and conversely, poor officers as well. I have also stated that since I have been here I do believe reserve officers bring a more diverse perspective to the table due to their civilian occupation which is often under appreciated by their active-duty brethren. There are a lot of prima-donnas from the service academies that I would never hire as a civilian employer. Even though they are highly intelligent, they do not have a real diverse background, have poor personnel management skills, do not understand the art of leadership well, and are very one dimensional. But each academy has great history, has produced some outstanding military leaders, and as the saying goes, never let progress stand in the way of tradition! But Mr. Ricks brings up some very interesting food for thought, which led to a very spirited discussion on power row this day.

LTC Waite at the MNF-I Engineer Conference

Day 166: Monday-April 20, 2009/Day 135-Iraq was the second and final day of the MNF-I Engineer Summit. As far as

I could tell yesterday went very well. I was back and forth from the seminar and my work area, and yes once again, I had to spring into action to put out a couple of small fires. After all, I am a fire chief so it was not really a problem! Hopefully today there would be a few less of these fires and once again I would move back and forth from the SOC to the seminar. Tomorrow I would be back to my normal battle rhythm.

Not much in the news today, but some of the Iraqi youth over here are getting on board and figuring out how to lead a better life instead of being recruited for an insurgent group, which are really nothing more than heavily armed gangs. Once the youth of this country get things figured out then Iraq will truly have something going on. One such young man, who is a Basrah council employee at 15 years of age stated that:

> Forty-five days ago I got a job in the council and I am very happy. Thank God, my life will now change. I get $200 each month and this is a good foundation for building my life. Everything is going to develop here. When the society strengthens, trade will come, so will investments and agriculture.
> **-MNFI, 20 APR 09:** *UK Guardian*

Good for this young man. But as we notice what is occurring around the world, given the current world economy, there is crime all over the globe. It is not only here in Iraq and Afghanistan, it is also in Mexico and on the high seas off the coast of Africa. Much of what is going on in Iraq now is not the twisted holy war born in the mind of al-Qaeda leadership, it is the misguided militias that continue to pop up that are only interested in getting a piece of the pie by attaining more wealth and/or power by whatever means necessary. Whether it is siphoning off oil from a pipeline to sell on the black market, or stealing electricity allotments from one of the eighteen provinces, it is all motivated by simple greed. Much like the drug cartels in Mexico and South America, the pirates of Somalia, or organized crime in the US, it is all about greed. It is not about overthrowing a corrupt and evil regime; it is not about religion; it is not about being a suppressed minority; it is about greed. This axiom is ageless and as old as man himself. This truism in human nature will never change or evolve. It is interesting to note that some experts believe that as economies wane the crime rate increases, while other experts believe this not to be the case (a simple Google search will enlighten you further). I would tend to believe that as more people become unemployed, cannot pay bills or feed their families, they will become more desperate. And as I know from my job as a fire chief, desperate people do desperate things. Many of the recent tragedies we have observed have been linked to people being laid off of work, that have lost their job, and see no way out of their economically challenged situation. Some of these people will simply commit suicide; some will take a gun(s) and kill others; and some will kill their entire family before taking their own life. Draw your own conclusions on this topic. Next time you see a mass casualty incident look for the underlying theme. Watch carefully as the economy worsens and then improves just how the crime rates correspond to this ebb and flow. It is all quite interesting but also so very, very tragic.

Day 167: Tuesday-April 21, 2009/Day 136-Iraq was a day with much news to report. To begin with, I had my first battle in Iraq last night. Yes, it was me, my broom, and an Iraqi critter of some sort! It must have been a few weeks ago that I had heard a noise that sounded like something rummaging around under my bed. So of course, I loaded my 9 mm pistol and began the hunt! No, just kidding. Do you think I am stupid? But I did grab my trusty broom and started moving boxes around not knowing what my adversary might be. It could be a snake, a large camel spider, or perhaps a rat! Well nothing was found during my search that day. But a few days later as I was looking in a gym bag in my wall closet, which I only keep open at night, I found some little chew marks on some packages of gum. Well, it has been quiet the last few weeks, but the other day as I lay on my bed watching a little television I noticed something move out of the corner of my eye. It moved quickly and with great stealth. I did not want to kill the little varmint so I grabbed my broom, opened my door to provide an escape route, and was hoping to simply usher the little pest out the door. As I searched further, this plan failed as my unwanted roommate was now gone as quickly as it had come.

After some more searching I found nothing and thought that I might need some Decon to take care of this intruder, so yesterday I bought some and placed it in a few key places around my room. Then, last night I decided to begin to arrange my second footlocker for its trip home. Tomorrow I would send home the first of three footlockers and we shall see how this drawn-out process goes. If you recall, last November I bought two nice camouflage duffel bags on wheels at Ft. McCoy, WI right after we were mobilized. If you refer back to the first twenty days you can refresh your memory and see that I started quite the trend as the PX sold out of them because people liked mine so much. I believe they were around $60-$70 each. Well, I moved the one I had on top of the other, and as I moved this bag I noticed the one on the floor had some Styrofoam hanging out of it. As I reached down to move it I heard some rustling inside the bag and knew the culprit was close. Game on! I grabbed the broom and poked the bag a few times and all of a sudden this frightening creature emerged ready for battle! He may have been little, but this little field mouse looked awfully tough! Once again, to keep PETA off of my back, and because I always give little creatures the opportunity to survive, I opened the door to the hallway once again. This time, the little critter took off and the chase was on. Around the footlocker, over the carpet, and around the frig he went before finally selecting his escape route right out the door. So I shut the door after I noticed the little fella had scrambled next door right outside of the GRC Sergeant Major's door! Well, even SGMs need some company every once in awhile!

So I began to check the damage in my second bag and this little bastard had had a field day. He/she actually had a pretty nice little condo! While I had no food in the bag, this little thing had crapped and pissed over about everything. It looked like a mini frat house. The only thing missing were the beer cans! So I brushed the turds off, cleaned off what I could, and then transferred everything from this bag into my remaining one. This little *Animal House* was shot! So to the dumpster it went. I thoroughly cleaned my room and then rearranged things a bit to get my remaining bag off of the floor. As I headed to the dumpster I noticed the little guy still sitting between the SGMs door and the rug sitting outside. I kicked the rug slightly and got *'Mighty Mouse'* moving down the hallway, which probably seemed like *The Longest Yard* to this little guy. Unfortunately, the door to the free world outside was closed so he ducked under the stairwell at the end of the hallway. Hopefully he would survive and find a new home.

I do have a little gap under my door, which was perhaps the point of entry for my little buddy. I believe I have the situation now rectified and there does not appear to be any other openings in the room that may lead to a repeat performance. So my little roomie had been evicted and I hoped I would not get another! I could not state with any certainty who won this battle. I was able to get 'Mighty Mouse' out and relocated without killing him/her, but he/she cost me a nice $60-$70 duffel bag. So perhaps we can call it a stalemate.

In real war news today, General Odierno was up north on Sunday to try and calm the situation between the Government of Iraq (GoI) and the Kurds **(MNFI, 21 APR 09: *al-Sharqiya*)**. As I have reported to you, this situation is a potential powder keg in the north, along with the al-Qaeda cells in Mosul. The US was supportive of the Kurds during the last Gulf War to keep Saddam from taking revenge upon them. This is where the northern no fly zone came into existence. The Kurds in return assisted the US in fighting against al-Qaeda in this war. But there is still great tension between the GoI and the Kurds. It this situation explodes, an already fragile government will begin to destabilize and the progress that has been painstakingly made over the past few years, and come at a very high cost, may begin to unravel. We are all keeping a vigilant eye on this situation and we will soon find out how far the political process has evolved in Iraq. Hopefully this dispute will be settled diplomatically and by violence.

Related to al-Qaeda, many here believe they are reeling and becoming desperate to remain relevant in Iraq. This is further validated by the fact that al-Qaeda in Iraq (AQI) is now recruiting children as suicide bombers **(MNFI, 21 APR 09: *AFP*)**. This is certainly a sad statement about AQI and I would agree that it certainly seems like an act of desperation.

An Iraqi officer commented on this desperate state:
This is a new method of recruiting child-soldiers and the method of someone who is losing the war. Al-Qaeda in Iraq's infrastructure is being destroyed and their leaders are being arrested and killed. They are sending a message that 'We are still here, but we have to rely on every method to carry out our terrorist operations.'

-MNFI, 21 APR 09: *AFP*

While this latest tactic is very pathetic and an act of desperation, keep in mind that there are other militias and groups still struggling for power here. So even if al-Qaeda is on its 'last gasp' as Secretary Gates recently stated and then retracted, there are still other security concerns for coalition and Iraqi Security Forces to be concerned with.

I Skyped Shari this AM and she said that being with four women for three days in Chicago was quite interesting. The human dynamics involved were very impressive I am certain.

Epic battle with a desert warrior!

||| ||| ||| ||| Day 168: Wednesday-April 22, 2009/Day 137-Iraq was off to a quick start. I decided to change my normal routine for today as any good soldier would do so as not to become too predictable. Of course, today was a no BUA day so I had time to experiment a bit. I began the day by Skyping Shari and we talked for about 45 minutes. My stepson, her oldest son, Matt is getting married soon and Shari will be assisting with the wedding plans. Of course this is a good thing because it will help keep her distracted. Shari's youngest son Josh is also getting married next year, but he wants to wait until I return to have the wedding. That is a very nice gesture, and oh by the way, he wants the wedding at our house! It is a beautiful place that Shari and I just love and that is fine by me. We entertain there often anyway, so what's another big function!

Then I was off to eat breakfast for the third time since I have been here! I have not been impressed with the breakfasts here and normally do not eat the morning meal anyway. Yes, I know, some people believe it is the most important meal of the day and gets your day off to a great start and provides you with more energy, blah, blah, blah! Horse hockey! Just as everything else, each individual is unique and as such people are affected differently by various stimuli. I do not believe that not being a breakfast person has affected me adversely in the least. I have a great deal of energy, which my wife probably would not like me to have anymore of, if you know what I mean☺! I am a big coffee drinker as well and I drink it all day long without adverse affects. I have no problem sleeping and I take no medication. In fact, I am the only one in my family that does not have hypertension. Anyway, I had an omelet, which was quite tasty, an English muffin, a container of OJ, and a hot cup of coffee. All in all, it was the best breakfast I have had here so far.

Then, I was off to mail my footlocker. I thought that perhaps by getting to the post office when it opened at 0730 that I could avoid the crowd and get in and out quickly. Boy was I right on the money! I had all the paperwork filled out properly; the young man inside named Mario did a fine job of inspecting and repacking the footlocker; it was then weighed and sealed; and then I paid the postage and that was that. The only glitch was that it was a bit oversized so that cost me an extra $42! This was a bit unexpected, and increased the total postage to $86. You can send up to 70 pounds in a box or footlocker, and this one weighed in at a hefty 60.5 pounds. But despite this minor issue, it is still worth the money to send this stuff home versus trying to carry it all, or send home on a ship that takes quite a long time to receive as we found out on the trip over here. I had actually thought this little venture may cost me over $100, which would still have been worth it. So this process was a piece of cake and I would have one more similarly sized footlocker to ship home probably next month, then my smaller Storm Case probably a few months before we leave for good. I may also need to send a few more boxes home, or get another Gorilla footlocker from the PX to ship home, but these are not oversized so the postage will be more reasonably priced.

So I was on a good roll to start today and everything was just clicking. Then off to the palace to begin my work day and my man Sean was still here! His farewell was last night so we all thought he would not be in this morning. But it was good to see him and he told me that he may need me to drive him to the airport this afternoon as Randy (MAJ Staab) lost his car keys yesterday! Not good! This would be fine with me as I would have a chance to say goodbye and get out of the SOC early all in one fell swoop!

The rest of the day was rather uneventful day as I prepared to be the backup for the GRD BUA briefing tomorrow as SFC Maltes would resume her role as the primary briefer after an elongated 4-day pass!

At about 1600 I took Sean over to BIAP. After he checked in we went over to my number one rated dining facility at Sather Air Base. The food was again very good and fresh, and the atmosphere quiet and comfortable. Even though the regular utensils were missing this time, so we had to lower our standards and use plastic ware, this place will maintain its number one status. After a fine meal, I walked Sean back over to the terminal and we unloaded his luggage. He would now just have to wait for about an hour to find out what time his flight would be leaving and then he would leave Baghdad far behind. He would most likely fly into Kuwait, stage there for a few days, and await transportation back to the US. Sean will be missed, but like everyone else who leaves this place, he is glad to be heading home. I do not blame him in the least. He is heading back to Ft. Irwin, California for a few months before heading to Ft. Gordon, Georgia for several months to attend a military specialty school. His family, currently in Korea, will join him in Georgia. So long Sean. You did a great job here and served your country proudly. I wish you the best. And so another one of 'power row' has departed and Sean is not being replaced as we continue to drawdown here. Such is life as a liaison officer. The only positive from this is that now my Rice

Krispie treats will last considerably longer as Sean certainly had a serious addiction for them!

'Power Row'
MAJ Randy Staab, MAJ Sean Song, LTC Mitch Waite, & LTC Bob Pritchard
(MNSTC-I LNO) (MNC-I LNO) (GRD LNO) (CENTCOM LNO)

Day 169: Thursday-April 23, 2009/Day 138-Iraq was the fourth Thursday of the month so you know what that means? Yes, it is GRD BUA briefing day. SFC Maltes would be ready to roll from the New Embassy Compound and I would be in position as the backup in case anything went awry.

It is warming up here as today the weather forecast calls for 99 degrees! I like the heat, but I have never been in heat like I will experience here when it approaches 120-130 degrees. I do not care how you describe it, or if it is a dry heat, that is just friggin' **HOT**!

In news today General Odierno met with the Iraqi Deputy Prime Minister recently to discuss the mechanisms needed to be in place if the Government of Iraq wants to request US troops to remain in certain cities within the country after the 30th of June of this year **(MNFI, 23 APR 09: *al-Sharqiya, KUNA, Radio Dijla*)**. This is most likely to occur in cities such as Mosul, perhaps Baqubah, and maybe even Baghdad.

The new US Ambassador to Iraq is Christopher Hill. He has finally been confirmed by the US Senate **(MNFI, 23 APR**

09: *NY Times, AP, Reuters, AFP, CNN, BBC*). He replaces Ryan Crocker who left in February. So once again, here we have a case of bureaucracy rearing its ugly head. We are still at war in this country and in a very vulnerable state right now. We know from experience and history that while the military has done an outstanding job of adjusting during the course of this war, it cannot achieve success by itself. There must be a strong diplomatic link in place, yet we have now gone over two months without a head diplomat in Iraq! General Odierno has done an excellent job of keeping things held together, but he should not be expected to have to be in charge of both military and diplomatic efforts. This is the kind of poor strategy, lack of vision, and proactivity that has cost this country many lives and billions of dollars. Not sure which administration to blame on this one, but yet another dropping of the ball here in Iraq!

And from our new ambassador, who received some votes of dissent because of his lack of experience in the Middle East:

> The real problem in the region for Iraq remains its ancient neighbor, Iran. The US believes, and the Iraqis definitely believe, that Iran needs to respect Iraqi sovereignty.
>
> **-MNFI, 23 APR 09:** *AFP*

This is a pretty well known fact. It most certainly is not the only issue facing this country, but Iranian influence is a major source of concern for both the US and Iraq. And as has been stated on many occasions, if the US pulls out too soon and the Iraqis are not quite ready to pick up the ball and run with it in all areas of government, any void that remains could be filled by Iran. Such a situation could easily lead to instability. General Odierno commented that he believes Ambassador Hill is very well qualified and an excellent diplomat. But the problem that everyone has failed to answer with any certainty here is when is it too soon for us to leave, and when will the Iraqi government be strong and ready enough? That is the billion dollar question right now.

I also had the opportunity to watch General Petraeus on the Pentagon Channel last night **(Pentagon Channel, 22 APR 09, 1900 GMT+3)**. The show was taped from the previous day at the John Kennedy School of Government at Harvard University in Cambridge, MA. General Petraeus spoke about the situation in Iraq and also in Afghanistan and Pakistan. General Petraeus does possess a Ph.D. from Princeton and was very articulate in selecting his words. He is certainly much different than most General Officers we have seen in the recent past and hopefully a glimpse of Generals to come. It was a very impressive presentation and I was pleased to have the opportunity to view it. We need many more General Petraeus' in the Army in order to move this tradition-laden and resistant-to-change culture progressively forward.

Happy 101st Birthday USAR

Day 170: Friday-April 24, 2009/Day 139-Iraq was my 52nd birthday. Certainly, not one noteworthy by number perhaps, but most assuredly it will be remembered by location. Now, how could I celebrate number 52 without my honey being here; without any real beer; without any birthday cake; without much of anything! I will let you know how this turned out tomorrow.

It is getting warm here as it was 100 degrees yesterday according to our Ugandan guards, and it is only April! It is also very dusty and when it suspends in the air it looks a lot like fog. But unlike fog, this stuff gets over and in everything.

Well, on this day there were a few high profile attacks reported that resulted in many civilian deaths. As I reported a few days ago, often the number of deaths and injuries reported by the Iraqi media channels are different than those reported by US forces on the ground. This indeed may be yet another such case, but the Iraqis are reporting 78 civilian deaths in two separate suicide bombings **(Fox News, online version, 24 APR 09, 0800 GMT+3)**. It is also interesting to note that a report came out yesterday that the Iraqi National Police had captured a key figure from al-Qaeda in Iraq. His name is Abu Omar al-Baghdadi. US officials have not yet confirmed that this is true, as apparently his capture has been reported in the past by Iraqi sources only to be disproven later. Al-Baghdadi is reportedly associated with the Islamic State of Iraq. However, the plot thickens as Abu Ayyub Al-Masri, supposed military leader of the Islamic State of Iraq, is calling for Iraqi insurgent groups to unite and admits committing mistakes in the past **(MNFI, 24 APR 09: *al-Jazeera*)**. This seems like a plea from a desperate group of insurgents to me! As has been stated on many an occasion, while security continues to improve, there is absolutely no realistic way to completely eliminate insurgents in Iraq. But as the insurgency loses momentum, they must resort to higher profile attacks in order to prove that they are still relevant. The unfortunate part of this is that this violence often kills innocent civilians and this simply drives the populace further away from them. The insurgents do not seem to understand this concept. Coalition forces made mistakes early in this war by fueling the insurgency through their heavy-handed tactics before finally resorting to tried and true counterinsurgent techniques. The people are the center of gravity in a counterinsurgent war and when they are on your side, then and only then, can progress and success be achieved. While we have managed to figure this out, some of the insurgents still have not!

Of course, as the US redeploys troops from Iraq, the insurgents will continue to probe and infiltrate the ranks of the Iraqi Security Forces. The radical elements of Islam do not want a strong government in place or they will cease to be relevant. The unfortunate part of this inevitable process here is that violence is likely to increase as coalition forces redeploy **(MNFI, 24 APR 09: *AP*)**. The ability of the ISF and the Government of Iraq to withstand this violence and bring these people to justice, or kill them in the process, will write the final chapter on this war. If the government is successful, the future will look much brighter for the country of Iraq, and the hard fought efforts of coalition forces will be rewarded. If the insurgents are successful, then the country is likely to fall into a civil war with many factions struggling for power.

The other chapter in this struggle here is related to the tension in the north between the Arabs and the Kurds. As I have told you this is another test for the Government of Iraq (GoI). Both groups have their own armies, and the worst case scenario is that fighting commences between the two in the north. An Iraqi store owner summed things up this way:

> The situation is going to become even more intense if there isn't political agreement between Arabs and Kurds. There are armed groups like al-Qaeda in Iraq that will try and exploit these disputes.
> **-MNFI, 24 APR 09:** *Reuters*

If the situation erupts in the north, this will be a testament that the GoI is not stable enough to handle the country with diplomatic efforts. This would not bode well for its future. The inability for the GoI to incorporate all factions within its country under one impartial and fair governmental umbrella will lead this country down the path toward greater violence and possibly civil war. The current US government administration is very unlikely to commit any more troops or money to Iraq if this becomes the situation. This country has been given a golden opportunity and now it is up to them to decide what to do with it.

I finished my birthday by leaving work a little early and went back to my room to have a little Johnsonville garlic summer sausage, a few slices of good Wisconsin cheese, and a near beer. Last year Shari and I were in Green Bay for a draft party which was fun. Two years ago I was in Vegas with Shari, my mother, daughter, and her friend. We celebrated my 50th there. So this was a stark contrast and not all that excellent. But it was another landmark passed and another day closer to going home.

▌▌ ▌▌ Day 171: Saturday-April 25, 2009/Day 140-Iraq was one in which I learned a little about myself. Shari and I had a little disagreement about our financial situation. So in the past month, we have managed to disagree about in-laws and finances, the top two issues that usually lead to divorce! But we have worked through these issues and I have found that once again I have been very unfair to my lovely wife. She does not really realize what a spaz she has married! I have noticed in others that when people spend a long time alone during their lives they often pour all of their energy into their careers. This often equates to great success professionally, but a real train wreck when it comes to intimacy and social skills. I have known some people that have been brilliant professionally, but a mess on a personal level. While this is not quite an accurate description of my situation - it is pretty darn close! I spent 21 years between marriages, and while I dated and managed to get close to a few women, at the end of the day I was still all by my lonesome. This has allowed me to become very successful professionally, but I need to find balance with my personal life. It is very unfair to Shari, who is doing the very best she can with the hand she has been dealt, and I have not fully appreciated her efforts. From this severe brain cramp, I have learned that I need to grow in this area and must beg Shari to forgive me for being such a bone head. While you hope that you will not be affected by the events that have unfolded over here, it is times such as this that you realize you have been. I write this piece so that you better understand the sacrifices of our men and women in uniform. Many people know someone who has been deployed to a war zone. Many people may see the visible scars of war in our returning veterans. But what most people never see is the pain of being away from home for an extended period and the stress it takes on a spouse. This is what the general public does not see or realize, but these types of things can remain with a veteran for a lifetime. These are the kinds of things that soldiers, marines, sailors, airmen, and coastguardsmen must contend with while also doing their jobs far from home. I would love to someday see any politician that votes to send troops to war to be imbedded with a unit for the duration of their deployment. Then, and only then, will they truly understand the sacrifice of our troops. But Shari and I will get through this, our marriage will be stronger as a result, and I have learned a little about myself that I can grow from. For someone who understands so much about the topic of leadership, I sure have not handled some of my situations with Shari very maturely, or as someone befitting such knowledge. But I will become a better person as a result despite the pain of it all.

I also received birthday greetings from my daughter, my parents, my aunt, my in-laws, and some of the boys at the fire department which were all very nice on such a melancholy day. And just when you begin to feel sorry for yourself, you realize that it is not all that bad compared to many others around the world. Parents have lost sons and daughters over here; innocent Iraqis have been killed in a rash of recent suicide bombings; seven police officers have died in the line of duty in the past month; and so on. And the final statement I will make on this issue is that the more education you receive in life, the more you truly realize that you really don't know that much at all! Think about that for a moment.

We are still having more violence in Iraq as I alluded to yesterday. Rocket and mortar attacks are also on the rise, but despite all of this, Iraqi citizens are noticing a difference in atmospherics and this change is more noticeable by them because they have been here for the duration of the war. This particular resident stated that:

> For sure things are improving, and one can sense a political maturation and some kind of quietness. No one can say that Baghdad is peaceful, but I can say it's safer. Prime Minister Maliki and many other politicians are making new alliances based on ideology and political interests rather than sects and religion and this is a really, really great thing.
> **-MNFI, 25 APR 09: *BBC***

Could this be another sign of progress? Let us hope so, and I sincerely hope the Iraqi people can overcome all of this chaos and build a brighter future for the coming generations.

Day 172: Sunday-April 26, 2009/Day 141-Iraq was going to be a busy one. I was going to begin to finalize my second book on EMS (Emergency Medical Services) Leadership called ***The EMS Leadership Challenge – A Call To Action.*** This is a follow on to my first book which targeted fire service leaders that was printed by Jones and Bartlett Publishers back in October 2007 titled ***Fire Service Leadership – Theories & Practices***. I was hoping to get this book published soon after returning home from Iraq, but there was still some work to be done to finalize things, get it edited, get it packaged, formatted properly, etc.

I would then begin final prep for my operational law exam for Naval War College. This was the second block of six in the Joint Maritime Operations course. By finishing this exam today, I would be a little over a month ahead of schedule. I wanted to be through Block 3 and into Block 4 by the time I went on R&R so I would not need to do any work while at home. While this course is not slated to officially end until late October, I wanted to be done well before that time and then decide if I would continue on with my studies or get out of the military altogether. Much of my decision in this regard hinges on the promotion board results which should be published around June or July. Once again, knowing the system as it stands, I am not holding my breath on this one and Shari will be very pleased if I am not on this list. It will also make my decision much easier about my future in the military. Perhaps after 33 years and the hardship of this deployment, it is time to get out and move on.

Then it would be on to recon the car wash close to GRC. I spotted this place about two weeks ago and will see if it is open on Sunday as the old chariot is looking pretty dingy these days.

I also want to report that the daughter of my Deputy Chief, Katelin Kerkman, sent over three care packages as a school project for her fifth grade class at St. Vincent's elementary school in Wisconsin Rapids, WI. One box was filled with candy; one with items for soldiers such as soap, shaving cream, razors, etc.; and the last with school items. I placed the candy and toiletry boxes on a table for the troops yesterday and the items were gone in a matter of moments. It is like sharks to chum! The troops really love this stuff. The last box I was working with Kendal Smith, the GRC Public Affairs Officer, to see if we could get it out to an Iraqi grade school and distribute its items which included: 20 stuffed animals; 2 Nerf footballs; 2 notebooks; 4 folders; 4 coloring books; 42 colored markers; 30 colored pencils; 40 regular pencils; some erasers; and 10 black pens. There should be just enough items for one class in Iraq courtesy of one class in the United States. I was hoping to get a photo of the distribution that I could send back to Katelin's class to document their generous efforts.

Secretary of State Hillary Clinton was in Baghdad yesterday. She is assessing the security situation and informed the Iraqi Foreign Minister that the new administration stands with Iraq, but there will be a limit **(MNFI, 26 APR 09:** *Washington Post, NY Times, AP, AFP, Reuters, SKY News, Fox News;* **JASG, 26 APR 09,** *Clinton tries to reassure Iraqis during visit,* **written by Matthew Lee; Stars & Stripes, 26 APR 09,** *Clinton: US won't abandon Iraq,* **written by Matthew Lee).** She also met with the new US Ambassador to Iraq Christopher Hill, the Chairman of the Joint Chiefs of Staff Admiral Mike Mullen, and General Odierno to discuss the security situation **(MNFI, 26 APR 09:** *al-Sharqiya*). It is interesting to note that her arrival comes less than a month after the President's. I believe the underlying theme here is this: 'We are still here to help...for now, but you cats have got to step it up diplomatically and get your security squared away because we will be leaving. We want you to succeed, but you have to want this more than we do'.

Today ended with a nice call home to the parents and a Skype to Shari. Yes, flowers will be in order and I will get them sent out today. And oh yes, by the way, I did get the car washed today. It was a little bit of a wait, but not bad and it cost me $10. This will be tax deductible and I won't be doing this often. But the car looks much better...for now!

St. Vincent's School in Wisconsin Rapids, WI
A generous gift from Katelin Kerkman's 5th grade class

Day 173: Monday-April 27, 2009/Day 142-Iraq was back to business. As I reported, Secretary Clinton's visit has drawn a lot of attention here. She not only met with the Foreign Minister on Saturday, she also met with Iraqi Prime Minister Maliki **(MNFI, 27 APR 09: *al-Sharqiya, KUNA*)**. As the diplomatic element in the DIME (Diplomacy, Information, Military, and Economic) model has been missing in action for much of the Bush administration, and there has been a gap in the diplomatic effort here between US Ambassadors Crocker and Hill, the diplomacy piece appears to be getting more attention in the new administration. As has been stated on many an occasion, the military instrument of power can be extremely successful in its operation, but it alone cannot win a war. All instruments must come into play, but diplomacy is what will ultimately close out a war, and it appears that the current administration understands this concept more clearly than the last.

This **'surge'** in diplomatic efforts is also being noticed by the Iraqis as an Iraqi lawmaker stated:

> Secretary Clinton met with representatives for civil society organizations in Iraq, as well as widows and orphans, which proves that the United States will play a new role in Iraq that is different from the previous military one; such as the role of supporting the political process in Iraq and the Iraqi government.
> **-MNFI, 27 APR 09: *al-Arabiya***

Secretary Clinton also held a town hall style meeting in Baghdad with about 100 Iraqis in attendance **(MNFI, 27 APR 09: *AP, McClatchy, AFP, UPI*)**. She informed this group that the US remains committed to assisting Iraq with its security, but will leave the resolution of political problems to the Iraqi government to solve. And solve them they must if they are ever to become a strong, stable government.

General Odierno also stated this day that the recent uptick in violence is related entirely to the "perception" of security. While the number of attacks has not risen dramatically, and the actual numbers validate this fact, they are higher profile in nature and have therefore drawn a lot of attention. But General Odierno went on to state that even though the numbers do not support a resurgence of al-Qaeda in Iraq, the "perception" by many Iraqis is that Iraqi Security Forces are unable to adequately protect them and this impedes governmental progression. As a result of these recent attacks, the ISF has demoted people; they have fired personnel; and they have created a task force to look into the reasons why security failed in these areas. So they are taking action to try and correct the situation, but because of the culture here and the infiltration of extremists into the ISF, these factors greatly impede progress. This is another area that the Iraqi government must shore up in order to move forward and be seen as a legitimate and stable entity.

There are a multitude of factors that are beginning to play out here in reference to the question General Petraeus asked long ago, which just happened to lead to a book by Linda Robinson with the same title - ***Tell Me How This Ends?*** How these variables are ultimately handled will go a long way in answering the General's question. A recent editorial suggested that Iraqis are beginning to reject terrorism because the attacks are directed primarily against innocent civilians. The statistical data bears this fact out, and the Iraqis know that al-Qaeda is attempting to incite sectarian violence and destabilize the government **(MNFI, 27 APR 09: *Asharq al-Awsat*)**. If this editorial is reflective of the attitude about terrorism from the majority of Iraqis, then al-Qaeda is in deep trouble. But Prime Minister Maliki is under extreme pressure concerning the reintegration of former Ba'ath Party members into the current government **(MNFI, 27 APR 09: *NY Times*)**. It has been suggested that the Prime Minister harbors great animosity toward anything related to the past and the Ba'ath Party, which has led to claims that perhaps he is not the man to lead Iraq into the future because of this bias toward reconciliation, which is imperative to bring the country together and stabilize the government. His ability to overcome this personal bias will have great impact upon the future of Iraq. And finally, we come full circle back to our friends in Iran. The Supreme Leader of Iran, Ayatollah Ali Khamenei, blames the US and Israel for the recent surge in violence within Iraq **(MNFI, 27 APR 09: *Reuters, AFP*)**. Secretary Clinton and General Odierno rejected these outrageous claims and it is more probable that it was Iranian influence that led to the uptick in violence. It is no secret that Iran hates Israel and vice versa. Because of this, Iran also hates the US because we are allies with Israel. It is also no secret that Iran hates the fact that the US is in Iraq, an Arab state, and will do anything to influence this situation. So there is a great deal of diplomacy that must occur in order for the situation to stabilize here in Iraq. How these critical variables are handled and ultimately play out, will go a long, long way in writing the final chapter on this war and in answering the question posed by General Petraeus.

Day 174: Tuesday-April 28, 2009/Day 143-Iraq started with a Skype to Shari. She received my flowers and loved the arrangement. I knew she would. I must put in a plug for **Blossoms and Bows** in Wisconsin Rapids who created and delivered the flowers. Because I have been a loyal customer they took care of the arrangement even though my debit card did not work for some reason. Betty, from the flower shop, sent me an email just telling me to call her to take care of the issue. How nice was that? Now that is outstanding customer service. I preach about this in my books, and have in this book as well. My two leadership books, ***Fire Service Leadership: Theories and Practices*** and ***The EMS Leadership Challenge: A Call To Action*** both have sections dedicated to the subject of customer service. And I can tell you that there are many organizations today that do not understand this concept. I am sure that because I am a returning customer, and not someone she did not know, that she could trust me to take care of the situation. With that stated, I am also a returning customer to many other businesses and have not received the same kind of customer service from them! So hats off to you Betty at Blossoms and Bows. Great customer service and I will spread the word. I guess I just did!

Now for the news of the day: General Odierno reiterated that it will be up to Prime Minister Maliki if US combat forces will remain in cities in the country after June 30th **(MNFI, 28 APR 09: *Khaleej Times*)**. Related to this tidbit, the spokesman for MNF-I, Major General David Perkins, stated that "Mosul is the one area where you may see US combat forces operating in the city. Mosul is sort of al-Qaeda in Iraq's last area where they have some maybe at least passive support" **(MNFI, 28 APR 09: *NY Times*)**. Then to top this off, an Iraqi National Police officer believes militant groups in Iraq are "re-energizing" **(MNFI, 28 APR 09: *TIME*)**. While he doubts that JAM (Jaysh al-Mahdi) will regain its former strength, they are prepared to fight if Muqtada Al-Sadr orders them to do so. Such are the dynamics here in Iraq. I believe that even after al-Qaeda is neutralized to a point where the Iraqi Security Forces can sustain security here, there are so many other variables that come into play that it will only take one small spark from any of these factions to start one large fire. US leadership is keen on the saying "the situation in Iraq is fragile and reversible" **(a phrase used often by General's Petraeus and Odierno)**. I fear that it will always be this way, even long after the US has left this country. That is just the way it is and it was quite unrealistic for our government to believe that they would be able to change an entire culture in a few short years. Talk about a mission doomed to fail! We have the same problem in the fire service trying to change a resistant culture on the tactics related to interior firefighting operations. It is just a tough nut to crack. I still think many criminal elements are just waiting for December 31st, 2011. Once all US forces leave Iraq, then we shall see how this all plays out. The situation will still be **"fragile and reversible"** and most likely will be for years to come. So the final chapter on **"how this ends"** may have to be written when US forces leave Iraq and the state of the country noted at that time. What happens after that is up to the Iraqis. And it is highly unlikely that if the situation worsens that the current US administration will commit any more funding or US troops to the effort. So in my humble opinion, after December 31st, 2011 one book will be finished and it will be time to begin an entirely new book.

    Day 175: Wednesday-April 29, 2009/Day 144-Iraq is a no BUA day so I will begin by outlining a new book I am now reading. The book is entitled – **Tell Me How This Ends?** written by Linda Robinson about General David Petraeus and his experience in the Iraq War. I am about 100 pages into the book and it is pretty good. It gives a detailed background on General Petraeus and his rise within the military community. The backdrop leading up to the surge is very interesting and this book, as well as the two written by Thomas Ricks that I have highlighted for you previously titled **Fiasco** and **The Gamble**, really provide the story behind the scenes that you may not, and most likely are not aware of.

    In news today, the Sahwa was the focal point. The Sahwa is just a different name for the Awakening Councils that were formed by tribes around Iraq to assist coalition forces to fight al-Qaeda. As I have noted before, many of these freedom fighters were to be incorporated into the Iraqi Security Forces and provided with employment. However, because of payments that were promised to these people by the Government of Iraq (GoI) and subsequently not paid, some of these resistance fighters have became disgruntled and may seek money from insurgent groups to again fight against coalition forces. Again, much like in the US, it is all about the money. The only groups that may not fall within this category are the radical extremists that hide behind their twisted religious beliefs. Yet, even they still need money to be successful! There have also been some recent events that have brought discredit upon the Sahwa, which General Odierno was quick to note that this is only a small faction and "99 percent of Sahwa continue to perform very well" **(MNFI, 29 APR 09: *al-Sharqiya*)**. A recent article from the United

Kingdom urges the GoI to reinforce allegiance with Sahwa members because they are the counterweight to AQI **(MNFI, 29 APR 09: *UK Guardian*)**. This is just another one of the major variables that the GoI must effectively address in order to succeed as a stable government.

Prime Minister Maliki is also still a big question mark here. Some see him as a leader incapable of overcoming personal bias and effectively reaching reconciliation with all major factions within Iraq. Others see him as a pawn for Iran. Still others view him as someone still evolving into an effective leader capable of leading Iraq toward stability. This is quite a wide spectrum of opinion and the jury is still out. He has stated that the US redeployment out of Iraq is final and that Iraqi Security Forces are capable of maintaining security despite recent attacks **(MNFI, 29 APR 09: *BBC*)**. This is just another of those huge variables whose outcome will not become fully known until US forces do leave Iraq for good. So the bottom-line is that it may still take several years for the answer to General Petraeus' question on how this will end to become fully transparent.

Day 176: Thursday-April 30, 2009/Day 145-Iraq began with the BUA. Our friends from SIGIR reported that Iraq's government has failed to provide ample funding for Iraqi Security Forces in regard to equipment and training **(MNFI, 30 APR 09: *AP*)**. This report also noted that the US military has established unrealistic training timetables for the ISF. One possible reason for this may be because the US timetable is always ahead of the Iraqis. This is simply a cultural difference and not that unusual. Another possible explanation is one that General Petraeus has outlined on several occasions related to the clocks that run in Washington compared to those in Baghdad. The Washington clock is always running fast because with soldiers continuing to die and money continuing to be spent, politicians are very impatient and expect swift results. On the other end, the Iraqis must overcome decades of autocratic rule and great sectarian friction that requires patience and persistence, and as a result, the clock here runs much slower. As General Petraeus has stated, he and General Odierno have tried to slow the clock in Washington by showing diplomats small signs of progress. The US Ambassador is working to get the Iraqi government to become more diverse, reconcile with the many factions within the country, and settle differences politically rather than through violence. This will also help to slow the Washington clock. You must remember that military engagements are heavily reliant upon political will, which is greatly influenced by public opinion. Casualties are always concerning, but the public establishes the level of acceptability in this area. When it becomes unacceptable, then politicians will get impatient and if they see no signs of sustainable progress, the clock will again speed up. This dichotomy becomes a real balancing act for military commanders. Military commanders that actually comprehend the type of conflict they are in must educate politicians that wars of counterinsurgency are long and progress is slow. There must be a clear and specific national strategy for military commanders to focus upon. Then, if certain benchmarks are not achieved within a realistic timeframe, then there will need to be a decision point by the military and political leadership on the next move. Counterinsurgent success is not like our victory in World War II when our troops returned home to throngs of cheering Americans parading down confetti-filled streets. It is also apparent that the US has a tendency to produce very unrealistic goals in its effort to achieve success. This has been a hard lesson to learn, and because of this fact expectations have been ratcheted back in both Iraq and Afghanistan. This is why it is imperative that the leadership at the highest levels within the government and the military must carefully examine all aspects of a situation before committing military forces. There must be clear and distinct strategic guidance that meshes all instruments of national power together (Diplomacy, Information, Military, and Economic) and incorporates accountability within each level. Then, in order for it to all work effectively, there must be effective leadership intertwined within the fabric at all levels. While it is easy to Monday morning quarterback just about any situation, it now becomes a matter of what we do with this information now that we are aware of it. If history tells us anything for those that pay careful attention to it, *we will forget all of these lessons within the next 15-20 years and be doomed to repeat our mistakes yet again!*

Day 177: Friday-May 1, 2009/Day 146-Iraq began another month. This would be a very significant month in my tour. I would reach the 50% mark around the 18th or 19th; I would reach Day 200 on May 24; Randy (MAJ Staab), the MNSTC-I LNO, would be leaving; my sister and stepson Josh would have birthdays, my in-laws, Dick and Kathy Kertis would have their 50th wedding anniversary; and I would begin the paperwork process for my R&R next month. So there was much to look forward to in the month of May.

In news today, the UK is finalizing its drawdown in Iraq. The end of April marked the closing out of the final chapter of combat operations for the Brits in Iraq. The UK prime Minister was here recently and stated that "today marks the closing chapter of the combat mission in Iraq" **(MNFI, 1 MAY 09: BBC, AP, SKY News & JASG, 1 MAY 09:** *British forces end combat operations in Iraq,* **written by Brian Murphy)**. The Brits have been a great ally and done a magnificent job here. They will truly be missed.

Great work blokes

There were more attacks in the country today and three more soldiers were killed. It appears as though al-Qaeda is attempting to make a statement through these recent attacks. A former Sadr City mayor stated that such attacks will "push us back to the sectarian violence. The Shi'a will be looking for revenge" **(MNCI, 1 MAY 09:** *McClatchy*).

MG David Perkins, MNF-I spokesman, stated that:
They are very emotionally charged targets. They are meant to go after a vulnerable aspect of society, to just literally kill as many innocent civilians as they randomly can. From al-Qaeda in Iraq's point of view, they are trying to generate a retribution attack and start ethno-sectarian violence up again. We have seen a couple of recent high-profile attacks, which obviously are a concern. But we don't think that's a fundamental shift.

-MNFI, 1 MAY 09: *AP*

So the situation is heating up over here and the fight is not yet finished. As has been stated time and time again, the situation is still very fragile and quite reversible. But it is time for the Iraqi Security Forces to step up, as well as the GoI, and take charge. General Odierno was also very agitated by the comments of the former Sadr City mayor because he believes his statements are unfounded and irresponsible. So you definitely observe both ends of the perception spectrum here. In a very dynamic and quickly evolving environment this is not all that unusual. Time will validate whose assessment is more accurate.

Today I also played **taxi driver** for a Captain that was down from GRN for a course at the palace (You talkin' to me? You talkin' to me? – Do you remember this movie?). Anyway, just another one of my many duties and yes indeed, it is unusual for a Lieutenant Colonel to be the driver for a Captain! Just the way it is here and really no big deal, plus it gets me out of the SOC for awhile. I will be playing this role for the next four days as Captain Tiffany Maraccini will be leaving tomorrow and then MAJ Sean Begley, one of my 416th TEC compadres will be over for a few days beginning on Monday. Just call me De Niro!

You talkin' to me?

▐▐ ▐▐ ▐▐ ▐▐ **Day 178: Saturday-May 2, 2009/Day 147-Iraq** began with no Internet in the barracks – still! It has been out now for the past three days so no Skyping Shari again! But I have called her from the palace so we have talked and stayed in touch.

It appears that the Iraqis really believe they are ready to take over security on 1 January 2012 **(MNFI, 2 MAY 09: al-Wakal, al-Mustaqila)**. I hope they are right, but even if they are not ready, I do not believe there will be anything that will keep US forces here after that date. I could be wrong, but all indications are that we will be out of here by then and then it is their country to run, ready or not!

General Odierno stated that al-Qaeda in Iraq is attempting to provoke sectarian violence with their recent round of high profile attacks. Secretary Gates verified that al-Qaeda in Iraq is responsible for the recent attacks here **(MNFI, 2 MAY 09: AFP)**. Not sure what their motive might be other than that they believe that over the next 32 months that Iraqi Security Forces will be strong enough to maintain security throughout the country. Otherwise, as al-Qaeda has proven to be very deliberate and patient in their planning, it would seem to make more sense that they simply wait until US forces leave at the end of 2011 and then make their move, unless they are worried that this may be too late. Perhaps they are a bit worried about their future here in Iraq!

As I have reported, many Iraqis are not falling for the antics of al-Qaeda and resorting back to sectarian violence. This is greatly impeding the efforts of al-Qaeda here. The Iraqi Ambassador to the US stated that:

> Sectarian violence in Iraq no longer exists, the instigators have no support, and there are still a number of political challenges facing the Iraqi government, but the important thing to notice is that these challenges are subject to dialogue away from the use of violence.
>
> -**MNFI, 2 MAY 09: *Aswat al-Iraq***

I am not certain I completely agree with his assessment, which I believe is a little optimistic, but clearly there are signs of sustained progress, movement away from violence, and the seeking of political solutions.

Finally today, to put this all of this information into perspective for you, it was reported that Baghdad is now safer than New Orleans! In the year 2008 there was a rate of 48 deaths per 100,000 residents in Baghdad, while New Orleans had a rate of 62 violent deaths per 100,000 residents **(MNFI, 2 MAY 09: *TIME*)**. So it appears that we are making more progress here in Iraq than we are in our own country! But as always, it all comes down to public perception.

Day 179: Sunday-May 3, 2009/Day 148-Iraq was a nice relaxing day. I slept in until 0700 this morning! That was nice and sorely needed. I made a pot of coffee and watched ***The Love Guru*** with Mike Myers. This was one of the movies that Sean (MAJ Song) left. Many people thought it was a stupid movie, but I enjoyed it. It certainly will not win any Oscars to be sure, but the silliness takes you away from the drudgery here for a few hours and that is always quite welcome. And I like Mike Myers and Jessica Alba so that made it worth watching. I do not think it was all that much different than the **Austin Powers** series which a lot of people liked. In fact, many of the same characters were in this movie, including mini-me!

I then did a little work on my Naval War College studies and watched a presentation of a taped lecture at the college in Newport, Rhode Island that focused upon the mission and capabilities of the United States Marine Corps. As a former marine, I was quite impressed with how well this service has adapted. Perhaps it is because it is smaller than its other three bigger sister services that fall under the Department of Defense umbrella (Army, Navy, & Air Force), but the USMC has a solid vision of its future and a good grasp on current activities. This branch of the armed forces decentralizes decision-making to very low levels within a unit and it has become very flexible and adaptable to its environment. This sounds an awful lot like my **ledocratic** organization I have mentioned to you and which I have implemented at our fire department. And while I have not yet done a thorough analysis on this organizational model to determine if it may work in much larger organizations, it has worked thus far in our smaller fire department, although the experiment continues. However, this organizational model seems as if it would work quite nicely in the USMC that has a size of about 200,000 personnel. The interesting question then becomes - would this same type of dynamic and flexible organizational model be effective in relation to larger organizations, such as the Army or the US government? This would make a great study for someone! I would postulate that with the right leadership in place, a **ledocracy** would work in any size organization, but of course, right now this is all simply theory.

I also heard from Sean (MAJ Song) who left us about 10 days ago. He made it back to Fort Irwin, CA alright and sent 'power row' some pictures he took before he left. I also called Shari today because Skype (Internet) is still down in the barracks. I had to inquire about its disposition with someone at GRC, and I was informed that someone was trying to download something they were not supposed to and that we basically had some sort of cyber attack! Due to these factors, the techno geeks simply shut everything down at GRC and are working diligently to correct the deficiency! The gentleman I spoke with was hopeful that the Internet would be back up and running by mid-week. We shall see! Back in November there was some other sort of mishap involving jump drives, and shortly thereafter, all jump drives were banned from military computers. The military still has not figured this one out! All of this makes it very inconvenient to move information from a classified to a non-classified source and vice versa. So either the cyber attackers are winning this phase of the war or our techno geeks are overpaid! Very frustrating!

It is also interesting to note that after General Odierno became very agitated by the comments of a former Sadr City mayor in relation to the recent attacks in Iraq which could lead toward an increase in sectarian violence, there is another story that moves 180 degrees **(MNFI, 3 MAY 09: *LA Times*)**. On Sundays, BUA slides are hung (inserted in a portal file on the MNF-I Web site) for people to read, but there is no formal briefing as there is on BUA days (Mondays, Tuesdays,

Thursdays, Fridays, and Saturdays). Anyway, a Baghdad resident stated that some Iraqis fear renewed sectarian violence after these recent bombings, but this resident believes the country has stabilized. Another resident believes the security gains made by coalition forces are still intact despite the recent uptick in violence, which is an indication of the last stand for the terrorists. While the timing is a bit odd on this story, it comes less than two days after General Odierno's comments! Can you say information operations? Information operations and winning the media war are very big in counterinsurgent campaigns. Information can help shape the battlefield and while the US was very poor at wielding this instrument of national power at the beginning of this war, it has gotten much better at it. And if information is power then we are beating the insurgents at this phase of the war as well. The question will then become, can the Iraqi government continue this information success after US forces leave? We shall see!

Day 180: Monday-May 4, 2009/Day 149-Iraq started off a bit differently than most. As I was driving to work and very close to the palace, the home stretch was blocked off for some reason today. So I had to quickly determine another route to the palace that I had not traveled previously. Being flexible and adaptable serves you well over here, and this is the same message I try and relate to my firefighters back in Wisconsin Rapids. This adaptability to change is an area that many in the fire service and the military have difficulty in identifying with. Few people like to move away from their normal routine or out of their comfort zone, but in dynamic environments such as the fire service and the military, if you are not flexible and adaptable, then you will fall behind the proverbial power curve very quickly!

There was much activity today after a fairly quiet weekend. General Odierno stated that the US is dedicated to the security agreement that is in place and remains committed to assisting Iraq to develop their institutions and security agencies **(MNFI, 4 MAY 09: *al-Arabiya*)**. All of this comes on the heels of Prime Minister Maliki's comments that Iraqi Security Forces will be ready to assume full command of security at the end of the security agreement **(31 DEC 2011)**, and a statement by Shi'a cleric Muqtada Al-Sadr that US troops need to be out of all Iraqi cities by 1 JUN 09. Actually, I believe the agreement calls for this to occur at the end of June 2009, but al-Sadr claims that if US troops are not out of cities by whatever the agreed upon date is, then he will direct his militia to again pick up arms against coalition forces. Such are the dynamics going on here, but it is interesting to note that al-Sadr just paid a visit to Turkey, and many people believe he is getting more heavily involved in the political realm, so his comments in relation to inciting violence send a mixed and confusing message. It is almost as if he were saying that if I do not get my way through the political process then I will resort back to violence! Not sure what kind of cleric sends such messages, but again, these are the dynamics afoot here.

The other issue I have been reporting to you of major significance is integration of Sahwa members into Iraqi Security Forces so these tribal members do not revert back to fighting coalition and ISF forces **(MNFI, 4 MAY 09: *World Tribune*)**. Paying those who have been integrated into the ISF already, and those that are still seeking such employment, has been a real issue but it is important to note that Iraq has just approved $350 million to reintegrate the nearly 100,000 Sahwa members and address this important inequity. So this news is very encouraging but still needs to be monitored closely.

Finally, the US is still doing damage control on the perceptions related to the recent violence here. MG Perkins, MNF-I spokesman, stated that al-Qaeda's recent attacks have attempted to incite sectarian violence, but this has not been effective. He goes on to state that while the recent attacks have been of a higher profile, the overall number of attacks is dramatically lower than one year ago **(MNFI, 4 MAY 09: *al-Sharqiya, Sot al-Iraq, Baghdad Times*)**. General Odierno follows these comments with:

> I believe the reaction of the Iraqi people and the reaction of the Iraqi government to recent terror attacks is sound: they did not direct any accusations and they understand these attacks. They know it is al-Qaeda in

Iraq, which is trying to ignite the fuse of sectarian violence. But Iraqis resist this and so do officials and they want to continue protecting the progress that was achieved.

-MNFI, 4 MAY 09: *al-Arabiya*

I finished off this Monday by watching **You Don't Mess With The Zohan** starring Adam Sandler. Now I just finished telling you about **The Love Guru** and the fact that it received terrible reviews and many people did not like it. But unlike television critics, I do not watch movies for their Oscar potential. I watch them because they entertain me, and everyone has different tastes. I prefer goofy comedies or action flicks to take me away from reality. I do not want to watch more reality shows as I am already living reality and prefer something a bit different! So while most people did not like **The Love Guru**, I thought it was alright as I mentioned previously. With that being stated, my stepson Jeremy took his girl friend and Drew to **You Don't Mess With The Zohan** at the local cinema and he said it was terrible and they walked out after just 15 minutes. But Sean left this movie here and I like Adam Sandler and movies such as **The Water Boy, Happy Gilmore, Mr. Deeds,** and even **Little Nicky**. But after watching this particular movie, I must admit that Jeremy was right on and it was really a stinker. So unlike **Happy Gilmore** or **The Water Boy** that I will watch over and over, this one was a one-time only viewing. Even though there were a few moments of levity in this movie, this was arguably his worst film. But even watching a bad movie is still better than being at work!

Iraqi boy with toys distributed by Gulf Region Division personnel

CHAPTER 11
400 DAYS – Days 181-200
May 5 – May 24, 2009

We are at war, and our security as a nation depends on winning that war.
--Secretary of State Condoleeza Rice
2005-2009
also served as National Security Advisor to President Bush
2001-2005

Day 181: Tuesday-May 5, 2009/Day 150-Iraq was another Red Letter Day of sorts. It was **Day 150** in Iraq, and in less than three weeks I would hit the halfway point of the 400, although it appears likely that our orders will conclude before we ever reach the magic 400. However, I am not changing the name of the book to 375 Days or something like that! The name will stand as my official orders to Iraq state for a period of 400 days. It is also Cinco de Mayo Day. Perhaps a non-alcoholic margarita is in order if they have them in the DFAC! It is also my sister's birthday today, but as a good brother, I will not reveal her age. Allyson, or Sis as we call her, is a very loving individual and a good human being. She is very sensitive and caring. She is a great sister and a good friend to Shari. They graduated from Lincoln High School in Wisconsin Rapids in the same class, but they really did not know one another in school. Lincoln is a big school with graduating classes approaching 600 students. And finally, my friend Randy Kubisiak, or Kubi as we call him, is celebrating his birthday today. I introduced you to Kubi long ago and he retired from the fire department about a year and half ago now. So there was much to celebrate today.

Back to the war today, the information operations campaign continues in earnest. MG David Perkins, MNF-I spokesman, stated that "terrorists may be accomplishing their task, which is to kill a lot of innocent civilians. But they are not accomplishing their purpose, which is to generate ethno-sectarian violence and chaos **(MNFI, 5 MAY 09:** *McClatchy***)**. Prime Minister Maliki states that:

> The violent days of the past will never return. These criminal acts are just trying to play on the sensitive chords of sectarianism but our security agencies have already infiltrated the terrorists organizations in their depths and will paralyze them using accurate intelligence rather than heavy weaponry.
> **-MNFI, 5 MAY 09:** *Aswat al-Iraq*

General Odierno took exception to this statement and stated that complacency and overconfidence can be a dangerous combination. General O prefers a more consistent and cautious approach versus such bold statements. Again, information operations is yet another work in progress over here.

I have already informed you of ***The Awakening*** that occurred in Al Anbar when the tribes in this western region of Iraq formed together, stopped fighting coalition forces, and began to assist coalition forces in the fight against al-Qaeda which was a major variable that caused this war to turn in our favor. As this movement was very successful, the Shaykhs in Anbar are now beginning a 'reconstruction awakening' and looking for $11 billion in foreign investments to fund reconstruction projects such as clean water, sewer, medical facilities, and other essential services **(MNFI, 5 MAY 09:** *LA Times***)**. If the Shaykhs are as successful with this venture as they were with the original awakening, then the future looks very bright for Al Anbar.

Finally, the Iraqi Army is having some difficulty in filling their ranks. Many of their fourteen divisions are short of personnel, especially at the officer level **(MNFI, 5 MAY 09: MNSTC-I update)**. This continues to be a work in progress but needs to be addressed with some urgency to fill the void that will be left when US forces leave in 32 months. There are approximately 50,000 positions yet to be filled to meet the required number of troops. The clock is ticking!

Happy Birthday Sis & Kubi

Day 182: Wednesday-May 6, 2009/Day 151-Iraq was off to a good start. I went to breakfast again this morning and had another omelet, with hash browns, a biscuit, and sausage. It was actually all very tasty, but I must tell you that the guys that do the serving are out of control! These guys are of all different nationalities, including Iraqis, and though very friendly, they do not understand the concept of portion control! They gave me twice as much as I could eat. Perhaps I need to be more specific when I order from now on. Typically you stand in a line (cafeteria style), move down the line, give the server your plate, and tell them what you want. I normally do not tell them how many items I want, but I guess I need to start doing so! Anyway, I could only eat about half of what they gave me and I hated to throw the rest away, but unfortunately I had to do so. Lesson learned!

I was then off to the post office to mail my second box. My first box and footlocker had arrived home with no issues. So today I would ship off another box, and in two weeks my second footlocker, as I continue to drawdown the equipment in my room and lessen the load to take home upon redeployment. Once again, getting to the post office as it opens is the best time to get there. So this mission was easily and quickly accomplished, and then it was off to the palace to start my work day.

I must also state that I talked to Shari yesterday on the phone from my workstation in the SOC as our Internet remains down in the barracks. She surprised me today with an announcement that she was going back to school! She does want this kept secret, but by the time this book comes out and people read this it will not be a secret any longer. But I am very proud of her for making this decision, which came right out of the blue. This certainly gives her something to fall back on should anything happen to me.

Anyway, to the news of the day: it appears that US forces will not remain in Iraqi cities past the June 30th date, or in Iraq past December 31st, 2011 **(MNFI, 6 MAY: *Reuters*; JASG, 5 MAY 09: *Iraq insists on US leaving cities by June 30*, written by Robert Reid; and Stars & Stripes, 5 MAY 09: *Iraq: No deadline extension for US*, written by Hamid Ahmed)**. I would say that not only is the Iraqi government pushing this issue, but also our own government. President Obama wants an end to the war here, and Iraq will need to step it up and take over ownership for their country. The time has come.

Following on the heels of this story is news that Iran will replace US influence in Iraq once we leave **(MNFI, 6 MAY 09: *Wall Street Journal*)**. This is what has been feared here and Iran is trying to position this influence within the political framework that has been established. However, at last check many of the Iranian backed candidates did not fare all that well in the provincial elections back in January! But there can be no doubt they will do their very best to influence the situation here as much as possible.

Finally, my pals from KBR are once again in the news **(Stars & Stripes, 5 MAY 09, *Top Pentagon auditor criticizes KBR support work*, source: Associated Press)**. The report criticizes KBR for their performance in relation to a

$31.7 billion combat support contract. The report goes on to state that the company, based out of Houston, failed to make certain that US taxpayers were getting the most for their money. Due to the lack of oversight, ineffective leadership, and the failure to incorporate accountability into the very fabric of many of these government approved private contracts, there can be no doubt that billions of dollars have been needlessly wasted. This is a sad report indeed, and the lesson to be learned in relation to other communities looking for that quick savings through privatization – **BUYER BEWARE!**

▍▍ ▍▍ Day 183: Thursday-May 7, 2009/Day 152-Iraq was another no Skype day! I fear that we are losing the cyberspace war here. Sure seems like our guys have great difficulty in figuring the counter strategy out to whatever is shutting the system down. They have yet to figure out a fix for the jump drive issue that began last November! It is pretty sad in my opinion.

I guess the editor from TIME magazine is now going to run the war from afar! It is amazing how many people actually believe they have better strategy than the people on the ground! It sure seems like we just went through this with the last administration. There certainly are a lot of people who believe they are the President, yet they never bothered to run for election! This is Monday morning quarterbacking at its finest and one of America's favorite past times. I have found this out first hand as a fire chief. And the bright idea bus is a mass transit system that exists in every community in the good ol' US of A! There are many times when everybody is an expert, has a better solution, but never runs for office or seeks a position in order to seek such change. We call these kinds of people snipers that prefer to sit back, remain uninvolved, and then bellyache about something when it is convenient for them to do so! Sound like anyone you know? But now we have a journalist offering opinion which I am sure will be considered at the highest levels of government. Not! I do not necessarily disagree with the base concept of the story that basically calls for a re-analysis of the 'Awakening' strategy because the Government of Iraq will not be able to sustain the salaries for all of its members **(MNFI, 7 MAY 09:** *TIME*). While this may or may not be correct, that will be for Iraq to figure out the solutions for. If you recall, many of the Awakening members were to be incorporated into the Iraqi Security Forces as basically a reward for assisting coalition forces in their fight against the insurgency. If this does not occur, these same people may once again turn their arms against coalition forces. I have also mentioned recently that the Iraqi Security Forces are woefully short of people. So I do not believe it is simply a numbers game. Ultimately however, the payment of such membership would have to be determined by the Government of Iraq and they will need to prioritize where this challenge falls in their long 'TO DO' list toward stability. The hardliners from this movement are those that need to be dealt with differently than the majority of these people who are simply trying to provide for their families. It is an interesting debate to be sure.

Today I think we will branch off in a little different direction. As I may have mentioned previously, but in no great depth, there is a lot of corruption within the country of Iraq. There are many people, including governmental and security force officials that have taken bribes; criminals who have siphoned off oil from the massive pipelines that dot the country and sell it on the black market; electrical sub-station workers who have either taken money for reallocating power to unauthorized areas or have been threatened with death or harm to their families; etc. A representative from the Commission on Public Integrity stated that "the reason for the massive corruption in Iraq is the belief by the corrupt that they are shielded from prosecution by the protection afforded to them by their political parties and sects" **(MNFI, 7 MAY 09: NY Times)**. So while everyone here knows there is corruption, it then becomes a matter of what is considered acceptable corruption by Iraqi society. It is a bit out of control at the moment over here, but do not kid yourself into believing that we do not have a good deal of corruption in our country! You read about it or see it on television almost every day, and a society will never be able to rid itself completely of corruption, so it becomes then a matter of what that society deems an acceptable level. I believe we have established that level in the

US, but such is not yet the case here. This too will come in time if the country continues to move forward and progress.

All aboard the good idea bus!

▌▌ ▌▌ ▌▌ ▌▌ **Day 184: Friday-May 8, 2009/Day 153-Iraq** was another day in Iraq. I feel that the story that must be told so that all who read this understand the trials and tribulations of not only the soldiers over here, but also of the families left behind. Even though this book will approach the New York City phone directory in size by the time it is published, it still will never truly do justice in articulating the realities of having someone deployed for months on end unless you have been through the process yourself. This experience is just very difficult to explain to someone if they have never been through it. Many of my guys at the fire station complain about going away to the National Fire Academy in Emmitsburg, MD for two weeks. Try being away from home for about a year!

There was a nice article in yesterday's *Stars & Stripes* about a unit deploying from North Carolina. While this is the second deployment for this National Guard unit, being deployed does not get any easier the more times you do it! The story outlines how difficult a deployment is for the families **(Stars & Stripes, 7 MAY 09:** *It takes everything I got to keep it inside: War may be waning, but families still struggling with deployments,* **written by Kevin Maurer and Mitch Weiss)**. This is the same experience that Shari has been going through, and in her 29 plus years of life, she has never had to go through this before. Nor have I for that matter, and even though I am proud to serve my country and understand that what we are doing here is important, it still sucks! Not sure how else to say it. I also understand the military, especially the Army, understands the strains placed upon its warriors and their families, but I am not so sure they have viable solutions to correct such an inequity. Through growing the force and longer dwell times at home they are making an effort to address the problem, but I do not believe this is enough. We shall see how effective it will be in time.

Map of Baghdad area (note the location of Adhamiya)

Even though my situation is not all that great, I will not complain about where I am stationed, and I have absolutely no desire to go outside of the 'wire' unless necessary. Certainly the situation here in Iraq is much better than it has ever been. But it is still a war zone and there are still people in this country that mean to do harm to anyone in a military uniform. My mother did not raise 'no fool'! I am perfectly content to stay right here on VBC.

The current book I am reading is called **Tell Me How This Ends?** written by Linda Robinson and is primarily about General David Petraeus and the surge. The book began a little slowly but is picking up speed. I am about 200 pages into it now and last evening I read about Task Force 1-26 (1st Battalion, 26th Infantry Regiment) that was stationed at Forward Operating Base Apache in Adhamiya. The soldiers of this unit saw the real horrors of war and lost many friends and teammates. If it was not al-Qaeda attacking them, it was the Jaysh al-Mahdi (JAM). This unit lost several members in single events, including huge explosions that flipped over 35 ton vehicles! This book goes into graphic detail, and as a paramedic I have seen some very bad scenes including multiple fatalities, but not when these fatalities were your friends. And at my rank of Lieutenant Colonel I could have easily been a Battalion Commander in charge of such a unit, and I cannot even fathom losing this many soldiers under my command. This does not imply that their commanders were not competent or caring. They were just in a very bad place and given a very difficult job. Many of the soldiers killed were in their early twenties and had young families. There is no amount of training that can prepare you for such horrors and because of this, those soldiers that picked up the body parts of their friends, and held the hands of the fallen as they died were also casualties of war. These are the real horrors of war that most people will never understand. It is very difficult to lose so many promising young men and women who had their whole lives in front of them. And this is just one unit of the many who have their own stories to tell and horrors to deal with long after the war has ended. These are the scars you carry with you for life. You never forget. So if you get nothing else from this book, I hope you remember and appreciate the extreme sacrifices of not only the fine men and women who have died or witnessed the horrors of war, but also the families that must cope with the loss of a loved one, or the aftermath of such life-altering events forever seared into the minds of those who return home. And because I feel so passionate about this message I will offer a salute here to not only the brave warriors of Task Force 1-26, but all of the men and women who have fought so bravely in Iraq and Afghanistan, and in all of America's previous military engagements. The real tragedy in all of this occurs when Americans forget such sacrifice. Many Americans that have never been affected by this war will remember these sacrifices a couple of times a year, typically on Memorial Day and Veterans Day. But for those that have been affected by war, whether in conflicts past or present, they remember the sacrifice each and every day! And to steal a quote from General Odierno who concludes each BUA with this little prayer: "May God Bless all of our soldiers, marines, sailors, airmen, coastguardsmen, and civilians." And I would add, their families as well.

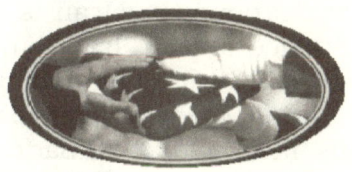

Day 185: Saturday-May 9, 2009/Day 154-Iraq was another day much like the previous 153 in Iraq! The weather is consistently in the mid-90s now and climbing. The past week has been warm and very dusty.

Last night I got to watch General Odierno's press conference at the Pentagon. Yes, he left Iraq around Wednesday and will return sometime in the next week to ten days **(Pentagon Channel, 8 MAY 09, 2000 GMT+3)**. Of course, I get to see and hear him during each BUA in Iraq so I have already heard everything of which he spoke, as he provided the media with this same message. You could tell he was a bit nervous as he played with the podium speakers a little and his glass of water, but I thought he did a good job of providing a consistent message. Bottom-line: the situation in Iraq is improving steadily despite some recent attacks perpetrated by al-Qaeda. Even though improving, the situation is still quite fragile and there is much work to be done, especially on the political front. But if current trends continue, there is no reason to believe that the Iraqi Security Forces cannot handle the situation here and all US troops should be out of Iraq by the end of 2011. That is the message, although General O did hedge his bet a bit by stating that any significant and unforeseen situation within the next month or next year could alter this assessment. The national elections at the end of this year or early next year will be a huge benchmark for Iraq and its political progression. The Minister of Foreign Affairs stated that Iraq's upcoming national elections "will be the most important elections held in this country since its establishment" because it will reveal the extent of Iraq's political progress **(MNFI, 8 MAY 09:** *al-Sharqiya*). I thought I just said that!

There are also indications that Iran is backing off a bit in regard to their influence within Iraq. And there has been rumor that Muqtada Al-Sadr has achieved the status of Ayatollah **(http://www.wisegeek.com/what-is-an-ayatollah.htm)**. This means that Al-Sadr has been elevated from a Shi'a cleric to the status of Ayatollah because of his diligent studies of the Qur'an. His father, who was murdered by Saddam Hussein's henchmen, was also an Ayatollah. Al-Sadr is also the person in command of the Shi'a insurgent group known as the Jaysh al-Mahdi (JAM) that have fought against collation forces in the past. One of his spokesman stated, "We of the Jaysh al-Mahdi have laid down our arms. We are certainly not going to use them against Iraqi soldiers. But the rebellion will go on" **(MNFI, 8 MAY 09: *Asia Times*)**. Hopefully the translation is that the rebellion will continue on through the political process and not via violence. But keeping this group under control is also very important for the stability of the Iraqi government.

So overall I thought the press conference went fairly well, but General O is a military guy and he gets paid to make military decisions. The talk of politics is a necessity at his level, but he is certainly not as comfortable dealing with these types of issues, and he should not be expected to be. This is why we have a President, a Secretary of Defense, a Secretary of State, a US Ambassador to Iraq, etc. Political matters are for this group to sort out and influence. Thus endeth the lesson!

There is much news to catch up on as signs of progress continue to emerge here in Iraq. If you recall, awhile ago I mentioned that there is a referendum this summer whereby all Iraqis will have an opportunity to vote on the Iraq-US security agreement **(MNFI, 9 MAY 09, 0730: *al-Sabaah*)**. If this referendum does come to fruition, the Iraqi populace may have a vote as to whether or not they want the US to leave the country before the agreed upon date of December 31, 2011. Dare I say that this sounds kinda like a democratic process here! And if I understand the process, although I am unsure how binding this vote may be, if the Iraqi people vote the US out of the country then we have a year from the vote to basically leave. This means that we will need to be out of Iraq almost 18 months sooner than expected. This will certainly throw a real big monkey wrench into all of the planning that went into the drawdown to get us down to 50,000 troops by August 2010 and to 0 by December 31, 2011! These same planners would no doubt have some serious cardiac complications should they need to shift this time schedule up almost 18 months!

Another sign of progress comes from a concrete barrier supplier in Baghdad. This gentleman states that because of the security improvements there has been far less demand for the big ten foot high concrete barriers called Texas T-Walls. This gentleman goes on to state:

> I used to install thousands of barriers for the American Army. I set them up in the most dangerous streets and neighborhoods. The demand for barriers has vanished lately with the improvement in the security situation. The work has been halted at some factories, and others have shut down completely.
> **-MNFI, 9 MAY 09, 0730:** *AFP*

This is good news, but it may mean more people on the unemployment line! Even progress has its downside.

Day 186: Sunday-May 10, 2009/Day 155-Iraq

was Mother's Day in Iraq. Because there is no BUA on Sundays, and because it is Mother's Day, I will take a break from war news today and rather, talk about my mom. To begin this section, I want to wish not only my mom a very Happy Mother's Day, but all mothers around the world, especially those with children or spouses serving in Iraq or Afghanistan.

I guess it could be said that I get my brains from my mother. She is a very intelligent woman who married a suave and debonair dude named Donald (I will talk about dad on Father's Day of course). She was born in Iowa, but basically grew up in upper Michigan. Her father, my grandpa Cecil, worked for the Nekoosa-Edwards paper industry as a forester. Her mother Iva was a stay at home mother and raised the three children. My mom was a very good student in school and enjoyed music. She learned to play the piano at a young age and enjoyed singing. She married my father Donald in Milwaukee in 1956. A little over a year later, at Columbia Hospital in Milwaukee, a young boy named Mitchell Rowell Waite was born. Little did anyone know at that time that this young lad would grow up to be a famous author. Alright, well if you are reading this book then let's just go with an author and take the famous off. Someday perhaps! Anyway, mother went on to graduate from nursing school and worked the majority of her career in Wisconsin Rapids at Riverview Hospital, and subsequently Riverwood Clinic. Mother did her best to get her children interested in music, but alas, her two boys became jocks and her only daughter did not head down that road either.

But my mom did her best to raise her children and instill the solid values and moral aptitude that we possess today. My grandmother assisted with raising the three little Waite desperados and I remember my childhood very fondly. While we may not have been the perfect family, and we like all other families had our challenges to deal with, I could not have asked for a better childhood or for better parents. And being over here in Iraq and seeing the poor kids and the lives they must endure in order to simply survive in this war torn country, it makes me appreciate my own childhood that much more. So on this day, Mother, I wish you love and thank you for all that you have done for me and continue to do.

Happy Mother's Day.

▌▌ ▌▌▌ ▌▌▌ ▌▌ ▌▌ ▌▌ ▌▌ ▌▌ ▌▌ ▌▌ **Day 187: Monday-May 11, 2009/Day 156-Iraq** was back to work, but not after finally Skyping Shari after about a week's absence. It was great to see her, and her sister Tina was over visiting when I called. They had taken their parents and my parents out for brunch on Mother's Day. My sister Allyson was up from Madison as well and apparently a good time was had by all. So this was all very good news from the home front.

Australia is ending their military mission in Iraq **(MNFI, 11 MAY 09: *al-Furat, KUNA, Radio NAWA*)**. They too have been here since 2003 and much like the British have been very good allies. Shari and I decided we are heading to Australia for a vacation when I return. We are forgoing the trip to Hawaii, which we will take at some other point in time. I left it up to her to decide where we would go and she chose Australia, which is also a destination I have never been to and have on my Bucket List. Yes, I started my list long ago and hopefully will have much time left on the old life clock to get most of the list accomplished. No, climbing Mount Everest is not on my list!

House Speaker Nancy Pelosi paid Baghdad a visit on Mother's Day! She apparently was over here to discuss bilateral economic relations, US troop redeployments, and Iraq-US strategic framework implementation strategies **(MNFI, 11 MAY 09: *AP, AFP*)**. Now I have to ask the question, what does it cost the American taxpayer for all of these US officials to fly over to Baghdad for a few hour visit, including all the planning that goes into such a trip and the security surrounding it! More research for me to do, but I just find this all very odd that President Obama is talking of sacrifice by all Americans in this time of economic depression, yet in the past few months we have had the President here, Secretary of State Clinton, and the House Speaker, not to mention that Secretary of Defense Gates was in Afghanistan this past week. And General Odierno just flew back to Washington! You probably all recall the lesson in poor decision-making a few weeks ago when someone on the 'bright idea bus' thought flying a large jet over New York City would be a cool thing to do! Not only did this event get massacred in the press, and rightfully so, but it cost the guy (perhaps the fall guy) who had the bright idea to lose his seat on the bus! Oh by the way, this little faux pas cost the American taxpayer thousands of dollars (approximately **$328,835** to be exact **(www.msnbc.msn.com/id/30643376/)**. So you can only imagine how much all of these little trips to Baghdad are costing taxpayers. Know I realize the discussions these politicians are having are quite important, but could these same people not plan better and come over as a delegation, or perhaps use secure teleconferencing to talk as we do over here, or even Skype! Yes, I am sure Skype is not secure, but the Tandberg system that we use in the military is a secure site and is a more sophisticated type of Skype. Just a question!

G'day Mate!

Day 188: Tuesday-May 12, 2009/Day 157-Iraq was a day full of bad news. Yesterday we had what we call a **'blue on blue'** incident. A soldier at a stress clinic right here on VBC shot and killed five other troops **(Stars & Stripes, 12 MAY 09, *US soldier kills 5 others in Iraq*, written by Jeff Schogol & JASG, 12 MAY 09, *American kills 5 soldiers at clinic in Iraq*, written by Robert Reid)**. This is a great tragedy and a very sad day for us all. It is bad enough that there are dangers that await personnel when outside the wire, or in what we call the **'Red Zone.'** But when you have such an incident within the relative safety of this base it is very disturbing. Details of the incident are still unclear and I am certain will be forthcoming in the days ahead, but the bottom-line is that five US troops are dead, killed at the hands of a US soldier.

I will be looking for how many deployments the individual who did the shooting has completed. I will be interested to hear what steps the military had taken to assist this individual. Then, if he has been diagnosed with PTSD, were there any warning signs that he was a danger to others? Every military member here carries a weapon and ammunition. While on base, almost everyone is to be in a **'Green Status'** which basically equates to a Barney Fife (for those of you old enough to remember Don Knotts on the Andy Griffith Show of yesteryear). You have a weapon, but it is not loaded, nor is there a round in the chamber. Your ammunition is on your person, but separated from your weapon. There are also clearing barrels all over this base which means that before you come back on base or enter any building, you are to clear your weapon to ensure there is no ammunition in it. The only people who are exempt from this **'Green Status'** are certain positions such as the Military Police or security personnel for high-ranking officers. So it will be interesting to follow this tragic story as the details unfold, and no doubt, the Army will take some knee-jerk reaction to the incident as they always do. However, I must state, as I did several weeks ago when I highlighted the suicide prevention program, the military is trying to find viable, long-term solutions. It does appear that they are hiring more mental health professionals to screen military personnel. They do screen soldiers more thoroughly before redeployment. They are talking of shorter deployments (9 months as opposed to 12) and longer dwell times (time back home). With a larger Army there should also be fewer deployments. So all of this is positive news and in my humble opinion, viable, long-term corrections to a serious problem. But it may still not be enough, but it is a good start. It is very difficult to foresee an incident such as we had here yesterday and so it is difficult to comment too much on corrective action until more facts about the incident are revealed. But no matter how you analyze this, there are five more soldiers who will not be going home and continuing on with life.

The other big news from yesterday was that the highest ranking officer in Afghanistan has been removed from command **(Pentagon Channel, Secretary of Defense/ Chairman Joint Chiefs of Staff news conference, 11 MAY 09, 2100 GMT+3; JASG, 12 MAY 09:** *Gates replacing top US Commander in Afghanistan*: **CNN; & Stars & Stripes, 12 MAY 09,** *McKiernan out of Afghanistan command,*

written by Leo Shane III). General David McKiernan has been relieved of command. The exact nature that has necessitated such a move is still not entirely clear. You could tell that both Secretary Gates and Admiral Mullen were uncomfortable discussing any details about the situation even though pressed by reporters. Perhaps it was resistance to change or the inability to adapt that necessitated such a change. I have remarked about this before and this may be the ultimate reason for the move. But it is highly unusual for a four-star General to be removed from command, especially during a time of war, when he is only about half way through his tour of duty. On the other hand I give credit to whom ever made the decision if the cause was just. We have seen far too many times where such a move has not been made and the wrong strategy pursued. The tragedy of this is that it can and has resulted in many lives lost and billions of dollars wasted. So I am sure the facts associated with this significant event will come out eventually and given the people and the personalities involved, it would appear to be the right call.

Finally today, there is a new FRAGO out that is basically calling for a stand down on all travel around Iraq due to detainee abuse photos that are pending release to the public sometime in the near future. Yes, these photos were taken over five years ago and for whatever reason are now being released to the public. Now the problem with all of this is that the abuse took place over five years ago and the perpetrators have been tried and punished. The Bush administration admits to signing documents that may have perpetrated such events, which has somehow morphed into the misnomer that this basically exonerates members of that unit. I do not care if it was supposedly authorized or not, there are still things that you have to question as a competent leader, and there were still too many things involved to simply lay the blame off on a memo from the White House. And in the case of Abu Ghraib there is plenty of blame to pass around. Not sure why the photos are being released other than because of a push by the ACLU (American Civil Liberties Union), but when this occurs it usually brings an uptick in violence and puts troops at risk. Because of this anticipated surge in violence due to this incredibly poor decision to release such photos at this time, this FRAGO is being sent out calling for a stand down by all military members with no movement into the **Red Zone** for a specified period until the atmospherics inform our intelligence community that it is relatively safe to venture back outside of the wire. I am certain that the people that perpetrated this entire misguided affair several years ago never realized that they would be putting their fellow troops in harm's way yet again, even five years later! It's like the gift that keeps on giving!

▌▌ ▌▌ **Day 189: Wednesday-May 13, 2009/Day 158-Iraq** was off to a good start. I Skyped Shari this AM and we had a nice long chat since it is no BUA Wednesday. She was holding baby Sydney who was over for the night. Whether she likes to admit it or not, she enjoys being a grandmother, although she looks far too young to be one! But by the time we were done talking about the day's activities and our trip to Australia, Sydney was asleep. Yes, how ironic is it that we are going to Sydney, Australia in December and we have a granddaughter by that name. Must be karma!

Anyway, today I will again comment on the tragic shooting of five people here on VBC. As the reports begin to trickle out and we begin to separate fact from fiction, I do believe the at least one of the victims was a sailor on the medical staff at the stress clinic where the shooting occurred. And as tragic as this is, and bearing in mind that there are just things in life that cannot be explained or foreseen, the risk manager inside of me cannot help but wonder if it was the young man who did the shooting that is completely to blame, or did the system in place once again let our troops down? Now to clarify, I hope this book is not coming off as being written by someone who dislikes the Army. That could not be further from the truth. I love the military and have spent over 34 years, the better part of my life, in the armed forces (Army & Marine Corps). But what does drive me batty is when I see a system as huge as the Army not functioning as it should. I have clearly outlined throughout this book where I believe this big, cumbersome Goliath could make changes to improve its systems and processes. In fact, I am encouraged to see that there are several things in the works within the military community to address these key and very sensitive issues. But because of the very bureaucratic, resistant-to-change organization that the Army is, this type of significant change comes ever so slowly. Now once again, as is a side effect of such a large organization, they are forced to react to a serious situation as opposed to being more proactive and foreseeing such a circumstance. So one must ask, could this tragedy have been avoided? You may think this is again Monday morning quarterbacking, but I look at this from a risk management perspective and truly believe that exercising vision can enable leaders to foresee such occurrences before they occur and then implement procedures to avoid them. In this particular case we do not yet know all of the facts, and it sounds like many of the current procedures in place were followed, yet we still have five dead military personnel! Perhaps nothing could have been done to avoid this tragedy, and I am certain it will be analyzed ad nauseam, but I will bet a six-pack of near beer that as a result of this shooting changes in procedures and policy will be made due to the scope of the incident. This is just human nature, and while there are no crystal balls out there that I am aware of that can foresee such incidents, a good risk management system can go a long way toward this end. However, in our society, it is just very difficult to request funding for something that you think might occur versus getting all the funding you need because we now have five dead troops! Again, this is just an inequity in human nature and this is how we typically and historically deal with such events. But changes after the fact do little to ease the pain of those families that have just lost someone.

Day 190: Thursday-May 14, 2009/Day 159-Iraq was GRD BUA day. SFC Maltes would be handling the duties this AM with me backing her up. I am not certain, but I do not believe the big guy is yet back from Washington. I watched the budget hearings for the Department of Defense in front of the House Armed Services Committee last evening **(Pentagon Channel, 13 MAY 09, 1700 GMT+3).** The hearing went pretty much the entire day and I felt sorry for Secretary Gates and Admiral Mullen as they had to navigate through all of the questions posed and the saber rattling from the politicians.

But they are old pros at this process and handled the situation with their normal calm and professionalism.

Well, I am sorry to state that I must go off once again on the systemic challenges facing the Army. It may seem that I picking on the Army throughout this book, but that is only because this is where my experience lies. I am not as familiar with the other armed services and I am sure they have their own unique challenges. I would also postulate that many of the issues found in the Army can be applied to all branches of the military, but perhaps not all. And it should be noted that the Army is the largest service by far which makes the situation more problematic. So I will begin once again with the tragic events that unfolded a few days ago when a soldier shot and killed five military personnel here on VBC. Details are beginning to emerge and once again it appears the system may have failed its soldiers. The soldier who did the shooting was on his third deployment. This is certainly not unusual given the fact that the war here has gone on over six years and the war in Afghanistan has gone on over seven. But mental health professionals know that "non-commissioned officers on their third and fourth deployments are more than twice as likely to have mental health problems as NCOs serving on their first deployment" **(Stars & Stripes, 13 MAY 09, *GI charged with murder of 5 after shooting in Iraq*, written by Steve Mraz & JASG, 13 MAY 09, *Soldier charged in comrades deaths*, source: AP)**. So this tells me that this information is out there, yet little was done to address it. In all fairness, I have reported that the Army is making organizational changes to shorten deployment time, increase dwell time at home, and grow the force in order to decrease the number of deployments. But all of this comes too late to help these five US troops, and it is too early to determine if these steps will even adequately address the situation at hand. But you would also think that knowing this information would have resulted in different steps being instituted to assist this troubled individual, which by all indications here on the ground did not occur. People around SGT John Russell (the shooter) knew he needed intervention well before the shooting, but these warning signs, as they often are, went unheeded **(Stars & Stripes, 14 MAY 09,** *Lives lost: Grieving a war zone tragedy*, **written by Geoff Ziezulewicz)**. When all the facts shake out in relation to this tragic incident it may be determined that there was simply nothing that could have been done differently to prevent such an occurrence. However, I believe otherwise, and I would venture to guess that there will be policy changes, new procedures put in place, and accelerated funding to assist troops with such mental health issues. But this is the typical reactive nature of the military as I have mentioned to you before. Greater proactivity and preventative measures are what should be occurring. But there is still that cultural stigma associated with mental health issues in the military that is an incredibly steep hill to climb and which there are no easy or quick fixes for. But the continued propensity for reacting as opposed to exercising sound risk management techniques is still a mystery to me.

In other news today, it now appears that there are some high ranking officials in Iraq that might formally request US forces to stay in cities such as Mosul and Baghdad past the June 30th deadline **(MNFI, 14 MAY 09:** *Khabaar*). Some people still believe the ISF are not yet ready to take on this mission by themselves in these more violent of cities in Iraq. This issue has gone back and forth with many comments from both US and Iraq officials. I guess we will know for sure on July 1st!

It now sounds as though President Obama may decide to stop the release of the detainee abuse photos I mentioned yesterday. His excuse for preventing the release of such photos is that he "strongly believes that the release of these photos would only serve the purpose of inflaming the theaters of war, jeopardizing US forces" **(MNFI, 14 MAY 09:** *Washington Post, AP, AFP*). He must have read my notes from yesterday! Good call, Mr. President. Stick that in your pipe and smoke it ACLU! There will be a better time and place to become transparent in relation to this situation that we already know about and which occurred over five years ago. It is not worth the life of one coalition member to release them at this moment in time. And if the ACLU does not like the decision, tough bananas! Perhaps if these same people that are belly-aching about it come over here and go out on patrol with our troops they would develop a much better appreciation why releasing the photos now is not a good idea. But I doubt that anyone from that organization will take me up on such an offer!

Finally today, the situation in Afghanistan continues to develop. Again, as I am in Iraq, this news will dominate this book, but I will occasionally take a peek to the east just to keep you abreast of the situation there. It does sound like the move to replace General McKiernan was the right call. LTGs McChrystal and Rodriguez appear incredibly capable and qualified for the situation in Afghanistan **(Stars & Stripes, 13 MAY 09,** *Petraeus: New US leaders in Afghanistan 'exceptional officers'*, **source: AP).** They have the necessary background and experience and will bring a fresh perspective to the deteriorating situation there. And when all is said and done, it may prove out that General McKiernan did nothing wrong while in command, he was just an old warrior that was trapped in a conventional warfare mindset. This happens very frequently and it may turn out that the reason he was replaced and General Odierno was not, was because General Odierno adapted to the situation here as I have highlighted to you before, and he has become very proficient in the type of fight we are engaged in whereas General McKiernan may not have been able to adapt.

It also appears that the Taliban is playing the wrong hand in Afghanistan and making similar mistakes to al-Qaeda in Iraq, which helped to turn the tide here. There extreme ideology and brutality toward the populace will turn this 'center of gravity' away from them. They had always been a little smarter than was al-Qaeda in Iraq in this regard, but this appears to be changing, and General Petraeus was very quick to identify this shift in Afghanistan. Stay tuned on this one as the tide may be changing to our east with a change in leadership and this strategic error on the part of the Taliban **(CENTCOM, 13 MAY 09, media highlights, source:** *CNN, UPI*).

Day 191: Friday-May 15, 2009/Day 160-Iraq was a day I decided I would be a little more unpredictable in nature. One of the things you learn quickly over here is that you want to vary your pattern so the enemy cannot predict your movements and habits as easily. Some of these variances are already built into the schedule with no BUA days or late BUA days, etc. But I decided to break the monotony of Groundhog Day and become less predictable by varying my pattern. So today I came in much earlier than normal for a Friday. I would then go to the Sports Oasis for lunch and dinner as opposed to the Café De Fleury. It is the little things that will break the monotony and drive any pursuing insurgent crazy!

I also want to get back to my report on the shootings of a few days ago here on VBC. More information is trickling out, and just to validate some of my claims on certain issues that I have informed you on, a mental health expert from the behavioral mental health department of the University of North Carolina states that military mental health services are getting better and they are diligently trying to improve such services, but the military culture still sees getting help as a sign of weakness **(Stars & Stripes, 14 MAY 09, *They broke him*, written by Steve Mraz).** As I have stated previously, this is the same type of culture found in the fire service that I am also very familiar with. There are no easy answers, nor any quick fixes. Putting more command emphasis upon this taboo topic and more funding toward it, as I have suggested, will assist in the process. But I will tell you that it will take effective leadership to adequately rectify this process, in addition to a serious cultural change within the military that unfortunately, I just do not see occurring anytime soon. I have heard too many comments from senior officers, and witnessed for myself, the nonchalant attitudes toward the 'softer' issues. This is very much like the suicide prevention challenges I mentioned several weeks ago. Comments such as "We don't have the time to add more training to the calendar," which translated into ineffective leadership-speak means "I really don't have the time or the desire to seriously attack this problem because the welfare of my soldiers is secondary to my mission!" Until the majority of military officers change their pathetic attitudes on these 'softer' issues it is unlikely to change very quickly, or at all. The problem is that such issues may be spoken of at the highest levels of government and within the military, but there is no oversight or accountability by supervisors to ensure meaningful action occurs. When

Generals start calling these officers on such issues and demand to see what they have done by making it a more performance-based assessment, and included such actions on their OER and tie it to promotional criteria, then we might start seeing a lot less lip service and a lot more action. This of course makes the large assumption that these same General Officers understand what is necessary to correct this significant problem! There also needs to be more training for officers and NCOs on these 'softer' issues in training curriculum at all levels, which there currently is not. The macho-image and testosterone-laden culture of the military is difficult to overcome and it will take years, patience, persistence, and a change in leadership attitudes to overcome such an obstacle. And you do not have to take my word for this, ask others outside of the military, and Dr. Goodale from the University of North Carolina just validated my theory on this. Furthermore, what a lot of senior leaders that I am around here at Al Faw palace do not seem to grasp in the least is that leadership is not about the rank on your uniform or how much lip service you provide to an issue. It is about action and taking the necessary steps, whatever they may entail, to adequately address and impact an issue. This takes strong and effective leadership and quite frankly, I have not witnessed this with any great regularity for sensitive issues that affect soldiers. If the topic deals with warfighting, or operational planning to defeat al-Qaeda, then these guys are all over that sort of challenge, which is another reason why the American military is as foreboding a force as it is. But conversely, when it comes to the 'softer,' more sensitive side of personnel issues, these same warriors are not nearly as adept at dealing with them.

In other news, while the US and Iraqi governments believe that al-Qaeda will continue to be degraded to a point where they can be effectively dealt with by ISF, the economy will play an important role in any timeline that is proposed **(MNFI, 15 MAY 09: *Christian Science Monitor*)**. I am not really certain if timelines and deadlines are a part of western culture or just an Americanism, but there are just some round pegs that simply do not fit into square holes! Politicians love timelines, especially when it furthers their own agenda or it is election time. Again, not sure this is what our founding fathers quite hand in mind, but it is what it is. But our commanders over here, including General Odierno who vigorously protested against any type of timeline, have always shied away from hard dates and prefer conditions-based criterion for making decisions **(CENTCOM, 13 MAY, media highlights, source: *Foreign Policy*)**. This concept simply has not flown in Washington, and much like Vietnam, our politicians may ultimately determine the fate of Iraq. Most people may not fully realize, especially those not old enough to recall the details of the Vietnam War, but the military did not lose that war. As stated by Anthony Cordesman, Center for Strategic and International Studies, our young men and women defeated the Viet Cong and forced them north to stop its offensives. We actually won the Vietnam War militarily speaking **(Stars & Stripes, 14 MAY 09, *If we don't 'hold and build' we'll lose Iraq*, written by Anthony Cordesman)**. His point is that if we rush to the exits in Iraq as most people want us to, without a coherent exit strategy based on conditions and atmospherics on the ground, then we could once again fail to learn from past history, as seems to be the American way, and Iraq could easily collapse back into chaos and sectarian violence. So the poor economic climate that has gripped the entire world, not just the United States, makes it difficult for Iraq to continue to build its security forces. And if they cannot do this, they will not be able to adequately secure the country from extremists by the end of 2011 when the US is scheduled to leave Iraq by security agreement. So while I have been impressed with our new President thus far and many of the decisions he has made, I think a defining moment in his current administration and his legacy will come by December 31, 2011. Our President often refers to Abraham Lincoln and the lessons of the past. We shall see if he has paid attention to what occurred in Vietnam. But unlike Vietnam, whose strategic importance is still debatable, Iraq is of significant importance and to adhere to a hard timeline when conditions on the ground suggest something entirely different, then adjustments should be pursued. Unfortunately, this book will be completed and hopefully published well before this deadline, so it will be up to you to see how this all plays out.

▌▌ ▌▌ Day 192: Saturday-May 16, 2009/Day 161-Iraq was another one of those days that I should have worn **'the stupid shirt'.** I am always telling the wife after dealing with people that simply cannot comprehend the logic behind anything, that their ignorance reminds me of the movie *'The Sixth Sense'* with Bruce Willis and Haley Joel Osment. But as opposed to seeing **'dead people',** I see **'stupid people'.** Yes, **these people** are everywhere! So my paramedic class of 2007-2008 got me a shirt that expresses just this thought that I wear on special occasions. As I have been watching the comments based on the President's decision not to release photos of detainee abuse that occurred over five years ago, I feel I must once again put **'the stupid shirt'** on! Now, I give the President great credit for making such a decision, and to me he is exhibiting the adaptability that is necessary in the White House. While he may have stated something entirely different while on the campaign trail, he now has much more information on which to base his decision **(MNFI, 16 MAY 09: CNN & JASG, 15 MAY 09,** *Abuse photos put US in 'Double Catch 22',* **written by Cal Perry).** Yet many **'stupid people'** want to hold him to whatever he stated months ago. I guess it makes for good political fodder for the Republicans, or good television commentary for those that oppose his administration. And many other politicians would have succumbed to such stupidity, and for whatever reason, they would have maintained their stance on a position just because! That is a tough way to try and run a country wouldn't you say? And here is what the President stated on the issue:

> I want to emphasize that these photos that were requested in this case are not particularly sensational. Any abuse of detainees is unacceptable. It is against our values. It endangers our security. It will not be tolerated. But the most direct consequence of releasing

the photos, I believe, would be to further inflame anti-American opinion and to put our troops in greater danger.

-MNFI, 15 MAY 09: *AP*

I would venture to guess that the people that agree with the releasing of such photos, even though it may endanger troops over here, are people that do not have a son or daughter in Iraq or Afghanistan. The **stupid shirt** is going on today!

Day 193: Sunday-May 17, 2009/Day 162-Iraq was a day to work on refining my EMS leadership book and finish my third assignment for Naval War College. I was hoping to finish refining my second leadership book by June so I can have one of my finest paramedics read it over to give it a sanity check, and then it would be off to the editor, and then hopefully by August/September it would be off to the publisher. This book that you are reading of course cannot be completed until I am off of active duty, which should be sometime in late November or early December. But I could do some preliminary editing and refining of the book on my own so it is close to being ready to go by the time my orders conclude. Then, on the trip to Australia, I will make final adjustments so I can get the book to the editor in January 2010 and hopefully to the publisher by February. This is plan anyway, and yes, I already cleared this with the wife!

Speaking of the wife, there was a very good editorial in *Stars & Stripes* recently concerning the inability of the Army to adequately support families of deployed members **(Stars & Stripes, 13 MAY 09,** *Army must better support extended family***, written by Kristina Kaufmann).** This is just another area that receives a lot of lip service but not a lot of action. Family readiness groups do not receive the funding they need, and the volunteers who support such groups are getting burnt out from all of the deployments. Now I will state that my unit, the 416th Theater Engineer Command, has done a good job of staying in touch with Shari. But because we live four hours away from the reserve center, it is difficult for her to attend any meetings or volunteer to assist. But whether these groups are from active duty or reserve units, the end result is the same. Perhaps no one knew these wars against terrorism (sorry Mr. President, but this is what they are no matter what name we place on it) would go on for this long, but again, where is the planning and vision that would have led us to conclude that this type of predicament would result? As I have mentioned to you previously, when it comes to planning for military operations the Army is all over it. But when it comes to 'softer' topics such as mental healthcare, or better supporting family readiness groups, these same people are very much lost in the fog of war! Perhaps the military's own unique culture is to blame, or perhaps it is because these topics are not intertwined into any leadership curriculum that I am aware of. But anticipating needs that result from war, especially protracted wars, is something the Army specifically, and the military in general, really need much work on. The military is big on the teachings of operational science and art, but it continually falters in understanding the finer nuances of the art of leadership. Without doubt the United States military machine is the greatest the world has seen to date, but it has many challenges it needs to address if it wants to remain on this lofty perch. And better taking care of its own people, who have done everything asked of them, should be a higher priority on the list of political and military leaders. I would conclude this by stating that the sacrifice of our military members and their families has been great. It is just unfortunate that this same degree of sacrifice is not being reciprocated by high ranking officials in both the government and the military. If this issue were reduced to a risk-benefit ratio, no insurance company in their proper mind would touch it!

Day 194: Monday-May 18, 2009/Day 163-Iraq was the halfway point of my Iraq adventure according to my calculations. We have two calculation devices that keep us abreast of our time served and time remaining here in Iraq. The more politically correct graph appears below. This is based on us departing near the end of October, the end date which appears in print in a FRAGO I recently observed. The other calculation device shows a good looking young woman in a bikini lying on a beach in the prone position. She displays a very fine caboose in this photo and this particular display is called a morale donut! Same data, just different graphics **(refer to Post Word)**. And while this young woman is nice to look at, she cannot hold a candle to my honey! Indeed honey!

How about some news for the day? There was an explosion reported by my boys at GRD in the IZ this AM. It does not appear that anyone was injured, but that is hitting close to home. There is still a lot of consternation concerning the withdrawal of troops from Iraqi cities by the end of June. Sadr City is right across the Tigris from the IZ and is still a hotbed of insurgent activity. One of the community leaders of this particular area commented that "This is the most dangerous decision being made. We will lose the security. The insurgents will come back. I will be the first one targeted" **(MNFI, 18 MAY 09: *Washington Post*)**. This is typically the tactic insurgents use when security leaves an area. They simply filter back into the neighborhoods just like cockroaches. A Company Commander in this area stated his feelings on leaving the area by saying, "it frustrates me that we have to leave. We're on the precipice. We can make this work. We just need a little more time. I've lost a lot of buddies here. I want to see it work" **(MNFI, 18 MAY 09: *Washington Post*)**. Remember what I said a few days ago about timelines? This is just another case in point. I do believe that there is some wiggle room in the agreement on both sides, but I am not sure the politics will allow US troops to remain in certain areas, and the side effect of keeping troops in the cities could be worse than adhering to terms of the agreement itself. Such is the fine balancing act taking place in Iraq. But it is readily apparent that some leaders in these communities want a US presence for a bit longer than the end of June, and the 'warrior spirit' is alive and well in our troops to see this thing through. Unfortunately, this level of decision-making is out of our hands. We shall now see what the politicians are made of in relation to this very sensitive issue.

There is still much clamoring going on about the President's decision to not release the detainee photos. Senator Lindsey Graham, a Republican from South Carolina, praised the President's decision by stating:

> Not releasing the detainee abuse photos says a lot about how President Obama makes decisions. He stood up to his political base and made a decision. Changing one's mind is a strength, not a weakness. He's realized the difference between being a candidate and being Commander in Chief.
> **-MNFI, 18 MAY 09: *LA Times***

It should be noted the Senator Graham is a military reservist and has served in both Iraq and Afghanistan. He understands the implications associated with the release of such photos. Now if he could just get the ACLU and their supporters to understand!

My friend Mike Girone always sends me very interesting articles to read. Yesterday he sent me an article called ***Burying Military Reputations*** by Ralph Peters. In this particular article Mr. Peters states that he supports Secretary Gates decision to change commanders in Afghanistan. But he is also very complimentary of General McKiernan, the commander being replaced. He further believes that General McKiernan was a conventional-style thinker that was thrust into a type of war he was not trained for, whereas LTG McChrystal is very adept at this type of warfare. Mr. Peters goes on to state that General McKiernan was given the wrong playbook for Afghanistan. The counterinsurgency manual that was used as a base document for progress in Iraq is not the right formula for Afghanistan. Mr. Peters states that General Petraeus quickly moved past the prescriptions of the counterinsurgent manual he helped to create while in Iraq, but this same success is unlikely to translate over to the Afghanistan theater of operation. Mr. Peters believes that if

General Petraeus allows LTGs McChrystal and Rodriguez to do what is necessary in Afghanistan, which may entail throwing away the counterinsurgent manual, then there is a chance of achieving success. But if General Petraeus does not allow them to operate as they deem necessary, then there may be problems and lessen the likelihood of success. Mr. Peters believes that General Petraeus is the key to the Afghanistan War as the CENTCOM Commander, and time will tell if he is flexible and nimble enough of mind to change gears and fight a different style of war than he did in Iraq. This situation gets back to the root of the Army style of leadership and its very culture. Most commanders could not change gears as is being suggested and as I have outlined to you previously. But General Petraeus is a cut above most officers as he has displayed on many occasions. It will be interesting to see if he can rise above the very culture that produced him!

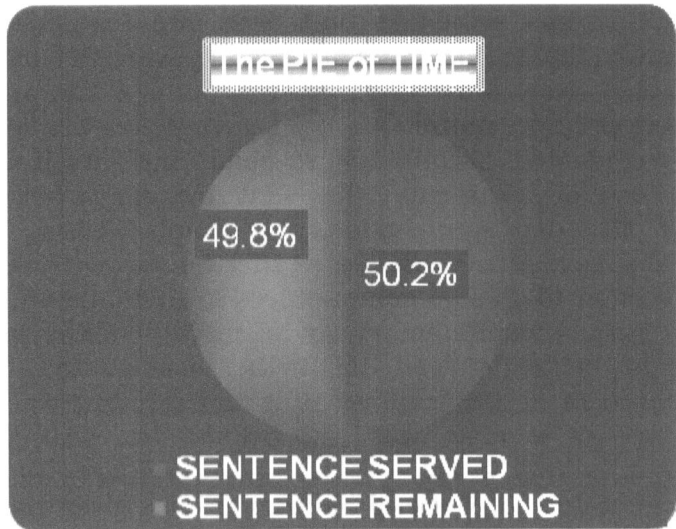

Day 195: Tuesday-May 19, 2009/Day 164-Iraq well a few days ago I began to have some teeth pain! Yes, I said teeth. Not a sharp pain, but a dull ache. I thought perhaps this was the same type of issue I had last summer when I was encouraged to try Sensodyne as a toothpaste. I thought this was simply a subtle way of saying that I was getting old! But the Sensodyne worked and the dull ache went away. Since I

have been deployed I have used Crest, but went back to Sensodyne a few days ago when the pain began. Unfortunately, it did not work this time so I knew something else was wrong. But like most normal human beings, I do not particularly care for trips to see the dentist. My dentist back home, Dr. Jerry Voelker, is a great guy and a good dentist, but that does not mean I enjoy seeing his bright, smiling face in his office! I mentioned my predicament to Shari who immediately asked if this was in the book! I told her it was not, but if I went to the dentist then it would find its way in. So here we are. Yesterday was miserable as it does hurt to eat and any pressure on the right upper molar in question is very painful. So I simply made my way to the dentist on base and got right in. This office was small, but quite high tech. I have never seen anything quite like this so it was a unique experience worth telling you about. A young Captain took me into a room and sat me in 'the chair.' You know, 'the chair' with the nasty drill, the suction tube, and the bright light! I described my current situation to him in detail and then he took his tools of torture and went to work. Actually, all he did was tap on about three teeth in the area in question with a metal instrument. When he hit the right tooth I gripped the chair with great vigor and he knew he had found the right location. His assistant then came in, stuck a device in my mouth, took a snapshot of the affected area with some other device, although I do not believe it was the normal radiation-type gizmo. Within 30 seconds the x-ray, or photo, or image, or whatever it is called appeared on a computer screen in front of me. The young Captain looked at it, and then used his computer mouse to gain greater detail on the tooth in question. I asked him what the verdict was and his response: some decay underneath the filling. Excellent! I have had root canals before and was hoping this was not the problem. For the most part I have pretty healthy teeth and have not had many oral health issues in my 52 years, except for the occasional foot-in-mouth problem that afflicts millions! As a young child I was struck in the mouth with one of those big, wooden swings at a beach park that knocked out one tooth, but that has been about it. My memories of military dentists are further tainted by an event about 31 years ago at Quantico, Virginia. As a young marine I had to have my wisdom teeth taken out. I can already hear Shari saying, well that it explains it! Anyway, I vividly recall two young Navy Lieutenants, one holding me in the chair since it did not come equipped with a safety belt, and the other literally standing on the arm of 'the chair' with a pair of pliers and using great force to extract the stubborn enamel nuggets. Now you understand my hesitation in visiting dentists! I do not have a fear or a phobia, just a healthy reluctance when it comes to visiting the dentist. Anyway, Friday I have an appointment to get the situation rectified, so it will be a long week of eating soft foods, liquids, and popping Advil. I will give you the outcome after my appointment.

Much news today beginning with the fact that Israel's Prime Minister Binyamin Netanyahu is in the US to discuss peace negotiations with President Obama **(CENTCOM, 18 MAY 09: *Al Jazeera, Al Arabiya, Khaleej Times*)**. It sounds as if Israel is serious about this and may finally agree upon a two-state solution. If it does happen on President Obama's watch, it will certainly be a huge feather in his cap. However, with this being stated, President Clinton worked very hard to achieve the same end but was not successful in the effort. The difference now may be because of the timing and the current atmospherics within Iran. As I have told you, Iran hates Israel and vice versa. Israel may believe that a two-state solution is more palatable at this point in time than a nuclear-capable Iran! So they are now using the US as the third party to assist with this possible peace accord, but also to pressure Iran on limiting their nuclear capability. Each situation would appear dependent upon the other for success. This situation could be extremely historic in nature, but also inherently important to Israel's well-being. This story most assuredly bears watching.

The other big news is that tensions continue to mount in the north here in Iraq **(MNFI, 19 MAY 09: *Washington Post* & MNFI and CENTCOM, 18 MAY 09: *NY Times*)**. I really am unsure how this will play out, but right now it does not look promising. The Kurds basically control three provinces in the north as I have reported to you. While they state that they want to be a part of the new Iraq, they have great distrust of the current Iraqi governmental leaders and do not believe their interests will be represented fairly. The Kurds have been adept at carving out their own infrastructure and economy, but I am really not certain a two-state solution will benefit either the Kurds or the rest of Iraq. But the Kurds are not recognizing the authority of the newest Arab governor in Ninawa and this is adding to the tension. Many of our own commanders are very worried about this situation escalating to the north and realize if violence breaks out there we will have a major problem. The US is attempting to play mediator at this point and find political solutions to ease the tensions, but this does not appear to be working too well. Prime Minister Maliki needs to step up in this situation because if this blows up, it will highlight his inability to adeptly negotiate peace between sectarian factions, and it will signal that Iraq is not nearly as stable as some may believe it to be, which will have great impact upon foreign investors. Prime Minister Maliki claims:

> We have political stability in Iraq, but we need to convince the other countries that Iraq is stable. Iraq can only be called stable when the political process is stable, thus companies risking their money in investments are investing in a stable country and will not be subjected to a political volatility.
> **-MNFI, 19 MAY 09: *al-Iraqiya***

The tension in the north is apparently also fueled by some disputed territory and this situation also bears watching closely.

Finally this busy day, it appears that the national elections will be held on January 30th, 2010 **(MNFI, 19 MAY 09: *AP, Reuters, AFP*)**. These elections will be huge in regard to the future of this country. This in fact may signal in which direction this country will head. If one party dominates these elections it may lead to sectarian division and progress may come to a screeching halt. If there is a good mixture of representation, this will signal political progress and a positive sign for the country's future. The good news is that I should be home about two and half months by the time the elections take place. But the elections in Iran are also coming up this summer and they will be extremely significant to Iraq, Israel, and the United States. All of these variables are why the year 2009 is being called one of great transition, not only in Iraq, but throughout the Middle East. This year could be looked back upon as one of the most pivotal in this war when all is said and done. You will need to judge for yourself.

Day 196: Wednesday-May 20, 2009/Day 165-Iraq began by watching Secretary of Defense Robert Gates on 60 Minutes **(AFN, 20 MAY 09, tape delay of 60 Minutes show of 17 MAY 09)**. Katie Couric interviewed him while on his recent trip to Afghanistan. Secretary Gates is very low key, but knows how to get things done. He has challenged conventional military thinking in order to provide troops with the resources they need today versus the weapons systems that might be effective for the notional war of tomorrow. Many politicians would not challenge the US military machine on such an issue, but he has. He has also recently relieved a good commander in Afghanistan simply because it needed to be done to change the mindset and direction of that war. This type of change is also rarely enacted at such a high level in the midst of a war and halfway through a good commander's tour of duty.

Secretary Gates feels the loss of each US military member on his watch, and he takes this very personally. He visits Arlington cemetery frequently. He is extremely hesitant to draw comparisons to the two Presidents he has served under in his current position, even though reporters push him to do so. He is not the typical Washingtonian bureaucrat, and avoids the mainstream circles. This is probably why I like this guy! In fact, he doesn't even like Washington that much. When asked about his job, because of all the pressures associated with it, and the continued loss of troops in two wars, he told Katie Couric that he feels it his duty to serve his country, but that he really does not like his job that much. It was a very impressive and insightful interview in my opinion, and I would further add that if we had more guys like Robert Gates in Washington our country would be a lot further ahead than it currently is. I hope he remains in office for a long time.

Well, today was send home footlocker number 2 day! The first arrived safely a few weeks ago, and I will inspect it and the contents inside for damage when I return home on R&R in July. But I again followed the same MO (modus operandi) as I did when I mailed the first footlocker and got to the post office when it opened. I was once again the first one there and got in and out in less than 30 minutes. The process went very smoothly and I must state, in all fairness, that I believe most of the workers at the post office are employees of the much maligned KBR organization. I have spewed enough venom their way concerning their fire protection and electrical code shortcomings that I now need to spread a little love in their direction on this one. I definitely maximized my load plan this time around. As I may have reported to you previously, 70 pounds is the maximum weight you can send in one box. This particular footlocker is large and already oversized, which will cost more to ship, but ironically it weighed more than the last one I sent home and cost less to send! The woman at the counter told me that the rates for oversized boxes just went down. I guess the cost of stamps went up, but the costs of some other things were reduced! The postal gods work in mysterious ways! Anyway, I knew this box would be close on the 70 pounds. In fact, when I placed it on the scale it read a hefty 75! So I took out a few items and got it down to 71 pounds. One more adjustment and it weighed in at a solid 69 pounds. So off it goes and it should be waiting for me upon my arrival home on R&R in July. I will probably send a box of my books home in two weeks and one more footlocker before I go home on leave. If I time this movement of goods correctly, all of these items should either be at home before I arrive or while I am on leave. That way I can check everything over and file any damage claims at that time if necessary to do so.

 I also hit 4868 on Solitaire on my little hand-held pocket computer today, which just cracked the all-time top ten scoring list so I was off to a very good start this day, which was expected to climb to a toasty 102 degrees today! Now I believe I have already told you that I like the heat. I have been to Phoenix on several occasions and prefer the dry heat over the more humid east coast weather. But I also have never visited Phoenix in the summer when it gets to 115-120 degrees. And while it is a 'dry' heat here, the mercury is rising a little each day and is expected to reach 108 the next four days. The Ugandan guards have a little digital thermometer at their entry point leading to the palace and it read 46.2 degrees Celsius yesterday at about 1600 (4:00 pm Baghdad time for you civilians) while sitting in the sun. Doing the conversion, this equates to about 115 degrees Fahrenheit, which is about how warm it actually felt. No matter how you cut it, that is **HOT**. Last year in July/August I think it peaked at about 130 degrees here. We have all heard the saying 'hotter than hell,' but since I have never been there nor do I have any plans to ever visit the place, I do not know how hot that actually is. But on a visit to Phoenix in 2000 with my daughter Lexie, it got unseasonably warm the latter part of April that particular year and the temperature reached into triple digits on a few days. I asked the manager at the hotel what it was like during the summer in Phoenix and he told me that if you turn on your oven and crank up the heat, then stick your head into the oven for a few moments, you will get the idea! OK, that is indeed **HOT**. But I can report to you that at 115 in the sun here right now, it reminds me a lot of grilling brats at my home back in Wisconsin, when you get the Charbroil Infrared Gas Grill nice and hot, and then you lift the cover to put your boiled in beer brats on the grill, and you are rudely met with that initial blast of heat that knocks you back on your heels for a moment. That is what this weather reminds me of, except it is not a momentary blast here! I am having a difficult time comprehending it becoming another 30 degrees warmer in July or August! No wonder why everyone moves slower over here. If you do anything quickly you are drenched in sweat!

The Coalition Café reopened yesterday after being closed for six weeks for renovations. I scanned the online menu for this DFAC and did not spot Pop's Burgers on the Wednesday menu! Boy I hope these did not go the way of the dinosaur, but just as a few other little morale boosters around here have vanished since I arrived, this may be the next casualty.

I will also begin to add a new feature concerning the rise in temperature over here as it becomes much hotter. This little **HOT**-meter will appear occasionally to keep you apprised of the rise in temperature as we now approach the oppressive heat of summer. The **HOT**-meter reading will include the forecasted temperature followed by the temperature reading on the digital thermometer at the guard shack leading into the palace. This digital thermometer sits out in the open so the temp I report will normally be at about 1400 when I return to work from my work out and lunch each day, and when the thermometer is typically sitting in the sun.

102/115

▌▌▌ ▌▌▌ ▌▌▌ ▌▌▌ **Day 197: Thursday-May 21, 2009/Day 166-Iraq** began with more talk of democracy. As I have reported to you, there is a great deal of corruption in Iraq. It is at a level that is not socially acceptable so there have been inquiries throughout the ministries in an attempt to cut back on the high incidence of corruption **(MNFI, 21 MAY: *McClatchy*)**. Many people view this as another sign of progress and democracy at work.

I suspect that there will be something that will occur with Iran in the near future that will either prove to be extremely positive or negative. There is simply too much saber rattling going on right now to believe otherwise. The Pentagon recently denounced accusations by Iran's Supreme Leader Ayatollah Ali Khamenei that the US is training terrorists in Iraqi Kurdistan. The Pentagon shot back stating that this type of statement was grossly inaccurate and damaging to future US-Iran relations, but the US has proof that Iran continues to train insurgents and supply weapons that surface in Iraq **(MNFI, 21 MAY 09: *AP, Reuters, AFP, VOA*)**. With the Iranian elections looming on the horizon next month, and the fact that the Israeli Prime Minister just visited Washington because he is more worried about Iran's nuclear capability than a two-state solution, all of these variables tell my gut that an ill-wind is blowing across the hot desert and something big is going to happen. And after watching the Pentagon Channel last night and Secretary Gates testifying before the House Armed

Services Committee about the military budget, it was clearly evident that we need to be prepared to deal with Iran on a military level **(Pentagon Channel, 20 MAY, 2000 GMT+3)**. This situation will bear close monitoring because it carries with it very serious consequences.

Finally today, there is still some fall out in relation to the release, or lack thereof, in regard to the detainee photos. Senators Lieberman and Graham are introducing legislation to establish an official procedure for the blocking of the release of such photos that could possibly endanger the lives of US personnel serving abroad **(MNFI, 21 MAY 09: *The Weekly Standard*)**. Senator Lieberman went on to state:

> President Barrack Obama made a bold decision that will protect our troops in Iraq, Afghanistan, and elsewhere and make it easier for them to carry out the missions that we have asked them to do.
> **-MNFI, 21 MAY 09: *The Weekly Standard***

This seems like a good move and a necessary one. I am certain the ACLU and their supporters will not appreciate this legislation, but again, chances are none of these people have a child or spouse serving in Iraq or Afghanistan and do not comprehend the ramifications of such a release. This just goes to show you that when people really have no clue about their actions, there are those times when others must provide that clue to them whether they like it or not! Time to go put on **'the stupid shirt'**!

106/122

Day 198: Friday-May 22, 2009/Day 167-Iraq was stepson Josh' 23rd birthday. Josh is studying to be a chef at Le Cordon Bleu in Minneapolis. Josh is the rock star of the family. He is very laid-back, all tatted up, has long hair, and is a very cool dude indeed. He would soon be graduating school and looking for a job as a chef. He would also be marrying his long-time girlfriend Kelly sometime next year, at our house if you recall! In return, Shari and I will make him cook for us on a few occasions to see what he has learned.

Today was also a day that I have both looked forward to and dreaded at the same time. Today is visit the dentist day! Yes, I have been eating strictly using my left side teeth, which has not made the right side too happy, except for the fact that I have no pain there when I do not use this side. Der! Hopefully today, the dental techniques will match the technology I observed on Monday and this will be a quick and relatively painless procedure. At least I hope so. I have a pretty high pain tolerance, but I still do not enjoy this type of activity. I would rather watch Jeff Gordon keep making left hand turns all day than go to the dentist (sorry honey!).

My buddies from KBR were back in the news yesterday **(Stars & Stripes, 21 MAY 09, *KBR got $83M in bonuses despite woes*, written by Kimberly Hefling)**. There are questions being raised in the halls of Congress as to how this could have happened! Please! It is the system in which it is allowed to operate. This is bureaucracy at its finest. Communication is slow and the system cumbersome. I have written papers on the topic of organizational design and mentioned it in this book on several occasions, but our government and military really need to look at a design change. These kinds of issues continue to occur with greater frequency because the archaic system we have in place in a fast-paced world is simply outdated. Why can no one else see this? And I would state that Iraq would probably not fare well with more of an adhocracy at this point in its evolvement, because a bureaucratic structure is a better fit for a fledgling democracy which needs a slower pace and greater structure. But for a more sophisticated democracy that has been in place for decades, that we supposedly are, a change is more likely to be successful. What do they have to lose? And oh by the way, I told you previously that KBR also is the contract service for fire protection over here, and after all of their 51 fire department locations were inspected, only 18 passed **(MNFI, 21 MAY 09: *TF SAFE*)**! Again, readiness levels, training skills, and equipment are all inspected during this evaluation. The average score for all 51 was a meager 78.7%! Again, as a fire chief myself, I find all of this very disturbing. Congress is now aware of the situation and moving slowly to rectify the issue, if they ever can! I must say it again – bureaucracy at its zenith of incompetence!

Now let us turn to another of my pet peeves. A recent study of federal workers revealed that management style is more vital to workers than are pay and benefits **(Stars & Stripes, 21 MAY 09,** *Survey: Management style vital to federal workers,* **written by Steve Vogel)**. After reading the article, I believe it was a mistake made by the article's author to confuse management with leadership. The two are not one in the same. The article goes on to identify leadership as being very important to workers, as well as communication within an organization. These are two of the primary components of a **ledocracy**! Once again however, these kinds of surveys and the exact same findings have been out and published for many years. Remember what I said about repeating history? It all boils down to effective leadership, and as I have informed you of previously, I have not seen a lot of it here. And because the Army specifically, and the military in general, is a big albatross which is slow to move and change direction, poor organizational design is still simply an excuse to mask ineffective leadership. I would even go on to state that there is more leadership found currently at the junior levels than in more senior positions. Younger officers and NCOs grew up in an era that required quicker thinking in a faster-paced society. These are also the same people out in the streets in Iraq and Afghanistan who need to make good, sound, and rapid decisions. It seems to me that those senior officers more removed from the fight, given the design of the organization in which they operate, are not following suit. So then one must ask him/herself, is the real problem the individuals involved, or the system in which they operate? I suggest it is both, but more a byproduct of an antiquated system. Real leaders work the system and find avenues to get things done in a timely manner. Conversely, those who are not effective leaders, but carry the rank, hide behind the rules and regulations and offer them as excuses as to why something cannot get done. I truly believe it is time for a change. But I will say I have seen some glimmers of hope as the Marine Corps empowers their junior NCOs, calling them 'Strategic Corporals,' and the Army allows young Captains and NCOs to make decisions on the ground whose consequences can carry serious strategic implications. These kinds of techniques simply need to be utilized on a much broader scale throughout the military and government. If we really want to stay ahead of our adversaries in the 21st Century, we need to make this organizational and cultural shift. There are many people who believe our government had the intelligence to identify the fact that we would suffer a catastrophic event on 9/11 of 2001. The problem was that this information never made it to the right people with decision-making authority. This same theory has been mentioned in relation to Pearl Harbor. So is it our intelligence community that failed us in both life-altering events, or is it the system in which they exist?

Happy Birthday Josh

Day 199: Saturday-May 23, 2009/Day 168-Iraq was my in-laws' 50th wedding anniversary. Dick and Kathy Kertis are great in-laws and I am sorry I would be missing this special day.

I also began with little pain in the tooth in question. Yes, my visit to the dentist yesterday went quite well. As I walked the **'Green Mile'** from my vehicle to the dentist's office in 110 degree heat, I began to recall all the sounds and smells of past visits. I recall the pain associated with the injecting of a needle into the gums (just a little pinch!) to numb the area and then the smell of burning enamel wafting into your nostrils as your jaw vibrates violently from the dentist's drill like a jackhammer on concrete. Well, I guess dentistry has come a long way indeed! I had to ask this young Captain named Walters if he even used a needle to numb the area! Indeed he did, but I felt nothing, not even a pinch. He took about an hour to replace the filling and even the material used these days is much different than that of yesteryear. Captain Walters explained that the old filling material was more like concrete and served dentists and their patients very well for years. But like concrete, it does deteriorate after time. The new stuff is more like glass and adheres to enamel much better, is much stronger, and will last longer. The filling is likely to outlast me! So he and his assistant did a good job, and of the four appointments they had scheduled yesterday, I was the only one to show! Perhaps the others were paralyzed by the memories of past dental experiences and could not muster up enough courage to walk the **'Green Mile'**! And while I will not state that I enjoyed the experience, or look forward to my next visit, it was certainly the most enjoyable visit I have ever had to a dentist's office.

The Long Walk

Light on the news today, but old KBR is of course denying the claims against them **(Stars & Stripes, 22 MAY 09, *KBR chief defends company's electrical work done in Iraq*, source: AP).** Since their contracts are worth millions of dollars, I am not surprised. Now, I would say that there is a modicum of truth in their rebuttal. I had heard that in regard to their questionable electrical work that they were not being held to the US electrical code standards, but rather the British code which more closely resembles that of Iraq. However, the fact that the buildings in question were never properly grounded, a basic premise of good and safe electrical work no matter what code you might be using, directly contributed to the electrocutions. I have also seen the reports and statistics firsthand, and although more of my experience lies in that of a firefighter and fire investigator and not an electrician, KBRs story will boil down to the fact that they were paid a lot of money and provided very shoddy work. This may be a result of subcontracting the work with little to no oversight. But the fact of the matter is that this American-based private contracting organization failed to watch out for the safety of the military members in Iraq. To me, this type of inept leadership found within this company is criminal in nature, and I hope all of their practices are examined with great scrutiny. End of story.

Taking a peek east for a moment, Admiral Mike Mullen, Chairman of the Joint Chiefs of Staff, states:

> The war in Iraq has taught us things about counterinsurgency warfare we might never have discovered otherwise. We will be smarter now in Afghanistan and more successful, not in spite of Iraq, but because of it.
>
> **-MNFI, 23 MAY 09: *KUNA***

While I would agree in part with the Admiral that many lessons of success are transferable to Afghanistan, there are also many nuances between the two countries that attempting to simply cut and past the Iraq strategy and apply it to Afghanistan would be a real strategic mistake. But with LTGs McChrystal and Rodriguez taking command, both of whom have the background and experience in that theater of war, this is unlikely to occur. The only way such a grave error will manifest itself in a mutant form would be if Washington tries to force these new commanders to fight the war differently than they see necessary. And yes, this happened in Vietnam and it happened early in this war. Politicians need to stick to diplomacy and the military to what they do best—fight and win wars. These two instruments of power should never cross over into the others realm, but they must mesh together and work closely with one another in order to achieve success while limiting the unnecessary loss of life and wasteful spending. We shall see if America learned anything.

Secretary Gates, along with pretty much everyone else in Washington, and the military for that matter, really have no idea if the new strategy and new leadership in Afghanistan will yield any progress. Many experts believe these moves will lead to slow progress, but many other experts do not feel such progress will be sustainable or long lasting. So indeed, this whole thing is one big crap shoot. Because of these variables, Secretary Gates is calling for a review on the tactics that will be employed sometime next year **(Stars & Stripes, 22 MAY 09, *Gates: Afghanistan tactics should be reviewed next year*, written by Anne Gearan).** This seems to be a prudent move, but the truth needs to come out, and unlike the early years in the Iraq War where many senior officials painted a rosy picture of a failing situation, this lack of candor only leads to more loss of life and wasteful spending. So a realistic assessment is what is necessary. Only then can critical decisions be made, including leaving the country altogether. However, in so doing, this may lead to future dire consequences. The strategy being employed in Afghanistan is a much higher stakes poker game than anything you will ever see in Vegas on ESPN 2!

Happy 50th Dick & Kathy

Day 200: Sunday-May 24, 2009/Day 169-Iraq was the official halfway point in my deployment and it began with some Naval War College studying. I was reviewing a very interesting section called the Failed State Index **(www.fundforpeace.org)**. This particular index is conducted by the Fund for Peace group using a Conflict Assessment System Tool (CAST) using 12 social, economic, political, and military indicators; 5 core state institutions including the military, police, civil service, system of justice, and leadership; and then sprinkling in the idiosyncrasies of each country. In 2008, this group rated 177 states. The higher the score - the worse the conditions within the state. To provide a little more perspective, the scale is as follows: 90-120 Alert mode; 60-89.9 Warning mode; 30-59.9 Monitoring mode; and 0-29.9 Sustainable mode.

So if you had to guess what country was the number one failed state in this index what would your guess be? If you guessed Iraq, you would be close, but incorrect. Think of pirates, and I am not referring to Captain Jack Sparrow! Yes, it is Somalia with a score of 114.2! But let me also tell you that Iraq has improved from the previous year, but is still number 5 with a score of 110.6; Afghanistan is better believe it or not at number 7 with a 105.4; and Pakistan is still better at number 9 with a 103.8. So to put this into perspective for you, our country is intimately involved with three countries in the top ten! Talk about a huge challenge and a bad draw!

By comparison, mirror, mirror on the wall, what country would you guess is number 177, or the most stable of them all? If you guessed the United States, you are absolutely incorrect! In fact, we came in at number 161 with a score of 32.8. Believe or not, Norway was the best at number 177 with a score of 16.8, followed closely by her Scandinavian sisters of Finland (18.4) and Sweden (19.8). And our friends to the north, Canada, ranks better at number 167 with a score of 26.3. I find this all very interesting. So if there are 16 countries ranked higher than the US, where are all these countries in fighting the war on terrorism or helping humanity?

Mirror, mirror on the wall, who is the greatest failed state of them all?

This Memorial Day weekend has been very melancholy thus far. I started off by watching another of Sean's films that he left behind. I sure hope he did not pay much for these things, or he has a weird taste in cinema, but the movie **Southland Tales** with The Rock, who now prefers to go by the name of Dwayne Johnson, also starring Sarah Michelle Gellar and Seann William Scott, was one of the most bizarre movies I have ever seen. Apparently it is classified as sci-fi drama. It was indeed a stinker! Sorry Rock...er I mean Dwayne!

But then I watched perhaps the saddest film I have ever seen, but also one of the best. I am not afraid to tell you that there have been movies that I have watched in the past that brought a tear to my eye and a lump in my throat. You know, the **Old Yeller** or **Ghost** kinds of movies. But not only did I tear up on several occasions while watching this movie, when the movie was over I actually cried and I could not stop for a few moments. I really do not know the reasons why it affected me as it did, but it had a very profound impact upon me. And what makes this even more amazing is the fact that I was working on this book as I watched it, so it did not even have my complete and undivided attention as the others where I became misty eyed had. This movie was released in February 2009 and stars Kevin Bacon. It was shown today on the Armed Forces Network right after **We Were Soldiers**, which also brought a tear to my eye even though I have seen this movie several times and is one of my favorites. But the movie I speak of is called **Taking Chance**. It was about a Marine Corps Lieutenant Colonel who had the honor of escorting a young 19-year old marine home to Wyoming after he had been killed in action just outside of Baghdad in April 2004. The entire movie is about this young hero's journey home and the people he affected along the way. It is deeply moving and an incredible film that is based upon a true story. This same film could have been about any of the 4299 (as of today) soldiers, marines, sailors, or airmen and it would have had about the same impact. And you need not be a member of the military to be moved by this film. You need not even have to be an American, simply a human being that understands the high cost of war. There have been numerous civilians and many other military members from other countries around the world that understand the high cost of war and they would appreciate this film as well. I may never really understand the reasons why this film moved me so deeply, and I never met or knew the young 19-year-old Chance Phelps, but I know I will never forget his story and will never forget his sacrifice to this country.

We must never forget the sacrifice

CHAPTER 12
400 DAYS - Days 201-220
May 25 - June 13, 2009

US Presidential candidate Barack Obama talks about 16 months. That, we think, would be the right timeframe for a withdrawal, with the possibility of slight changes.
-- Iraqi Prime Minister Nouri al-Maliki
2006-present

Day 201: Monday-May 25, 2009/Day 170-Iraq was Memorial Day in Iraq. It was business as usual on this holiday for most Americans not engaged in war. There will be parties, cookouts, family get-togethers, and many other activities across America this day. While Americans enjoy this day, others will be standing watch in Iraq and Afghanistan, and other areas around the globe, to ensure that our families can celebrate this day without fear of attack from a foreign invader. Today is a day when all Americans should remember the sacrifices of the many, both past and present, for it is due to those who have served and made the ultimate sacrifice that we can celebrate such a holiday in a land of freedom. This same argument holds true for the families of those military members who have served or are currently serving, and especially for those who have lost a loved one in the course of protecting our freedoms. People who do not believe in war or its purpose are idealists and I only wish their dream for peace throughout the world could become reality. But in the years that man has roamed the earth, he has never been at peace with his fellow man, and it is unlikely that this will ever occur.

While few people enjoy war as it depletes resources and robs life from men and women in uniform well before their time on earth should be finished, it is unfortunately a necessary evil. Because these idealists do not realize that without our military and the wars it must wage, there are other people around the world who seek to do everything in their power to destroy our way of life. Some may do this because of jealousy of our freedoms we enjoy. Others may do it under the guise of religious fanaticism. But whatever the cause, without our brave men and women in uniform, the rights and freedoms that millions of Americans enjoy this day may otherwise just be wishful thinking. So I hope the people who have never served their country, or somehow never been affected by any war, take time from their festivities to observe a moment of silence for those who have served, those who have fallen, and their families that have supported their efforts. This is what Memorial Day should be all about.

> We cherish, too, the poppy red,
> that grows on fields where valor led.
> It seems to signal to the skies,
> that blood of heroes never dies.
> **-Moina Michael, 1915**

This Memorial Day ended on an incredibly sad note. MG Eyre, our Commanding General for GRD, was on his way back from visiting a water treatment plant in Fallujah this afternoon when his convoy was hit by an IED. Three people were killed (one coalition force member and two civilians). The coalition member was one of GRD's own. Two security personnel were also wounded in this attack. MG Eyre was unhurt. I will report more on this as details become available in the days ahead. Please keep the families of those who died on this Memorial Day in your prayers.

Day 202: Tuesday-May 26, 2009/Day 171-Iraq began with an interesting article forwarded to us by our operations director. The story was about General Odierno, the hero and was written by Ralph Peters, a retired Army Lieutenant Colonel, who is a freelance writer. It is interesting to note how people put a different spin on the same story. Now I have already highlighted why General Odierno is a case study in evolvement and will be remembered in American history as one its greatest Generals. I have also outlined what another writer has mentioned about the heavy-handed tactics he was used while Commander of the 4th Infantry Division early in the war. Many people viewed these tactics as actually fueling the insurgency as opposed to quelling it. But General O was not alone in this more aggressive tactic, while Major General Petraeus (in 2003) successfully used a classic counterinsurgent philosophy in Mosul during this same period as Commander of the 101st Airborne Division. Mr. Peters stated that he wrote his article to correct a grave error. In his opinion, General Odierno showed the insurgents who was boss and he was correct in his application of aggressive tactics. Mr. Peters then takes a swipe at the tactics he called 'soft' that were applied by General Petraeus and then he goes on to blame these tactics for the current situation in areas where it was used. I find all of this very interesting, and this is a topic even the most prolific experts on the Iraqi War will disagree upon. And while I do not possess the expertise to sway the argument in either direction, I can make a few observations.

Mr. Peters is a former infantry officer. He is a product of the US Army system, which system's glaring deficiencies I have highlighted throughout this book. Most infantrymen would believe that you should always be a hard-charger and that kicking ass and taking names later is always the best strategy! Mr. Peters also retired from the military before the events of 9/11 and joined the military after I did, so he missed the Vietnam experience as well. He also never attended War College **(http://en.wikipedia.org/wiki/Ralph_Peters)**. So his expertise to make such claims is questionable at best. But even with these facts thrown on the table, I am not saying his opinion is incorrect, and I am sure it is one shared by many. But if his opinion is correct, it is interesting to note that

General Odierno has evolved and changed his tactics during the course of this war. He has also been the MNC-I Commander here in charge of all military operations, and is now the MNF-I Commander in charge of all operations within Iraq. In these positions he has not stopped kicking insurgent ass, but he has stopped using the same tactics that he used early in the war. In fact, the tactics he is employing now are similar to the ones used by General Petraeus back in Mosul in 2003! So Mr. Peters' argument is simply not holding up to closer scrutiny in my mind. This is why examining every possible angle on a topic is important before drawing conclusions. And as someone who is in Iraq, hears the comments from General Odierno every day, and observes the actions taking place on the ground throughout Iraq, none of this supports Mr. Peters viewpoint, other than General Odierno being a great American and an outstanding officer.

I have also just begun to read a book written by Lieutenant General Ricardo Sanchez, who was the first coalition force commander in the Iraq War, called **Wiser In Battle**. He has been chastised in almost every book I have read as being very inexperienced when he was named the commander here, and he was put in place over more senior generals to be a governmental puppet for the Bush administration. He has been blamed on many accounts for his actions, or inactions, during the first 15 months of this war, including the Abu Ghraib detainee abuse scandal. I believe LTG Sanchez has written this book to provide his perspective. I am only about 25 pages into the book, but already he disputes the allegations against him and basically blames the Bush administration for the grave errors made early in this war. At least this part of the story appears to be consistent on all accounts thus far. But LTG Sanchez also blames the Bush administration for having no Phase IV to the war plan, which is basically what actions are to be taken after combat operations cease. Now it has been documented by other sources that the CENTCOM Commander, General Tommy Franks, was responsible for the creation of such a plan, one that never existed. General Franks even states that his mission was simply to remove Saddam Hussein from power and occupy Baghdad. And this he did successfully. So one would have to dig deeper to determine if such a post combat operations plan is the military's responsibility, or that of the government. From the training I have had, it is the military's responsibility to produce such a plan based on the guidance from key government officials, such as the President, Secretary of State, Secretary of Defense, the National Security Advisor, etc. So there seems to be plenty of blame to pass around on this grave strategic error that certainly was paid for through the lives of American troops. I would conclude from all of this that the highest-ranking officials of this country, both in the military and the government, fumbled the ball on this one.

▌▌ ▌▌ Day 203: Wednesday-May 27, 2009/Day 172-Iraq was a no BUA day and since I rambled on so much yesterday, I will catch you up on a few notes. I will first state that Iraq is a very interesting place. At one point it probably was a very exotic place to visit, and perhaps can be again someday. But years of repressive rule have hurt this country to the point that they may not know peace and political compromise without violence for many years to come, if at all. And from my observations and in my humble opinion, this country's posture will still be very fragile when we depart at the end of 2011, if this indeed occurs as scheduled. That is just the way it is here. Fragile and reversible are terms that have been used frequently by military leaders to inform US government officials of the conditions on the ground in Iraq. These terms will be used for years to come when describing Iraq's situation. And while little signs of progress provide hope for a brighter future, it is the incidents that happen every few days here, and which claimed the life of one of GRDs own yesterday, that tell you that it is a constant struggle and seems like for every step forward, it is one step back. An Iraqi lawmaker talked of sectarian violence that could tear this country apart if it becomes widespread:

> The Shi'a have limited patience, but by not responding to recent sectarian bombings they have proved that they want to prevent Iraq from getting dragged back to the sectarian conflict. If this does happen, it would

burn everything. It would take the political process back to zero.
-MNFI, 26 MAY 09: *Washington Times*

This statement reeks of fragility and the patience for not retaliating could end at any time. And in order to move forward the Iraqi government needs to accomplish several tasks, such as addressing the widespread corruption which undermines the legitimacy of the government in the eyes of the people; strengthen the rule of law process by building more prisons, as well as educating, retaining, and protecting judges and attorneys to prosecute and punish criminals and terrorists; continue to rebuild Iraq's infrastructure through better electrical service, potable water capability, sewage removal and treatment, the building of educational institutions and medical facilities; and better securing the populace from violence. These are all tall challenges, and they will not be adequately addressed by the end of 2011. But baby steps are occurring and Prime Minister Maliki is stepping up efforts to address corruption as he just accepted the resignation of the Minister of Trade **(MNFI, 26 MAY 09:** *AP***)**. This is a start in this forward movement toward legitimacy, but there is a long road ahead.

My good-looking bride also got braces today. I saw them during our Skype session and they are clear on top and of the heavy metal variety on the bottom. She has such a big, beautiful smile already that it will be hard to improve upon perfection, but she feels it necessary. I believe she said she needs them on for a year. And although my teeth could probably use braces, I am already 52 years old and a bit more practical as I do not like anything foreign on my teeth unless I put it there! We have already documented my experience with dentists!

Day 204: Thursday-May 28, 2009/Day 173-Iraq was once again GRD BUA day. I would again be the backup at the palace. I would have two more such briefings until R&R. Then SFC Maltes was going to be on her own. She would be taking her R&R after I returned and then I would be the primary briefer until she got back. So we just switched off on the duties and each had our part to accomplish in this process. If my calculations were correct, I only had about eight more of these briefings left to do!

I also had to visit my friend Mr. Barber yesterday. I am not sure if they are getting better at cutting hair or I am just getting used to them, but my barber did a good job again yesterday. Looks like my bald spot continues to grow, but such is life and the aging process. Life would probably be easier without hair anyway! Then you could save on shampoo, hairbrushes, and haircuts! It wouldn't get in your face, no need to worry about bed-head hair in the morning, and it wouldn't fall into your food! Oh yes, life without hair. I'm just preparing mentality for that eventuality in case you haven't picked up on that yet!

Not much in the way of media news today, but we lost another soldier last evening with three more wounded. The final statistics are not yet in for May 2009, but it will have been a bad month, the worst of this year, and the most since I have been here. This includes GRD Navy Commander Duane G. Wolfe. Commander Wolfe, 54, was a reservist from Los Osos, California and only had five more weeks left in his tour of duty here. Commander Wolfe worked at Vandenberg Air Force Base as a civilian deputy commander for the 30th Space Wing Mission Support Group. CDR Wolfe was the Commander for the Gulf Region's Central District office at Al Anbar. Commander Wolfe leaves behind a wife and three children.

One of the others killed in Monday's explosion was a member of the Department of State, Terry Barnich. Terry, 56, was from Chicago and I had just met him at our Engineer Summit last month as I was tasked to escort him into the palace. He was one of the speakers at this conference as the Deputy Director for the Iraqi Transition Assistance Office (ITAO). Terry was scheduled to return home at the end of June.

The third individual killed in Monday's attack was Dr. Maged Hussein. Dr. Hussein, 43, hailed from Egypt and worked for the United States Army Corps of Engineers in the Jacksonville district office. Dr. Hussein leaves behind a wife and 5-year old daughter.

Rest in peace, you who have tried to make a difference here. May you find your reward in heaven and may your extreme sacrifice never be forgotten.

▌▐ ▌▐ ▌▐ ▌▐ **Day 205: Friday-May 29, 2009/Day 174-Iraq** was full of news. To begin with, I mentioned a few days ago how Iraq is cracking down on corruption throughout its government. As counterinsurgency experts will tell you, when a government is toppled, a vacuum is created that is rapidly filled by many interested parties. These various elements can include another type of government, a military regime, religious fanatics, or all of the above who then wage a colossal battle for power and control. This is what has happened in Iraq, and the fledgling government led by Prime Minister Nouri al-Maliki continues to search for avenues toward legitimacy. But al-Qaeda, other militias, Muqtada Al-Sadr, and other interested groups have all tried to seize such power and control and delegitimize the Maliki led government. Without a legitimate government that can protect the populace and provide essential services, distrust will manifest itself within the populace, which then allows these other interested parties to continue to operate. Another side effect of the massive effort to build a legitimate government is corruption. But one step toward progress and stability is now beginning to occur on a much broader scale in Iraq. The Minister of Trade has just resigned due to rumors of high level corruption, and the Head of Iraq's Commission on Public Integrity said that 997 arrest warrants have been issued based on other allegations of corruption **(MNFI, 29 MAY 09:** *Reuters, AFP, BBC* **& JASG, 29 MAY 09:** *BBC*). Of these, 53 are warrants for senior level officials. Another step toward progress!

In a bit of an odd story, a young Iraqi daredevil is seeking permission to skydive over Baghdad! I am sure he is looking for permission, as opposed to just doing it, so our guys don't blast him out of the sky! But this guy is a bit of an idealist and perhaps a little premature in his remarks, but I sure like his attitude. He states that:

> Diving over a city that has suffered from war is recognized in the skydiving community as a symbol that the war is over. Because I want to say to all the world that we are now in peace, and it's not war anymore.
> **-MNFI, 29 MAY 09: *UK Guardian***

And while symbolism is very big over here, perhaps this young guy should inform the insurgents that the war is over. Not sure they have been issued the memo on that one yet!

The Army Chief of Staff, General George Casey, has stated that he is planning for a decade of deployments for troops in Iraq and Afghanistan! General Casey states this is necessary because global trends are moving in the wrong direction **(Stars & Stripes, 29 MAY 09, *Casey plans for a decade of deployment*, written by Tom Curley).** I am unsure if General Casey cleared his commentary through the Pentagon or the Obama administration, but I am not sure they would concur with his assessment. There can be no doubt that counterinsurgent wars are very long and protracted affairs. History has told us this time and time again. But there is currently a plan in place to leave Iraq by the end of 2011. That date would put this war at almost the nine-year mark. And in case people have forgotten, we have been at war in Afghanistan longer than we have in Iraq. That war is going on its eighth year already and there is no end yet in sight! So General Casey's comments probably will not play well with the Secretary of Defense or the Obama administration. Nor is it good news for troops that have already made extreme sacrifices, as have their families. I guess if he is going to throw commentary like that out there for all to hear, then I hope he has a solid recruiting and retention plan as well. And I hope Congress is planning on spending a lot more on the Army budget for growing the force and doing a better job at protecting those in harm's way. If history tells me anything, I highly doubt that this type of strategic planning has occurred at these high levels! If the US is going to be the lead for all conflicts in relation to the growing list of failed states around the globe, our leadership better have a much better strategy for doing so than they have displayed thus far. Otherwise, as some have already predicted, we are headed to the ruination of the Army. much as we witnessed after the last protracted war veiled in ambiguity – Vietnam!

▍▍ ▍▍ **Day 206: Saturday-May 30, 2009/Day 175-Iraq** was another filled with baffling news. There was a recent article in *Stars & Stripes* that stated that our generals are finding solutions to the Army's suicide problem to be very elusive **(Stars & Stripes, 29 MAY 09, *Generals find suicide a frustrating enemy*, written by Ann Scott Tyson and Greg Jaffe).** This just validates my theory of serious systemic problems within the Army. You might believe this simply to be a side effect of the Army being such a monolithic bureaucracy that is resistive to change and slow to progress. I have already stated that this is a major issue within the Army and outlined the reasons for my statements. But after a piece I read that documents the facts surrounding several troops being deployed to Iraq and/or Afghanistan after being officially diagnosed with PTSD and on medication to treat such a disorder, it seems to me that we have some serious leadership issues to deal with besides the organizational challenges **(http://www.veteranstoday.com/modules.php?name=News &file=article&sid=3835).** Perhaps we can attribute this issue to the fact that the military is stretched too thin; or because more than 600,000 Americans have served multiple tours; or perhaps because 12% of our soldiers in Iraq and 17% in Afghanistan are taking medication to cope. And it is also interesting to note that only about 1% of our population serves in the military. And of this 1%, the Army and Marine Corps are carrying the majority of the burden in relation to the wars in Iraq and Afghanistan. So we ask a great deal of only a few and it has obviously taken its toll. So I do not find this as mystifying as the Generals do. Some possible ideas to mitigate or rectify such a large challenge include: a complete shift in national strategy that uses our troops more judiciously;

reinstituting some form of draft to spread the obligation to protect America throughout more of the populace; budgeting more money for a larger military; incorporating Navy and Air Force assets into counterinsurgent warfare more effectively; if a larger force, then better use of active duty and reserve forces and a reconfiguration of certain assets within each component; shorter rotation times; more mental health specialists; better and more focused training within the military at all levels; more effective leadership at higher levels; and finally, a change in military culture to overcome the stigma of mental illness. None of these possibilities of which I speak have easy solutions or quick ones that can be addressed quickly, even though time is not on our side. But lip service versus action, which I have observed far too often, does nothing to solve the problem but only creates more distrust between the troops and our government and military leaders. Perhaps because some politicians feel that because we are only tapping into 1% of our population it is not that big of an issue for them to solve. I am just not convinced that our leaders are that committed to solving this problem. There have been token indications of such an occurrence, but nothing that I would call sweeping change. Too bad, because our troops deserve a much better effort and one worthy of the sacrifices they have given for their country.

 Today was Randy's (MAJ Staab) last day in Iraq. Last evening we did Randy's farewell in the SOC with LTC Bob 'Maddog' Pritchard doing the honors. Randy gave a nice departing speech and had some very nice comments about his 'power row' comrades. Randy will be heading back to San Diego to gather up the family and then off to Monterrey, California to begin his studies for his master's degree in electrical engineering. We also performed our typical 'power row' ritual of going off for pizza before one of own leaves. So we enjoyed some North End pizza over in my neck of the woods (GRC) and I gave Randy the same parting gift that I gave to Ryan LaPorte and Sean Song. This is a Wisconsin Rapids Fire Department's Chief's coin. Coins are huge in the military and you often get one from a General Officer upon completion of a good tour of duty. You can buy others at the PX and can quickly accumulate quite a collection. I have such a collection at home. So a few years ago I decided to purchase some for the fire department and begin a new tradition there. I would hand out these coins to my firefighters on special occasions for going above and beyond their normal duties. So while Randy just received a coin from his Commanding General here, and one from an Australian SOC mate, I highly doubt that he has a coin from a fire department – until now. While there are many coins that may look like this one, this particular coin is unique. So it will be a nice and very unique addition to his collection. Randy has been a pleasure to work with and he will be missed. Job well done, Randy.

Power Row Salute

Day 207: Sunday-May 31, 2009/Day 176-Iraq was going to be a little different than my normal Sunday. Today I traveled to the IZ for another run with the GRD Commander MG Eyre and an ODP session. I guess it has already been about three months since our last one. Now I have already outlined the trials and tribulations concerning travel throughout this war torn country. Nothing comes easy or fast. So I wanted to spend my Sunday morning completing my normal tasks, such as laundry, paying bills, working on this book, etc., and then head to the IZ in the early afternoon, as our run with the CG was not until about 1630 (it takes about 45 minutes to go from VBC to the IZ). Oh, and by the way, it is only forecasted to be about 106 on Sunday! Should be a real fun run! Then, my plan was to spend the night and return early Monday morning to get back into my normal flow. As the old adage goes, 'the best laid plans of mice and men often go awry'. The plan of one mouse has already gone awry during my tour, as my battle with this little desert warrior has already been chronicled! Anyway, this is how I submitted my movement plan. Unfortunately, due to circumstances outside of my control, which occurs a lot here by the way, I would now be traveling over mid-morning and not return until late Monday afternoon! This is where adaptability and flexibility comes in, something that I am always preaching to my firefighters. So Plan B was now to get up early and accomplish as much as possible before my trek to the IZ. Once there, I would check in and get the key for my CHU and then continue on with my work until lunch. Then I would go to lunch with my old pal LTC Mike Girone (GRD G1). Then back to the CHU for more work before we would head off to do our run with the MG Eyre followed by our officer development session that evening. The next morning I would go over and hang out with the gang in the G3 (Operations) section, watch the Monday morning BUA, conduct some business and then get a little more work done before heading back to VBC. Once back at VBC I would get my work out in, albeit a little later than usual, go to chow, and then call it a day. Then back to my normal schedule on Tuesday. I will let you know Tuesday morning how Plan B turned out as I write this before the events of Sunday unfold.

In reality, even though this was not the original plan, it was nice to leave VBC once in awhile to break up the monotony. And going to the IZ was almost like a mini-vacation. People who have come from the IZ to VBC for a few days tell me the same thing. So it is all good!

Oh yes, we did go on our fun run today! When we left to go it was 108 degrees F. in the sun and we ran on asphalt, which made the temperature more in the neighborhood of 125-130 degrees! This was most certainly not the brightest idea our leadership came up with since I have been in Iraq, but everyone survived the desert death march and then headed back to our compound. The briefing we had for our officer development was not all that enlightening. Most of the information presented I had already heard or seen before, but we did have a lot of new faces at GRD that I am certain got much more from the session than did I. Then, there was a showing of the movie **300** which was non-mandatory attendance. I like this movie a lot and have seen it about six times. I also use portions of it to illustrate key points when I teach leadership classes. But I decided I did not need to see it for a seventh time and it was back to the CHU for some shuteye.

Day 208: Monday-June 1, 2009/Day 177-Iraq I awoke in a CHU in the IZ, much as I had my first night in Iraq. I turned on the news to start my day and observed that the situation in North Korea continues to get very interesting. North Korea may be using their nuclear program as a bargaining chip to get other goods and services in return for discussing non-proliferation of their program, and they may feel their timing is right because our military is stretched very thin right now. While there is some truth and logic to this if it is correct, we still have many other assets that are not tied up in Iraq or Afghanistan, especially Air Force and Navy power. The Army Chief of Staff General George Casey stated that the Army would need to shift gears in order to mount any military operations against North Korea **(Stars and Stripes, 30 MAY 09,** *Casey: Army would have to shift gears for North Korea battle,* **written by Jeff Schogol).** This is stating the obvious as a conflict with North Korea would be of a more conventional-style than the irregular warfare we are involved with in Iraq and Afghanistan. The problem the Army and Marine Corps will have is maintaining proficiency in conventional-style tactics while actively engaged in fighting a non-conventional war. It will not be easy given the current size of our military, but being able to remain proficient in both styles of warfare is a necessity, not a luxury. So unless the Army splits and has dedicated forces to only fight each style of warfare, which is not viable given its current size and structure, this dual role will include a lot of training and funding to move the Army ahead.

Before my ride back to VBC I hung out much of the day with the GRD G1 LTC Mike Girone. We had lunch at the MNSTC-I DFAC and talked of baseball, our current situation, and the way ahead. We always have very insightful conversations and it was nice to see some of my 416th TEC brethren on this trip as it has been a few months since I was last over in the IZ. The trip itself was really non-enlightening, but still nice to get away from the monotonous routine at VBC if only for a few days. And yes, Plan B worked out quite nicely and the ride back to VBC along Route Irish was uneventful, which is just the way I like it.

▌▌ ▌▌ Day 209: Tuesday-June 2, 2009/Day 178-Iraq was back to my normal schedule and the morning BUA. North Korea continues to rattle its saber. There are really only two options associated with their actions. One: they are using their nuclear capability as a bargaining chip to get other commodities for their country that they are badly in need of. They have exercised such strategy in the past and may be doing so now. Two: their leader is a nut case and looking for a fight. If this is the case, it will not be pretty. As I have stated previously, many Americans believe the US cannot react to such a threat because we are already spread too thin militarily due to the wars in Iraq and Afghanistan. While this is true on the Army and Marine Corps side, we still have great capability to unleash on North Korea if necessary to do so. The Air Force has these things called strategic bombers and fighter jets and the Navy has these big floating airports called aircraft carriers, not to mention a lot of missile capability. In addition, the Army and Marine Corps also have artillery and armor capability that they are really not utilizing to fight a counterinsurgent war. So if North Korea believes we are vulnerable right now because of a two- front war, this would be a large strategic miscalculation on their part. Plus, we have about 28,000 ground troops in the area and South Korea has a very capable and well-armed military as well.
(www.breitbart.com/article.php?id=D98G47401&show_article=1).

Plus, because numb nuts has already played many of his cards, the element of surprise for a ground invasion is gone! So to me, he is trying to play high stakes poker to get something in return, but he doesn't appear to be much of a poker player. But this situation still warrants our attention and preparedness.

I have mentioned on several occasions how many officers over here are struggling to embrace counterinsurgency operations after decades of training and preparing for conventional-style warfare. But there are many others that definitely understand effective counterinsurgent strategy. One such officer is LTC Ben Matthews who is stationed in Mosul, one of the last bastions of AQI in the country. LTC Matthews states:

> We haven't killed our way out of this insurgency. We have bought ourselves out with other means. Employment and money are my biggest weapons. It's like a free enterprise and trade thing I've got going on against the insurgents.
> **-MNFI, 2 JUN 09:** *Times of London*

Sounds a lot like America to me! But this is exactly how these types of wars are achieving success. The people are the center of gravity for both sides and whoever has the populace on their side has the advantage in a counterinsurgency war. There are still pockets around Iraq that sympathize with the insurgents, are intimidated by them, or remain neutral and help neither side. This is why it has been difficult to root out some of these insurgent cells. But as money talks here as it does everywhere else around the world, and employment reduces the recruiting pool for insurgents, this strategy has proven very effective here in Iraq. I am not certain how this strategy is being applied in Afghanistan, or if any success is being realized if they are. But it is clearly effective here, and if it is taking hold in Mosul then there is hope that we can root out these last strongholds of AQI and move Iraq forward.

As we drawdown here in Iraq, we are seeing the effects of this action in the SOC as well. Many people are leaving and not being replaced. We have felt the effects in **'power row'** as well. We are now at half strength. We have transitioned from four LNOs to two since I have been here (6 months). The two that remain after Randy's departure are the CENTCOM LNO LTC Bob 'Maddog' Pritchard and myself. I am now the senior member of power row! The only good news from this is that Bob and I have a lot more room! I also understand that when I am on R&R they will replace me with a Captain from the GRD

G3 section who will just be here part-time and do what is necessary before heading back to the new GRD digs just down the road from the palace. If this is what actually occurs, then it kind of makes you wonder if they really need me here in a fulltime capacity! Just another one of those unexplainable mysteries here in Iraq!

Day 210: Wednesday-June 3, 2009/Day 179-Iraq was post office day, as I would send home a box of books for my personal library. My second green footlocker arrived while I was talking to the wife on Monday. It only took about 13 days to arrive. Thus far, the two boxes and the two foot-lockers have been sent and arrived at their destination successfully. I will check the contents to ensure there has been no damage in transit when I get home on R&R, which is now less than a month away! I cannot wait and am certain I will not look forward to the return trip. But in order to write the final chapter in this book, it must be done! If my calculations are correct, it should only be about 146 more days until we begin to prepare to head out of here. Yee haw!

There is also much reporting through intelligence sources that AQI is plotting a major attack on the new embassy and VBC. They are also targeting a dining facility here on VBC, which would be a devastating attack because during meals times these DFACs are filled with many people. All of the DFACs have a protective overhead cover that can withstand direct rocket and mortar hits. There are also contracted security guards at each entrance that check and scan everyone coming into the DFAC. A suicide bomber getting into one of the facilities would be the worst-case scenario. So vigilance is the key and extra precaution is necessary in order to foil the dastardly plans of AQI. This is a very high stakes chess match that is counted in lives lost or saved. Yet another little tidbit I will not share with the family until they read this book.

Overhead cover for De Fleury DFAC on VBC

Our friends from Iran are drawing the ire of General Petraeus **(MNFI, 3 JUN 09: *al-Sharqiya, Asharq al-Aswat, al-Hayah*)**. General Petraeus criticized Iran for continuing to supply weapons and train and fund insurgents as they continue their proxy war against the US. This type of activity is doing very little to assist in our efforts to improve relations with Iran. In fact, the olive branch that President Obama has extended is being sawed off by the current Iranian leadership. Certainly, after years of turmoil and ill will between the countries, it will take a long time to improve relations, but it seems clear that until there is a change in Iranian leadership this will not be possible. The Iranian elections are this month and it appears as though the current President, Mahmoud Ahmadinejad, is the front runner and the choice of the Grand Ayatollah who carries great clout in Iran. So a different posture toward the US is unlikely without any changes in leadership within Iran. And of course, our relationship with Israel is also an important part of this equation. But it was not all that long ago that Iran was very US friendly when the Shah was in power. Perhaps this can be so once again, but it may take a long time. However, it does not appear at this point that Iran is ready to meet us in the middle so the tension between the two countries will remain.

As I left the palace last night I noticed the rotunda was blocked off. People here are getting ready for **The Colbert Report** to do a few live shows here from the palace! I believe he is coming sometime in the next week or so. I never really watched Stephen Colbert, whose show appears during the week on the Comedy Channel, until I got to Iraq, but he is very entertaining to watch. It should be an interesting week as preparations continue. Wonder who his main guests will be? Stay tuned and I will report more on the show later in the chapter.

▍▍ ▍▍ ▍▍ ▍▍ **Day 211: Thursday-June 4, 2009/Day 180-Iraq** began with news that President Obama is picking the Los Angeles Lakers in six games over the Orlando Magic for the National Basketball Association championship. I must state that even though President George W. Bush used to own a professional baseball team (Texas Rangers), he never came out and did the overt predicting as has President Obama. If his forecasting of sporting events is any indication of his ability to forecast economic recovery and foreign policy then we are in good hands. Thus far, the President has accurately picked North Carolina to win the NCAA men's basketball tournament and the Pittsburgh Steelers to win the Super Bowl. Now yes, both of these teams were the prohibitive favorites to win their respective championships, as are the Lakers, so he is not moving too far out onto that limb, but he is still batting 100%. Not bad for a President!

In the BUA today there was discussion of sticking to the agreement to have all US combat forces out of Iraqi cities by the end of this month **(MNFI, 4 JUN 09: *al-Iraqiya, al-Sharqiya, al-Jazeera* & CENTCOM, 3 JUN 09: *NY Times*)**. General Odierno stated that "I think it's time for us to move out of the cities. I think it's important that people understand we are going to stick to the security agreement that we've signed" **(MNFI, 4 JUN 09: *al-Jazeera*)**. So in places like Mosul and Baghdad, where there had been discussion of keeping US combat forces in the more violent areas of Iraq, it appears as though Iraqi Security Forces are ready to step forward and assume these duties. To me this is a good sign of Iraq's progression, but time will tell if this was a wise decision or just hopeful thinking. We will monitor this one closely.

I watched a little of LTG Stanley McChrystal's congressional hearing to confirm him as the Commander of Coalition Forces in Afghanistan as he replaces General David McKiernan **(Pentagon Channel, 2 JUN 09: 1800 GMT+3)**. LTG McChrystal believes the war in Afghanistan is still winnable, but it will take a concerted effort over the next 18-24 months before an accurate assessment can be made

(CENTCOM, 3 JUN 09: *AP, CSM, WSJ, France 24*). He goes on to state that the strategy that proved successful in Iraq will not translate to success in Afghanistan and that Pakistan is a key variable to any success in this area of operation. It is important to note that terminology is often very misleading. The term 'winning' should not be translated by the American public as a situation where US forces defeat and eliminate all insurgents in Iraq and Afghanistan and both governments stabilize and prosper as democracies. This was the original and very tainted perspective by many high level US officials, but I am not sure the term winning is the most appropriate term to use here. Success and sustained progress are perhaps better and more realistic measurements to use. If the coalition can neutralize the insurgents to the point that each country's security forces can handle them effectively; each country can protect its borders to stop foreign fighters and weapons from freely crossing into and out of its sovereign territory; each country's respective military is capable of self-protection from invading enemies; and each government can stabilize and work to provide its respective populace with essential services and employment opportunities; then success may be claimed. It should be clearly articulated that past perceptions about what constitutes winning a war, especially one of an irregular nature, are very different than what occurred in World War I, World War II, or the first Gulf War. So Americans, including many in the military, must better frame what success in Iraq and Afghanistan actually means.

Finally, it appears as though al-Qaeda is looking at the Mexican border to possible launch an attack into the US **(CENTCOM, 3 JUN 09: *Washington Times*)**. People must also bear in mind that insurgents do not need to win the war they simply just need to not lose it! As long as they exist to disrupt normalcy and put fear into the hearts and minds of the populace they are achieving success in their twisted minds. So people should not become complacent and believe that insurgents are only found over in Iraq and Afghanistan. They are all over the globe and that is why this is truly a global war on terrorism no matter what other terminology we may want to refer to it as. This is not just a US problem—it is a world issue. So our border security forces in the US and FBI need to remain vigilant that al-Qaeda is going to try and strike again in the US. Al Qaeda is like an aggressive cancer that you try and eradicate through radiation and chemo, and although you can slow it down and force it into remission you can never entirely get rid of it. And once you let down your guard, even slightly, it comes back with a vengeance. I do not mean to scare you with such news, but knowledge is power and you need to know the facts. This is a societal cancer we will be dealing with for decades to come. So preparedness is the key and keeping the pressure on these radical elements is a necessity in order to be successful. This is just the world in which we currently live.

109/124

Day 212: Friday-June 5, 2009/Day 181-Iraq began with no Internet in the barracks – again! Who knows what the problem might be this time! It is always something around here. People who complain about our services in the US should spend a few weeks over here and then they would appreciate what we have a lot more!

The pace of things has really slowed here recently. There are very few new projects to work on, just a continuation of many old ones. The days are just running together now and I cannot wait until my R&R in a few more weeks.

Well, yesterday President Obama gave a historic speech in Egypt. He is once again reaching out to the Muslim community **(MNFI, 5 JUN 09: *LA Times, AP, Reuters, AFP, PBS* & *JASG, 5 JUN 09: BBC*)**. As usual, depending upon who you listen to and what media outlet you watch, there are a variety of differing spins on the one speech! Overall, I would say that many people, on each side of the argument, believe this is a good start. The President was also careful not to use the word **terrorism** in his speech as many people outside of the Muslim community link the terms terrorism and Muslims together. These are the millions of ignorant people who really have no clue as to what is going on in the world or care to take the time to research it further. However, no matter how you cut it or how you say it, we are in a **global war on terrorism!** President Bush got this one right anyway. Terrorism attempts to instill fear in people for no particular purpose. Terrorism uses cowardly acts such as the recruiting of females, children, or the handicapped to become suicide bombers. Terrorism attempts to keep everything in a constant state of chaos. So we can call terrorism whatever we want, but to paraphrase an old adage that has recently morphed into something slightly different during the presidential campaign but carries the same meaning, 'You can put lipstick on a pig, but in the end, it is still but a pig.'

One important excerpt from the President's historic speech was:

> America has a dual responsibility: to help Iraq forge a better future and to leave Iraq to Iraqis. I have made it clear to the Iraqi people that we pursue no bases, and no claim on their territory or resources. Iraq's sovereignty is its own. We will honor our agreement with Iraq's democratically-elected government. We will help train its Security Forces and develop its economy. But we will support a secure and united Iraq as a partner, and never as a patron.
> **-MNFI, 5 JUN 09:** *AP*

Of course al-Qaeda officials were not impressed with the speech and Iran was lukewarm with the whole situation. They believe actions speak louder than words, but I am unsure what actions they expect the President to take. It is interesting however that the speech comes just before the elections in Iran. We shall see how this all plays out. And while al-Qaeda is evil, they are no dummies. They came out with a video denouncing the President's speech before he even presented it! Of course, al Qaeda really has no objective but to kill as many Americans as it can and they seek to continue to brainwash those who listen to such rhetoric. They will intimidate and coerce innocent people to do their dirty deeds for them. After reading several excerpts from a commission report on 9/11 for Naval War College, it is quite apparent that these people have twisted the word of Islam for their own evil purposes. This group of misfits is not poverty stricken or uneducated individuals. But because of the vacuum that is often created in many failed states, of which there are many, this vacuum is often filled with people who seek power through corruption, greed, or simply to control the masses. This is often found in countries without a stable government that can protect its populace and provide essential services. These are variables that often lead to an insurgency. This is why it is imperative that the Iraqi and Afghan governments stabilize and provide this security and service to their population. But even if this is accomplished in these two areas of the world, there are many more failed or quickly failing states out there.

Keep in mind the scariest part of this whole issue of terrorism. Al Qaeda seeks to prevent governments from stabilizing through the use of terrorism (sorry Mr. President, but it is what it is). Through chaos they have more control. Through chaos there is more corruption. Through chaos there is less stability. But the only reason they seek to attack Americans on our soil is not to overthrow our government or continue the chaos, they simply seek to kill Americans for the sake of killing. So those Americans who do not believe that we need to keep the pressure on these evil-doers have no sense of reality. The sad part of this is that we will continue to lose young men and women in uniform in the course of this fight. That happens in war. And this is not a holy war, nor a war of philosophies. It is not a war of religions or civilizations. It is basically a war of good versus evil. It is a war of those who embrace life versus those who value death. It is a war we cannot lose for the sake of all humanity. These extremists are a cancer to the fabric of global society, and while it would be nice to completely eradicate them through massive doses of chemotherapy, the best we can hope for is skillful surgery followed by point radiation to contain these radical elements so they cannot launch devastating attacks upon any country. If their cause were so just, and they really believed the crap they were spewing, then they would be the ones strapping on a suicide vest to their body and blowing themselves up instead of getting young children or the handicapped to do it for them.

But this is their cowardly and twisted version of whatever ideal they are fighting for, if they really even have one. And it should be clearly understood that this is not just America's fight, it is a global war on terrorism (sorry again, Mr. President). These fanatics can strike anywhere and prefer softer targets that are not prepared and offer little resistance. This is why it is hard to believe there is not more assistance from other countries in this war. It will be sad if these countries must learn the hard way and the price for turning a blind eye is paid for in lives of the innocent. But for America today, and its future tomorrow, we cannot afford to lose this battle.

Let's call terrorism something else!

Day 213: Saturday-June 6, 2009/Day 182-Iraq was the beginning of the weekend, such as it is here in Iraq. There really is no such thing as TGIF here, but Saturday's are normally a little slower and Sunday is a half-day and slower yet. TGITWEIIIHII may be a better acronym to use here. Now if you can figure this one out before I give it to you I will buy you a cold beer (real beer)! Survey says: *Thank God It's The Weekend Even If It Is Here In Iraq! How did you do?*

Today marks the 65th anniversary of D-Day. This would be another one of those days to remember those that fought so bravely on foreign soil. Without their sacrifices during this period in history, our world may have become a much different place. Thank you to all of our veterans, both current and past. We owe our freedom to you.

There is still much discussion of the President's speech from a few days ago in Egypt. It continues to get mixed reviews and if you wrote down who you thought would be encouraged by such a speech and those who would view it skeptically, you would probably have done much better than you just did with my new acronym! The hardliners on each side were not about to be swayed, but you hope such a speech can positively persuade many people that sit in the middle. But of course, and I cannot disagree with this point view, many people on both sides are waiting for actions to follow the words **(CENTCOM, 5 JUN 09:** *Asharq Alawsat***)**. So it is still too early to effectively evaluate the new US posture on the Middle East and the Muslim world.

Not surprisingly, our friend Muqtada Al Sadr, the Shi'a cleric, rejected the President's speech and stated that "the resistance will continue and the opposition will continue until US troops redeploy from Iraq and Afghanistan and end their support of the Israeli enemy" **(MNFI, 6 JUN 09:** *al-Malaf***)**. Well, I am with him on this one, and am all for redeploying but he should help us to rid his country of terrorism first, and then I think we can oblige his wishes!

Now this next little vignette will capture how divided people are on the President's speech, even between religious leaders.

> Muqtada Al Sadr's analysis was:
> The honeyed and flowery speeches express only one thing: America wants to adopt a different attitude in subduing the world and putting it under its control and globalization.
> **-MNFI, 6 JUN 09:** *USA Today*

This guy sure seems to have a lot of pent-up anger and appears to be a glass half empty sort of dude. On the other hand, here is a guy who is clearly a glass half-full kind of guy, Baghdad cleric Shaykh Muhannad Al-Moussawi:

> President Barack Obama expanded his calls for religious coexistence, and a departure from the extremism and insanity. I'm glad that he told the American people that Islam is a religion of tolerance, love, and peaceful coexistence, like other monotheistic faiths. His speech in Cairo broke new ground. We appreciated the overtures to Muslims he made today,

particularly his promise to withdraw from Iraq and his recognition of Palestinian's suffering.

-MNFI, 6 JUN 09: *DPA*

And finally, many Iraqis have mixed emotions on the President's speech. Many Iraqis appreciated his efforts to integrate the Muslim faith into his speech, while others felt his rhetoric was hollow **(MNFI, 6 JUN 09: *LA Times, FOX News, USA Today)*.** So again, it was very unlikely the President's speech was going to sway the hardliners. But if he moves those sitting in the middle of the issue, which is the majority, then I would have to believe it was well worth the effort. Now if he can follow up the speech with noticeable actions, then he really has made a historic impact. Stay tuned as this will be a story for days to come.

D-Day – June 6th, 1944

Day 214: Sunday-June 7, 2009/Day 183-Iraq was a nice quiet day. I got much accomplished in the morning as I worked on refining this book. I really do not know how this masterpiece will turn out in the end with all the color and the pictures, but as the author I am going to make the original manuscript something I can be very proud of. It will then be up to my editor and publisher to dash these dreams!

Shari was leaving for Florida today with son Josh. Shari is basically going along to keep Josh company. Josh has finished his studies at Le Cordon Bleu in Minneapolis and is now heading to Orlando to begin an internship at a restaurant there. Shari lived in Orlando for about nine years before she moved back to Wisconsin a few years ago because of little ol' me! Bond – James Bond! The rest is history. Anyway, Josh's fiancée Kelly is still in Orlando and that is where he started his culinary school before moving to Minneapolis last summer. Don't ask!

Bond – James Bond

Preparations continue for the taping of a few Colbert Report shows here in the palace **(Stars & Stripes, 6 JUN 09, *Colbert plans to broadcast shows from Baghdad*, written by Jake Coyle)**. The studio is pretty cool, but because of the dim lighting I am having difficulty getting a good picture of the set for you, but my pursuit will continue. This studio also occupies the entire rotunda so it is a bit of an inconvenience, but if the troops enjoy this experience then it is no big deal. This guy is pretty hilarious and I look forward to seeing the shows on television if I do not get a glimpse of a show while filming it live.

In a couple of other good news/bad news stories, I will begin with the bad news. As I am sure you recall all my misgivings concerning the Army's inability to adequately address the rising suicide issue, there was a story in *Stars & Stripes* about a soldier's parents who were complaining to the Army about an unlicensed psychologist inaccurately diagnosing their son's condition, and consequently, he committed suicide a few weeks later **(Stars & Stripes, 6 JUN 09, *IG says complaints about mental care unfounded*, written by Kimberly Hefling)**. Now, as I do not have all of the facts associated with this case, I cannot and will not speculate if this ruling is valid or just a way of CTA (Covering Their Ass). But I can tell you, after being in the system for over 34 years, it would not surprise me if the parent's had good reason to be suspicious! As I have stated time and time again, the Army is a huge bureaucracy that is in need of a systemic enema. To address a major issue they often use a media blitz to increase awareness and then unilaterally issue blanket policies. I have told you on several occasions that this is simply placing a band-aid on a gaping wound, and the one-size-fits-all mentality is not effective and a waste of valuable resources. Or, you may think of it as the little Dutch dude sticking his finger in a leak in the failing dam! I do believe the Army is doing their best to hire more healthcare professionals to address this huge problem, but I really do not know where they are at with these efforts at this point in time. For this young man however, this assistance comes too late and I can tell you as someone who took three years of psychology, the human mind is extremely complex and even the best of psychiatrists often have issues diagnosing problems. So having an unlicensed individual trying to do this, well, this is just another poor decision in a very large system that doesn't have the capacity to care. I say this because I have seen it often. Many administrators or General Officers will claim such statements are simply not true and many people may view this as sour grapes. In some cases this is very accurate, but my statement is not that these General Officers do not truly care about troops, it is the unforgiving system in which they operate that does not care. In his book **Wiser In Battle**, retired Lieutenant General Ricardo Sanchez mentions that he stopped by to visit a retired General he had known and who had been a mentor to him during his career. When he asked this General how he liked retirement after 30 plus years in the Army, he told LTG Sanchez that it was not worth a dam. He went on to state that the Army really does not care about individuals, especially when they get out. Now if this is an accurate portrayal, and this is from a General Officer, it certainly doesn't paint the organization as one that cares. The same holds true for many of the suicides that occur when troops leave the military. Often, the military just loses track of these warriors and that does not exhibit the characteristics of a caring organization to me! This kind of inside knowledge makes it very difficult for me to recruit people for the Army. And that is sad news indeed!

Plug that leak

Now the good news of the day is that LTG Jack Stultz, Chief of the Army Reserve, is quickly becoming my hero. He is working with Representative Joe Wilson from South Carolina to improve benefits for reservists **(Stars & Stripes, 6 JUN 09, *Improving returns for risk and sacrifice,* written by Tom Philpott)**. They are teaming to move Congress to approve two different options on retirement packages, which in my estimation are long overdue. To begin with, up until recently, reservists had to wait until age 60 to draw a pension from the government. This pension is based upon a point system that takes into account years of service, training, additional duty, and your rank. Then, on January 28, 2008 Congress passed a bill that allows a reservist to reduce the time from age 60 by three months for every 90 consecutive days served during a mobilization. This means that I will be able to reduce my time by about one year when my tour is complete. So that moves me down to age 59! The problem with this particular bill is that it is not retroactive, so you have many reservists who served post 9/11 in the wars in Iraq and Afghanistan, during the most tumultuous periods in these wars, and they get nothing! Remember what I just told you about this bureaucratic monster and its inability to care! This new bill seeks to make the bill of 2008 retroactive back to back to 9/11. This would be the fair thing to do and we will see if it is once again lip service, or if our elected officials really want to take care of those that have sacrificed so much.

The other option relates to years of service over 20. In this option, for every year a reservist stays beyond 20 years they can reduce their time until pension award by six months. I like this idea as well. There is likely to be a cap at age 55 otherwise this plan may be cost prohibitive. I had heard that this was the primary reason why similar bills had not been approved previously, but as our government bails out banks, auto industries, and pretty much everyone else who needs it, how about they step up and take care of its veterans? Yes, of course I am biased on this one. What did you think I was going to say? But if this ever really materializes, and it is pretty iffy as it stands anyway, the math looks like this in my situation: due to my mobilization (400 days) I will be down to age 59 in order to draw a pension; I have 31 good years toward retirement as we stand now; so that is 11 years over the 20; take 11 and divide by 2 equals 5.5 years; then minus 5.5 from 59 and presto, this gets me to 55! I think I will write my Congressional leaders on this one and tell them to support it. I told you this was good news!

Colbert arrives in Iraq

Day 215: Monday-June 8, 2009/Day 184-Iraq was just another Manic Monday (sorry Bangles). Only three more weeks until I begin my trek home for some R&R. I cannot wait!

There were a few tidbits in the news today. Again, you must understand that the war of words is also referred to as information operations. It is all about perception, not only here, but everywhere. We have reached an age of unprecedented political correctness whereby people must choose their words very carefully. To me, this is because the real message cannot be articulated properly, but that is just me! For instance, I have already explained the phrase "war on terrorism." The President does not want to use this phrase and even the military is now using the new term 'overseas contingency operations' versus the 'global war on terrorism', better known as GWOT. Now I believe the rationale is because people link the terms 'terrorism' and 'Muslim' together. I guess I never made that leap, but there must be some poll out there that suggests this is the case. Again, I do not care what we call it, but in my opinion it is most accurately described as a global war on terrorism! Remember the lipstick on the pig?

We now have a few more situations that bear watching here in Iraq. As you may recall, by security agreement all US combat forces are to be out of all Iraqi cities by the end of June 2009. Yes, that is within the next 23 days! The security will then be taken over by the Iraqi Security Forces. General Odierno also stated that it is important that everyone remember that the terms 'combat forces' and 'advisors and trainers' should not be used interchangeably. While all 'combat forces' will be out of the cities by June 30th, not all US forces will be. The Government of Iraq has requested that some US forces stay in a few of the cities to continue to assist and advise the Iraqis **(MNFI, 8 JUN 09: *Radio SAWA*)**. I am not sure if this means these forces will dress differently or not carry weapons! Otherwise, how would the average Iraqi citizen know the difference, or the insurgents? Got me on that one, but again, it is all about perception. However, to me the perception will be that we still have US forces in the cities unless they are incognito! I will let you know more on this one once I figure it out.

There is also another mystery that puzzles me over here. Many civilian contractors have been allowed to wear military uniforms while performing work over here for the last several years. These are the same uniforms our troops wear except instead of having rank and service affiliation on the uniform, they wear a DA Civilian patch or something similar. Many of these people do not wear the uniform well if you catch my drift. Many are overweight or have long hair, and in my opinion as a soldier, it just looks bad. I am really unsure of the rationale behind this move other than perhaps the military wanted to create the illusion of more troops than there actually were. Recently a FRAGO came out informing all civilians that they now could not wear the uniforms and they had 45 days to make other arrangements! OK, this to me is a good call, but not sure why they have 45 days as this does not coincide with the end of June or any other timeline of significance that I am aware of. But perhaps this is occurring now because the military does not want to create the perception that we have all these people in uniform over here as we try and reduce our footprint in Iraq. Oh yes, the ebb and flow of decision making. Keep in mind that we have roughly 135,000 troops in Iraq as of today and continue to drawdown. There is also about one contractor for every US troop as well and many of these people are wearing military uniforms. So I will let you draw your own conclusions on this one and I will let you know if I find out the rationale behind the decision.

Next we have Admiral Mullen who recently reported that AQI has been severely weakened and are now primarily confined to the areas in Baghdad and Mosul. He goes on to state that the ISF are capable of providing the necessary security and dismantling the terrorist organization **(MNFI, 8 JUN 09: *al-Arabiya*)**. Now I have to tell you, I believe I understand why the Admiral is making these statements. I believe he is trying to inform the American public and our elected officials that progress in Iraq continues. Again, it is all about perception. These statements probably go a long way in appeasing the public and Congress. If he stated that things are not going so well and violence is increasing, there would be more pressure from Congress to get out of Iraq sooner or stop the flow of funds for the effort. I understand this completely. However, as someone on the ground here, I can tell you that the situation is improving and AQI is a hurting unit right now. However, there are also many other issues to deal with over here besides AQI. But since AQI is linked to

9/11, and it is an acronym recognized and despised by both the American public and Congress, they are the primary focus. I get this point as well. But every time a high ranking US official makes such claims, whether true or not, there always seems to be an uptick in violence that ensues. Obviously, whatever group we are claiming to degrade or defeat they feel the need to prove otherwise. We saw this recently when Secretary Gates made such a statement, which he quickly recanted for the exact reasons I just laid out for you. Because of the statement, we did see a noticeable increase in violence. Even terrorists watch CNN! Most probably we will see this again after the Admiral's statement and the people that pay the price for this war of words are the troops that patrol the streets. Sometimes transparency can be dangerous! If these same people making such claims feel so confident about their statement, then let them come over for 12 months and patrol the streets in Baghdad or Mosul and drive the IED laden roadways! Not likely!

Finally today, one of my favorite writers is in the news again. Thomas Friedman, who writes for the *New York Times*, suggests that the US should work with the UN and European Union to prioritize efforts to unify Iraqi political discord. As someone who knows the Middle East very well, Mr. Friedman truly believes there is an opportunity to create a successful democracy here **(MNFI, 8 JUN 09: *NY Times*)**. I hope he is right, but if this is to be it must be accomplished politically, not militarily. I believe our current government understands this.

Stephen Colbert & General Odierno at Al Faw

Day 216: Tuesday-June 9, 2009/Day 185-Iraq was another day much like many others. Let's see, I would go to work early for the morning BUA; attend the post BUA huddle; coordinate and disseminate information; go workout and eat lunch; come back to the palace and accomplish more work and a few reports; call Shari; and then head back to the barracks late after the evening update; read the *Stars & Stripes*; read about 10-15 pages of whatever book I am reading at the time; perhaps watch a little Armed Forces Network; and then call it a day. Can you say – rut? I am sure I have mentioned this a time a two already, so this can be considered a recurring theme throughout the book, but as the days begin to run together here it reminds me of the Bill Murray movie **Groundhog Day**. The alarm goes off and you awake to the same day over and over and over! Oh well, it could be much worse compared to the brave souls patrolling the streets and either being shot at or worrying about IEDs. They may be in a rut as well, but it is a much more dangerous rut than the one I find myself in. Tip of the hat to these brave men and women.

As there was not much new news in the BUA today, I will report on a few items from last evening's CENTCOM CUB. Secretary of State Clinton had some tough words for North

Korea as the US is talking tough right back at old Kim Jong-il. She states that if North Korea launches a nuclear attack upon Israel, this would be followed by a swift and decisive retaliatory response against North Korea **(CENTCOM, 8 JUN 09: *Bloomberg News, UPI, Reuters, Al Jazeera*)**. Now let us contemplate the ramifications of all of this. Let us say that Mr. Kim Jong-who is mentally ill from North Korea launches a nuke at Israel. Certainly, this would be devastating and kill many innocent people. Then the US and/or Israel will launch an attack on North Korea in response. As we have a lot more nukes in our arsenal than does North Korea this would seem to be a suicide mission on their part and doom the entire country. Of course, such a cataclysmic event would also have an impact upon South Korea and Japan due to their proximity to North Korea. So any scenario involving nuclear weapons is never good, and yet another major reason why rogue states and terrorists cannot possess such weapons. It would certainly be a much safer world if no one had these types of weapons, but in the wrong hands they can change the world.

There are a couple of interesting notes to report just to our east, which is still in the CENTCOM area of operations. A Pakistani intelligence officer states that he has trained one of Taliban leaders and there is no possibility of overpowering them so we should talk with them rather than sacrificing more lives **(CENTCOM, 8 JUN 09: *London Times*)**. Now there may be two different ways to view such a statement. To begin with, the intelligence officer is correct in that it is virtually impossible to eliminate an insurgent force. Again, they do not need to win a war—they just simply cannot lose it. So as long as they exist, they have the ability to cause disruption. This we know from past insurgencies, the Russians unsuccessful efforts against the Taliban in the 1980s, and after seven plus years of war in Afghanistan. While history tells us that most insurgencies are not successful, it also whispers in our ear that they typically end in a negotiated settlement. This is most likely going to be the case in Afghanistan as well. The Taliban may carve out their little world somewhere in the country with those who chose to live under their extremist life style. The remainder of the country can be protected by security forces and exist under a less extreme form of government. This is basically a two-state solution and if it occurs, you would hope the two opposing forces could live in peaceful co-existence.

On the other hand, perhaps the Taliban is feeling the heat. The US is sending in more troops; the Afghan Security Forces are getting larger and more capable; and the Pakistanis are stepping up their efforts against the Taliban and securing their borders. In addition, some Pakistani tribesmen are now fighting against the Taliban for blowing up one of their mosques **(CENTCOM, 8 JUN 09: *DAWN, Al Jazeera, McClatchy Newspapers*)**. If the tribes rise up against the Taliban then it will be similar to the Sunni Awakening that occurred here in Iraq, as various tribes joined together and fought against al Qaeda. This watershed event, in conjunction with the US troop surge, and Muqtada Al Sadr telling his militia to work the political system as opposed to violence, were all important factors in turning the tide here. So all of this may be a bit of gamesmanship on the part of the Taliban, but in reality a negotiated settlement is probably the best course of action to pursue if we pay attention to history.

Finally, Iran has its big election this month **(CENTCOM, 8 JUN 09: *media highlights*)**. It appears that the two front runners are the incumbent President Mahmoud Ahmadinejad and challenger Mir Hossein Mousavi. This is an interesting dichotomy to watch as the incumbent is: anti-western; more of a hardliner; very women unfriendly; and basically status quo. He also appears to be the chosen one as far the Grand Ayatollah is concerned. While the Grand Ayatollah may be out in left field on many issues concerning the modern world, he still carries great weight in Iran. The main challenger is: more of reformist; pro-West; women friendly; youth friendly; and speaks of change in a struggling country. Haven't we heard this theme of change somewhere recently? While it is likely that Ahmadinejad will be re-elected, voter turnout especially among youth and women is the key if there is to be any chance of an upset.

Groundhog Day in Iraq?

Day 217: Wednesday-June 10, 2009/Day 186-Iraq was back to normal at the palace. All of the Colbert Report equipment had been packed up and taken away as they filmed they last show here in Iraq last night. I think it was wonderful that he came over here and performed. While we have had entertainers here many times before, I believe this is the first time anyone has ever filmed a series of shows here **(Stars & Stripes, 9 JUN 09, *Operation Iraqi Stephen*, written by Kim Gamel & JASG, 9 JUN 09, *In Iraq, Colbert does his shtick for the troops*: NY Times)**. So I may be incorrect on the exact nature of the historical value of the visit, but I do know that it was a first. So I applaud Mr. Colbert for his patriotism and taking time to come over here. I know from seeing the response from the troops that it was a huge success. While I never got to see any of the shows live due to my schedule, I do look forward to viewing them on television. That is if my cable is working! It went out again last night. Have I mentioned how much I look forward to leaving this place?

On a few other notes, we need to revisit a few stories I have droned on about previously. But as this book is intended to educate and enlighten those who read it, this will just validate and reinforce some past stories. To begin with, we go back to the contractors here in Iraq. Now I do not want to lump all contractors together, because I am sure there are some good ones, many that fit somewhere in the middle, and then some bad ones (the old Bell curve). But the bottom-line is that the US government really has exercised little control or oversight over these contractors. This war has been unprecedented in the use of civilian contractors and many people have gotten rich as a result **(Stars & Stripes, 9 JUN 09, *Report cites missteps in war contracts*, written by Richard Lardner)**. I have already told you on multiple occasions about the issues associated with one of the larger contractors over here – KBR. And oh by the way, they are mentioned in this article as well and are now drawing the attention and ire of Congress. About friggin' time! Anyway, the bottom-line to all of this takes us back to the serious lack of strategic planning before engaging in this war. This contractor fiasco (with all due respect to Mr. Ricks) is just another result of that failure to plan. As I always tell my firefighters, 'fail to plan, plan to fail.'

Next, we get back to the stress placed upon our troops and its side effects such as depression and suicide. While many people are still scratching their heads on how to impact these alarming trends, the questions continue to mount and the answers are few **(Stars & Stripes, 9 JUN 09, *Battalion's leaders look for stress*, written by Teri Weaver)**. The key point in this article so I do not rehash old information is that officers can become stressed and are often the last to seek assistance. This again relates right back to the stigma attached to the admission of a mental health issue in the military. Many officers believe admission of such a condition will limit their promotional chances and career opportunities. They see themselves as damaged goods, and this my friends, is directly related to culture within the military as I have mentioned to you before. Changing this culture, if it is even possible to do so, will take many years and a lot of resources. Much like a protracted war in a rapid satisfaction society, I doubt that the military and our government have the persistence and patience to provide enough funding and resources to adequately impact this alarming problem. And I say this not because there is no desire to fix the problem, I say this because of the bureaucratic structure in which the answers exist. This bureaucratic monster will absorb the issue to the point that nothing meaningful or long lasting will come out it. It reminds me of the old 1958 movie called **The Blob** with Steve McQueen where this blob of matter from outer space just absorbs everything in its path and nothing ever comes back out of it! Can you say – bureaucracy?

The Blob = Bureaucracy?

Day 218: Thursday-June 11, 2009/Day 187-Iraq was once again GRD BUA Day. So it was into the palace early to ensure the slides were ready to roll. After today, there was only one more of these to do before I left on R&R. But today was a bit different than past Thursdays as SFC Maltes called me in the SOC and informed me that she could not gain entry into the NEC for some reason so I would be the primary script reader for the BUA today! Well, this is why we have the system I suggested in place, and why I rehearse along with SFC Maltes in case of situations exactly like this. So I am ready to spring into action like the jungle cat upon his prey.

The Jungle Cat

Once again the cable was out in my room. Not sure if it is just my room or all of GRC, but I did heard **Everybody Loves Raymond** on in another room at 0530 this AM, which is a show that is on cable over here so I suspect it is once again just my room! Not sure what is going on with that but I will inquire with the cable guy at GRC – again!

In news today there was one item of note that goes back to the referendum that was to occur this summer concerning the presence of US forces in Iraq **(MNFI, 11 JUN 09: *NY Times, AP, AFP*)**. You will have to think way back when I mentioned this last, but now there seems to be some political posturing going on over here. I think the Iraqis are beginning to get the hang of this political thing! Anyway, there was supposed to be, and may in fact still be, a referendum concerning a vote by the Iraqi people to determine the fate of US forces in Iraq. I am unsure that this referendum is binding and would supersede the already agreed upon and signed security agreement between the US and Iraqi governments. If so, then I believe we have one year to leave Iraq once the vote is finalized if that is the wish of the Iraqi people. But as I understand the situation, this is the only vote on the ballot, and I know from my own experience back in the states as a polling station official that it costs a lot of money to do any election. I have heard this referendum election, with only this one item, would cost in the millions of dollars! I believe there is a debate going on as we speak that because of the poor economic climate here right now, such a vote for one item seems very wasteful. Others, obviously the anti-Westerners want to use the vote as political fodder and move troops out sooner than the security agreement calls for. If the referendum is binding and the Iraqis vote us out, then we would leave about 18 months **(June/July 2010)** earlier than the security agreement calls for **(December 2011)**. The plan that some lawmakers are pitching is to still have such a referendum, but tie it into the national election process in order to save money. That seems like a good compromise and a smart economic suggestion. If this does occur, and the vote goes against us at that time, then we will only be out of Iraq about a year earlier than projected. We shall watch this story as it develops.

Oh yes, and by the way, the briefing went well to the big guy and it may have been my best yet. Just call me Walter Cronkite Junior. Indeed!

And that's the way it is – June 11th, 2009

▌▌ ▌▌ **Day 219: Friday-June 12, 2009/Day 188-Iraq** was going to be a big day in Iran. There are two opposing parties that are completely opposite of one another. One offers the status quo – the other offers hope. One offers women an opportunity – the other seeks to keep them repressed. One offers youth a better future – the other seeks to keep them locked in the past. Two opposing views – one presidency at stake. There are two keys to victory for either candidate. The more liberal, pro-Western candidate needs people to get out and vote. He needs the youth and women votes in order to have a legitimate chance of winning. But perhaps the biggest wild card is the Grand Ayatollah. His voice has always carried great clout and because of his own personal perspectives, it appears as though he is supporting the status quo. So just judging from past history and how issues such as this unfold in the Middle East, I am going to go with the status quo. I think trying to overcome history and the Grand Ayatollah will be too steep a hill to climb this time around. But the winds of change, they are a blowin' in Iran and this will be an interesting race to watch.

Not much in the news today but Prime Minister Maliki is warning that violence will undoubtedly increase as US troops withdraw from Iraqi cities and the national elections draw closer **(MNFI, 12 JUN 09: *Wall Street Journal*)**. His line is that the Iraqi Security Forces can handle the responsibility. This statement may be debatable, but the bottom-line is that they are going to have to. It is better to find out the gaps in their security now while we are still here then to find out when we are stepping on the plane heading home! Violence will always be a part of the culture here. It is up the ISF to ensure that it is at least containable. We have much violence in America as well in case people may have forgotten, but by all appearances ours is being contained.

LTC Chuck Samaris called this AM and congratulated me on my performance of yesterday. Apparently I received rave reviews from COL Fuller on my Iraqi pronunciations. Yes, this is the same COL Fuller who made me audition for him when I first arrived in Iraq! But this was nice to hear since I did have to spring into action on short notice, and this also validated my personal assessment of my performance. And that's the way it is!

Well, Shari made it home from Florida all right. It was a quick trip but she has her youngest son now situated in Orlando. She also informed me she was going to have her nose pierced and get a tattoo on her ankle! Now Shari is a good-looking woman and already has a very nice tattoo on her back. Now, she wants one on her ankle, which will be a bracelet with the birthstones of her children on it. I asked her why she just didn't wear an ankle bracelet with charms and she said that was just silly! No sense in arguing on that one! But on our honeymoon it was I that mentioned something about getting her nose pierced. And so she did, after a little liquid courage to get up enough nerve! I thought it looked sexy. So after she had a few issues with the original piercing I was surprised she wanted to do this again. It will be quite difficult for her to get any sexier than she already is! Me, I have no piercings or tattoos on my bod after 52 years of existence! I will never say *never*, but let's just say neither of these things is high on my priority list, or on my bucket list.

No piercings or tattoos on my bucket list!

Day 220: Saturday-June 13, 2009/Day 189-Iraq began with interesting news from both war fronts. In Iraq, the situation continues to progress. As US troops continue to drawdown it appears that US diplomacy is increasing. This is the right formula to achieve success in a counterinsurgent war. As a result of this increase in diplomatic efforts, many of the military reconstruction projects around Iraq are being turned over to the Department of State **(MNFI, 13 JUN 09: *TIME*)**. Diplomacy is the instrument of national power that by all accounts was clearly lacking for the first few years of this war. It also appears as though it wasn't until about 2006 or 2007 that the meshing of the military and diplomatic efforts began to occur. I will state many more observations on this situation in the final chapter, but it is abundantly clear that a clash of personalities, lack of a coherent national strategy, and ineffective leadership at the highest levels all played a part in our early floundering here that was paid for in the lives of many US troops, innocent Iraqi civilians, and billions of taxpayer dollars. I have researched this topic adequately and observed enough during my tour that I am confident in making such statements. Such a waste! But success is now within the grasp of the Iraqi people. And Iraq's envoy to the US, Samir Al-Sumaydi, summed up the situation succinctly by stating that "when Iraq goes right, the pride should not be felt only in Iraq, but also in the United States" **(MNFI, 13 JUN 09: *KUNA*)**.

In Afghanistan, the situation is different but evolving. Because we took our eyes off of the ball for the same reasons we struggled early in Iraq, we must now go back into Afghanistan with a new purpose, a realistic strategy, and basically regain some of the momentum we had gained before we invaded Iraq. So General Petraeus is preparing the public and our government for the worst in Afghanistan **(JASG, 12 JUN 09, *Petraeus: Afghanistan attacks at high*: CNN & Stars & Stripes, 12 JUN 09, *Petraeus: 'Difficult times' in Afghan fight*, written by Kevin Baron).** We know the fight will be tough and we know we will see many more flag draped containers returning to US soil before all is said and done. But perhaps for the first time in seven years, there is new leadership that understands the fight we are in; a meshing of all instruments of national power; a proper resourcing for this theater of operations; more emphasis upon the Pakistan situation and their role in this particular war; a new resolve that this will not be a quick or easy fight in which to achieve success; and some increased economic leverage that the US can effectively utilize in this war. Why did it take us so long to reach this point? In my estimation, it all gets back to what I have told you all along – leadership. Having the right people in the right places to make the right decisions is sometimes more an art than a science. But perhaps the key point here is not about putting smart people into such high level positions, but rather, putting the people that have the wisdom to do what is right in those key positions. As we have seen all too often in both wars, intellect does not necessarily equate to wisdom!

To close out this chapter, I must tell you about a recent situation at the De Fleury DFAC that boggles the mind. So I am going to lunch, as usual, and I grab a plastic plate and head into the DFAC to peruse my food selections. I notice they are having Sara Lee cheesecake today. The De Fleury has this dessert about every three days or so. My normal MO since I have been here is to grab a piece when I can and put it in a Styrofoam food carrying container to take with me for consumption later. My grandma Grace loved Sara Lee cheesecake and every time I have a piece it reminds me of her. She was a wonderful woman who passed away December 9th, 2000 at the tender age of 92. I miss her very much. Now Gram had a sweet tooth, which must be hereditary because my dad is afflicted with the same condition! Anyway, cheesecake was not listed on their online menu and this would make the second day in a row they had this tasty treat which was unusual. But being the smart guy that I am, I decided to grab this opportunity at another piece, put it in a Styrofoam container, and take it back to my room. I normally cut the piece in half and have half one night and finish it off the next. But today would be a bit different. I went over to grab a Styrofoam container, as usual, and a young Specialist told me I could not take one. I asked him what he meant as I had been doing this same practice for over six months now. He said it was a new policy and just went into effect that day! I told him I was grabbing a container to carrying my prize desert back to my barracks. He said the new policy was one plate and you can eat your food at the DFAC, or a Styrofoam container and you can take the food with you, but one and only one choice! I must admit, I was a bit miffed and was very rude to the good Specialist. I went back to finish my meal and developed an alternative plan which still worked out quite nicely, but left the DFAC wondering what was up with this new policy? I could tell the poor Specialist was getting an earful from a lot of people. Was there a critical Styrofoam shortage? Was this part of a weight control program? Was the contractor trying to cut costs? Was there a litter problem on VBC? I just had to find out the answer as this inquisitive nature is in my DNA. Just ask Shari! As I thought about the entire bizarre issue, my biggest thought was how rude I was to this young Specialist who was just trying to do his job. It wasn't his policy. He was not to blame for this strange and unadvertised decision. So I felt I had to apologize the next time I saw him for my inexcusable actions. And I did not have to wait long as I saw him again the very next day and offered my sincerest apology. Then he proceeded to tell me that it was now OK to take two containers, as the policy had been reversed! At least for the time being! He then told me that the brainstorm, or brain fart, came from the Mayor's Cell, which are a group of military people who run certain sections of VBC much like a Mayor runs a community, hence the name. This particular day I did not need two containers, but I apologized to him again and told him to keep up the good work. Now I blame these dipsticks who came up with the idea in the first place, failed to notify anyone of such a change after all of these years using the same procedure, and then made these poor kids carry out their dirty work. Now that is poor leadership and a failure to plan. I will make it a point to check with one of these Mayors on VBC and find out why the change is being implemented. Perhaps it is for a very good reason, but their implementation of such a change has been terribly flawed. I will keep you informed of this incredibly important drama here on VBC that I shall call – Styrofoam-gate!

Cheesecake & the Styrofoam mystery

CHAPTER 13
400 DAYS – Days 221-254
June 14 – July 17, 2009

> *There is no action the American military can take that, by itself, can bring about success in Iraq. But there are actions that the US and Iraqi governments, working together, can and should take to increase the probability of avoiding disaster there, and increase the chance of success.*
> --Secretary of Defense Robert Gates
> 2006-present

Day 221: Sunday-June 14, 2009/Day 190-Iraq is a day to be remembered. To begin with, today is my beautiful wife's birthday. However, as a gentleman I will not reveal her age. Today she will receive flowers and I will bring her gift home with me when I go on R&R in a few more weeks – that is besides me of course!

Shari is the love of my life. Even though we do not see eye to eye on politics or some NASCAR drivers, we have so much in common. Shari has had a lot of tough times in her earlier years, but she has emerged from it all a much stronger person. While I affectionately call her a 'redneck' every so often, I would not change a thing about her. It is too bad we could not have found one another sooner than we did, but as the saying goes, 'better late than never.' So we plan on making the best of our time together as we approach the 'golden years'. And that is why being away for this long after just finding one another has been incredibly difficult. But this too shall pass and we will move on. Perhaps the greatest gift I can give to Shari is to get out of the military. Then we need not worry about another deployment and having to go through this all over again.

Today is also Flag Day and the Army's birthday. Now I must share a little irony with you, but it works out pretty well for me especially as my mind begins to fail! Shari's birthday falls on the Army birthday. I have spent 28 years in the Army (National Guard & Army Reserve) so it has been a big part of my adult life. Not that I would ever forget my beloved wife's birthday, but I have a few big reminders to help me just in case! In addition to this, we were married on November 10th, 2007. November 10th just happens to be the United States Marine Corps birthday! The USMC is where I started my military career over 34 years ago. Shari, and not me, picked our wedding date as our church was not available in October and November 10th just happened to be the next open date available! But again, another big reminder so I should never forget our anniversary. Believe me when I say that the irony of it all is not lost upon me!

As today is Sunday, I want to take some time to explain to you how difficult the balancing act is over here and that every word and every action has a consequence. I have already explained the situation with press releases or comments from high-ranking officials and how they cause difficult issues for our troops over here. So let me try and articulate this point more clearly.

The issue of releasing the detainee abuse photos of several years ago has caused great consternation within our country. I have already documented this so I will not expand on this point any further. But since this issue surfaced a few weeks ago I have accomplished a little research. I tried to go back into the archives to see what happened on the ground when the news of Abu Ghraib first broke back in 2004, but I could not locate any data. There is some information that the beheading of an American contractor was related to this photo release and the abuse, but this may only be loosely related. However, in early April of this year, after Secretary Gates stated that al Qaeda was basically degraded to a point that

ISF could handle the situation if there was an uptick in violence. I will not state that there is any direct correlation, but I am also not saying there is not. There was a significant increase in HPA (High Profile Attacks) in April 2009, of which most were in Baghdad and Mosul, which just happen to be the last strongholds of insurgent activity. Coincidence? Perhaps. The day after the Secretary's statement was released, six soldiers were killed in Mosul! Coincidence? Perhaps. So while it may be difficult to identify significant trends here, it bears watching. It is also interesting to note that shortly after Secretary of Gates made this statement he quickly recanted it.

Now in light of Admiral Mullen's comments, we will have to wait and see how this plays out, but yesterday there was a vehicle borne IED just to the south of Baghdad that killed 29 local nationals (LN) and injured another 53. Just east of that incident, 4 coalition forces were injured in an RPG attack. Coincidence? Perhaps. But my point here is that if these are not coincidence and these attacks are a direct retaliation as a result of high-level commentary, guess who is paying this price? It is not the people making the statements. It is the troops on the ground! And I know, or at least I hope this is correct, that both the Secretary of Defense and Chairman of the Joint Chiefs of Staff understand this cause and effect relationship. But the balancing act to keep elected officials and our public informed of the progress over here, tempered with the fact that any statements can lead to an increase in violence by insurgents to invalidate such claims, is a very, very fine line to be sure.

The day concluded on a high note. We had a very nice celebration in the rotunda at the palace for the Army's 234th birthday. There was music, an inspiring video, a speech by General Odierno, and birthday cake. It was all very well done.

Then I called Shari on her 29th plus birthday. She had a few Jägermeisters on board when I called so she was feeling good! But she had a nice birthday and she received my roses. No one had baked her a cake however so she was a little disappointed in that. But next year is a red-letter birthday for her and God willing, we will all be home to celebrate. Happy birthday, honey. I love you.

Happy birthday! Especially you, honey

Day 222: Monday-June 15, 2009/Day 191-Iraq is a day that I need to congratulate the Pittsburgh Penguins for winning the Stanley Cup this year. Not sure that the President picked them as I do not believe he is a big hockey fan, but I do not know this for certain! But the Steel City now has two major sports championships this year with the Penguins big victory and the Steelers Super Bowl win last February.

I also need to congratulate the Los Angeles Lakers on their 15th National Basketball Association championship. I guess the President was just a bit off on this one as the Lakers won the championship in five games instead of his predicted six. Still pretty good Mr. President!

It also appears as though Mahmoud Ahmadinejad will retain his Presidency in Iran, as I predicted **(JASG, 15 JUN 09:** *Ahmadinejad set for poll win,* **source: BBC)**. But, there are a lot of rumors concerning corruption at the polls and many people in Iran are not happy. Based upon the numbers given and the pre-election polls, there is quite a discrepancy that few experts can explain. So either there is something that does not pass the smell test in the election process, or the pollsters need to go back to the drawing board because they are clueless! So it looks like the status quo in good old Iran, but there looks like trouble a brewin' to our east. Sure doesn't look like a fair election process took place, but then we have had hanging chads and other irregularities in our system so we should not be too quick to rush to judgment. But if the Ayatollah basically selects who will serve as his puppet in Iran, why don't they just say so and call it what it is – a theocratic dictatorship. It certainly doesn't resemble a democratic and free election process to me. I don't think they were trying to simply put lipstick on this pig. They put a wig, a dress, high heels, and a tiara on this one in Iran and called it – free and fair elections! Good luck selling that one to the world. They can't even sell that one in Iran! But if the Ayatollah is calling the shots anyway, does it really matter who the President is? Perhaps it would, but more likely it would be about the same old position on issues, just a different face. Too bad for all those women and the youth in Iran who really wanted to see some change. But as is human nature, people that sit far away in other countries only see the few bad people that give that country a bad name. They do not see all of the struggles and tribulations of the masses that really want change. These are the people who are just looking for a brighter future and the ones I feel sorry for.

And in Israel, Prime Minister Binyamin Netanyahu addressed his country yesterday and spoke of a possible two-state solution! While still too early to celebrate this long awaited compromise, it does make one take pause and provides some hope in this area of the world **(JASG, 15 JUN 09, *Israel PM calls for demilitarized Palestinian State*, source: CNN)**. Of course, as I mentioned a few days ago after his visit to Washington, this outcome may only become possible because of Iran and their quest for nuclear power. And to validate this observation one only need watch the Prime Minister's news conference as he opened up such a historic announcement with a discussion of none other than – Iran! Think he has Iran on the old brain?

Finally today, one of my favorite writers was again in the news today as Thomas Friedman stated that democracy has spread outside of Iraq's borders. He gives old George W. kudos for his vision of democracy in the Middle East by stating that "President Bush had a simple idea, that the Arabs could be democratic. The presence of a US Army in Iraq created a sense that change was possible" **(MNFI, 15 JUN 09: *NY Times*)**. Has everyone been so wrong about President Bush? Was he a visionary? Well, not likely considering his administration couldn't think far enough ahead to develop a strategic plan once we reached Baghdad! But dependent upon how all of this turns out, long after we are all gone, the history books may look back upon President Bush very favorably. Time will tell.

Well of course the Iranian elections were fair!

▌▌▌ ▌▌▌ ▌▌▌ ▌▌ **Day 223: Tuesday-June 16, 2009/Day 192-Iraq** is one with lots of news. The situation is shaping up for the end of June when all US combat forces must be out of Iraqi cities. An Iraqi government spokesman stated that this will occur, but that there will be a few troops that remain in the cities as trainers and advisors **(MNFI, 16 JUN 09: al-Iraqiya, Aswat al-Iraq)**. General Odierno reinforces this message by stating that there will be a small number of advisors and trainers in some cities still working with Iraqi Security Forces **(MNFI, 16 JUN 09: AP)**.

Then, as US troops begin to leave and will ultimately depart Iraq for good in December 2011 at the latest, some other Arab countries are stepping up their assistance levels. Egypt may assume some of the security and training roles once the US departs, particularly in the areas of the Air Force, Navy, and Special Forces **(MNFI, 16 JUN 09: al-Sharqiya)**. This seems like a good move and a very smart one at that.

As security and stability continue to progress the Iraqis are turning their efforts to rebuilding the country. They are looking to build many hospitals and clinics to address healthcare. Prime Minister Maliki stated:

> Iraqis have been deprived of development these last, long years. We are determined to continue

reconstructing the country despite all hardships. Security and stability now reign in Iraq, despite the efforts of supporters of terrorism and sabotage.
-MNFI, 16 JUN 09: *DPA*

    Today I would also like to exhibit to you just how difficult some of the decisions are over here. The other day I explained how people in Washington must balance keeping our elected officials and the general public informed on the situation in Iraq versus force protection. But there are many other decisions over here that can affect a situation both negatively and positively. It is a very fine line indeed. Let us see how you would react to a particular incident. This kind of reminds me of the old thought provoking question in relation to the proverbial spaceship that: can carry only 20, one of which is an astronaut and the pilot; and oh yes, the earth is about to blow up! Who do you send to carry on mankind? Would your group include people with different specialties such as a scientist, a physician, or an engineer? Would it include people of different races, religions, and of differing gender? This little vignette may not be exactly the way I have laid it out, but you get the point.
    Now, let us say you are over here in Iraq and are the commander of a personal security detachment. Your number one job is to ensure your 'clients' reach their destination safely. In order to accomplish this mission, sometimes you drive without due regard and piss off many Iraqi drivers as you try and avoid areas where an IED may be hidden or a known area where small arms fire attacks have occurred recently. So the question becomes: do you drive with the flow of traffic to avoid making other Iraqi drivers mad, or do you ensure you do whatever is necessary in order to get your client to their destination safely? Hold on a minute, it is not quite that easy. Keep these variables in mind as you mull over your answer. There may be a total of four to five people in one vehicle that may be hit by an IED; however, you may piss off 20-30 Iraqis as you weave in and out of traffic; and these 20 or 30 Iraqis then tell another 2-3 people each; and now there are close to 100 pissed off Iraqis that are upset with the Americans. Now it must be stated that most Iraqis are good people and seek a better future. There are only a handful that

are bad apples, but because of the insurgent's ability to collect intelligence and then use it against coalition forces, they will use such incidents to recruit more insurgents and as the impetus to launch more attacks. I told you it was not that easy! Remember, saving face and getting vengeance is very important to the culture over here. As a result of the erratic driving habits, you **may** now have fueled the insurgency further and this **may** result on several attacks upon coalition forces. What would you do? Is it worth the risk? Again, these are the very complex and difficult decisions that must be made over here by 19 and 20 year olds. And you wonder why there is stress over here? So from the highest ranking US officials to the youngest private patrolling the streets, every action, every word has meaning, and one false slip of the tongue or one inappropriate reaction can lead to very dire consequences.

Who is going on a trip?

Day 224: Wednesday-June 17, 2009/Day 193-Iraq is one that I should outline a few more details on my experience here. After all, this book is supposed to allow you to see this war through my eyes. Obviously, it is difficult to highlight everything that goes on, but I hope I have and can continue to provide some insight into this experience. With this being stated, I have already highlighted my normal weekly battle rhythm (routine) for you. But outside of my normal work routine and the atmospherics around Iraq and the Middle East, I am not sure I have given you sufficient background. So, I will try and fill in some of the gaps today.

To begin with, I may have told you that my day typically begins at 0500. Then I make a pot of coffee, watch a little **CNN** to catch up on the world news (if the cable is working), and then Skype Shari. Then about 0600 it is time for personal hygiene and getting ready for work.

I leave the barracks about 0630 every day. Some days I may arrive at the palace a little later if there is no BUA or a later one, otherwise I get there about 0645 or so. Once at the palace I dive right into my normal routine for that particular day.

Then at around 1100 or so it is time to go workout (cardio routine on Mondays, Wednesdays, and Fridays and perfect pushups and sit-ups on Tuesdays, Thursdays, and Saturdays). After working out I try and cool down and relax a bit before going off to lunch. I will provide my normal weekly menu later in the chapter. During my little respite in the afternoon to break up an otherwise long day I may watch **The Simpsons, Everybody Loves Raymond**, or some more news. I may also take a power nap if I am really dragging. Then after lunch it is back to the palace for my normal duties until about 2000 and then back to the barracks. At the barracks I will often have a bowl of cereal and my half slice of cheesecake for dinner, read a little from whatever book I am enjoying at the time, and then perhaps watch a little television before heading off to see Mr. Sandman. As I may have told you, there are only so many cable channels to choose from over here and many carry the same exact programming! So your selections are very limited, but there are a few shows I like to watch during the week including **CSI, Chuck, and Bones**. Some others such as The Unit, Ghost Whisperer, and Supernatural I enjoy as well but they come on at a later hour when I am normally counting camels. And then the next day, I start the same basic routine all over again!

Today I also sent out footlocker number three, my prized Storm Case. It was stocked full of stuff and tipped out the scale at 68 pounds, so I had two left to spare! This footlocker was not oversized as the first two had been, so it was considerably cheaper to ship. Once again the process went smoothly with the people from KBR and this footlocker should beat me home or be there shortly after I arrive.

There was a nice little article in the *Stars and Stripes* about Stephen Colbert's trip over here **(Stars & Stripes, 16 JUN 09, *Colbert connected in Iraq by not playing it safe*, written by Linda Campbell)**. Mr. Colbert did not just blow in, do a show and blow out, he did a series of shows and even had his prized locks cut by General Odierno. I highly doubt that many other celebrities would have done the same. I watched one of his taped shows last night after he had returned from Iraq and the audience was very appreciative of his efforts to entertain the troops. So, as he often states on his show, 'a tip of the hat' to Stephen Colbert.

Taking a peek east, General McChrystal has taken command in Afghanistan. General McChrystal clearly understands the fight we are in. Now he must shape all actions in this theater of war around a coherent strategy **(Stars & Stripes, 16 JUN 09, *General McChrystal takes command in Afghanistan*, written by Jason Straziuso)**. There can be no doubt that this fight will be long, it will be tough, and there will be many more flag-draped containers returning to US soil before all is said and done. But when fighting a global insurgency hell-bent on destroying the infidels in the West, this is a fight that we must continue to wage and we must not lose. The future generations of America and the free world are depending on us. And this is a mission we do not take lightly, but it will be our public support and political will that will allow our efforts to succeed. Once this support ends, the insurgents have won. And as there are less than 1% of the American population waging this battle, it would be very easy for the other 99% to say, "Let's end this now." But this is when the insurgents will be able to train, plot at will, and carry out more dastardly attacks. We must continue to push them, hunt them down, and give them no rest and no quarter. In this quest we must not fail.

Armed Forces Network television

Day 225: Thursday-June 18, 2009/Day 194-Iraq is one that continues to focus upon June 30th. If you have been paying attention, this is the date when all US combat forces are to be out of all Iraqi cities. This message is being driven hard. Attached to this message is the fact that there will still be some US advisors and trainers in some of the cities. This of course is to inform the Iraqi people that they may still see some US troops in the cities, but they are only there in an advisory capacity. Prime Minister Maliki states that Iraqi Security Forces will not ask US forces to participate in any combat operations after the June 30th deadline. Apparently they will use us as a taxi service however to move their troops around as they "don't have any planes" **(MNFI, 17 JUN 09: AFP)**. I hope he is right, and he is trying to show the Iraqi people that he is in charge and sticking to the terms outlined in the security agreement. This will further legitimize the government and lead to greater stability. Sooner or later this had to occur and I hope they are up for the challenge. The good news is that we will not be that far away should they need any assistance.

There is also much news from our neighbors to the east – Iran. There appears to be great unrest and discontent related to the sham of an election process they just conducted. There also appears to be an underlying dichotomy between the old guard generation that is anti-West and autocratic, and the younger generation, including many women, who want a softer stance with the West, an improved economy, and a more liberal form of government **(JASG, 17 JUN 09, *Rival rallies in Iran*, source: BBC)**. This situation is worth keeping a close eye on as civil discontent continues. This may lead to a civil war within Iran. As I have stated on numerous occasions, it all comes back to leadership. People who try and misuse power and authority will ultimately fail. Just ask old Adolph? Oh, I guess he is dead. How about old Saddam? Oh, I guess he is dead too. Get the point? When the majority of people want a change for the betterment of their lives, those people that lie and cheat to keep them from realizing such a dream will be met with great resistance and revolution. There can be no doubt that the elections in Iran were unfair. I am not sure whom they are trying to fool on this one. They are not even good cheaters! So now they hide behind religious and military veils and insist the elections were fair so that they may remain status quo. There are obviously many Iranians who believe otherwise. So keep an eye on this as there is trouble a brewin' in old Iran. Again, due to greed and corruption it is the millions of innocents who suffer. Maybe Iran should look at a **ledocratic** form of government! Just a thought!

Day 226: Friday-June 19, 2009/Day 195-Iraq was haircut day once again. I have stretched my current cut to about three and half weeks and it is getting a little shaggy. But if my barber does a **'George Custer'** then I will have about 11-12 days for it to grow out before I go home on R&R. Then, with 15 days of R&R I will need a cut as soon as I return, or perhaps even in transit if I have time. After that haircut, with just over 100 days remaining upon my return to Iraq, I may need only about four more chop jobs. I do not say this disparagingly because some of the barbers here are actually quite good, but they still are not my regular barber, Todd Twait, back in good old Wisconsin Rapids.

Just a trim please, not a scalp job!

Dominating the news today is still our movement out of Iraqi cities by the end of June per the terms of the security agreement. For whatever reason, some of the insurgent groups want to increase attacks on coalition forces to claim they are forcing the withdrawal, as opposed to the US and Iraq adhering to the terms of the security agreement **(MNFI, 19 JUN 09: *Buratha News*)**. Again it is all about perception, but unless people have simply not been paying attention, the US has launched an aggressive information operations campaign to explain exactly what is occurring. Perhaps it is the insurgents who have not been paying attention!

There has developed a bond, a kinship of sorts, between Iraqi Security Forces and US forces. As of this date, Mosul and Baghdad are the biggest areas of concern in relation to insurgent activity. So I doubt that US forces will move too far out of these areas, but they will move out nonetheless. A Colonel from the ISF in Mosul sums up the situation this way:

> My soldiers will do their job but we still need the Americans. They are like our right hand, especially when it comes to air support and logistics. We need electronic-jamming devices because our vehicles don't have them. We also need air support because our soldiers feel more secure with helicopters above them, and it deters the enemy.
>
> **-MNFI, 19 JUN 09: *AP***

So we shall keep a very close eye on the atmospherics that develop after June 30th.

As we have talked of the turmoil in Iran the past few days, we will look to the Far East and our friends in North Korea. Old Kim Jong-il, to quote a famous chef, is 'kicking it up a notch' in regard to his rhetoric. There has been discussion of North Korea launching a rocket toward Hawaii as they continue nuclear testing **(Fox News, 18 JUN 09, Report: *N. Korea may fire ICBM toward Hawaii*: reported by Jennifer Griffin and AP)**. I would say that old 'dingle dorf' is dangerously close to getting his butt handed to him! He is not impressing China or Russia either, so he may be standing alone on this one. There sure is a great deal of discontent around the globe right now. What happened to peace on earth and good will toward man? Why must we always be in conflict? Sure seems like a flaw in our DNA, does it not?

Kickin' it up a notch

Day 227: Saturday-June 20, 2009/Day 196-Iraq was more of the same news in regard to the unrest in Iran and our withdrawal from Iraqi cities by the end of the month. Of course, the news of the US movement out of Iraqi cities has been beaten like a dead horse **(Security Agreement media information, 20 JUN 09: *al-Baghdadia*)**. So while some of the not-so-bright insurgent group leaders will still try and claim it was them that kicked us out of the cities, most Iraqis, unless they have no television, no radio, they cannot read the newspaper, or they cannot hear the talk in streets, know what the real deal is. I think the confusing part of the whole situation will occur with US trainers and advisors remaining in some of the cities after June 30th. They will still wear the uniform and carry weapons. So we shall see on that one.

The ISF continues to become very adept at taking control of their country. In Diyala Province, just north-northwest of Baghdad, the ISF arrested four members of an Iranian-backed insurgent group and seized a cache of weapons and equipment **(MNFI, 20 JUN 09: *al-Sharqiya*)**. Hopefully the little issue in Iran right now will inhibit this flow of funding and equipment into Iraq for a sustained period.

I continue to develop myself professionally and am now reading a book called ***Learning To Eat Soup With A Knife*** written by counterinsurgent expert John Nagl. It is more of a textbook than a novel, but it is right in my wheelhouse as it highlights the culture and organization of the US Army and some of its impediments to progress! Much of what I have read thus far in this book simply validates my own observations I have made throughout this book. I will talk more about Mr. Nagl's book in succeeding days, although I plan to finish it while traveling home on R&R, but the basic recipe for an insurgency is beginning to take shape in Iran. The government is now not viewed as legitimate in the eyes of many due to the latest election faux pas, especially among the college educated and female populations; the economy is poor and cannot take care of the populace; and there is great discontent and unrest. If smaller splinter groups begin to take up arms against the government and its security forces, then you may have the beginnings of a full-blown revolution. We shall see how much clout the recent words of the Supreme Leader will carry as he calls for peace, states the elections were just and fair (just an outright lie by a religious man), and tells those creating the unrest to cease and desist or there will be consequences. Ayatollah Khamenei clearly despises the West, a fact he does not try and hide in his statements, and since his selection for the Presidency is his own, what did you expect him to say? We shall see if the people of Iran will settle down and continue to live in this mock free society, or if they will say enough is enough. Stay tuned on this huge story over the next few weeks as we watch to see if it continues to pick up steam or simply fizzles out.

One fact you may not know is that Iraq is in the midst of a serious drought. This is the fourth straight year with drought conditions and the temperatures continue to climb and precipitation is missing in action. It has been a long, long time since I have seen any precipitation here. Because of this, the Government of Iraq is exercising its ability to govern by seeking terms with Turkey to provide more water from the Tigris and Euphrates to assist the farmers with irrigation and also aid hydroelectric power generation **(MNFI, 20 JUN 09: AFP)**. So it is good to see the relationship building between Iraq and Turkey as the government continues to evolve. Bad news for the insurgents!

The Sons of Iraq program that I have spoken about previously is still a work in progress **(MNFI, 20 JUN 09: unclassified SoI update)**. However, if the Iraqi government fumbles the ball on this one then things could go south in this country very quickly. If you recall, these people belong to the same tribes that fought against us earlier in the war. But because al-Qaeda was so brutal toward the tribes and would indiscriminately kill innocent people, this same group decided the coalition was the better of two evils and turned their weapons on al Qaeda. One of the key phases of counterinsurgency is incorporating the less radical elements of the insurgency into society. Most of these people just want to find work and provide for their families. So after al Qaeda was degraded considerably, coalition officials sought to get the Government of Iraq to provide jobs for these individuals. It is a good concept, but because many of these people are Sunni and the Prime Minister is Shi'a, it has caused great sectarian tension and is a real test to determine if the government is really legitimate and ready to consider the welfare of all the people of Iraq, or if it simply is seeking to take care of its own and damn the rest! So this is a very big deal, and with the payment issues and this sectarian influence, this is still a work in progress. However, if these Sons of Iraq get pissed off, they may revert back to the insurgency and pick up arms against the coalition and the government, which could quickly destabilize the situation here. So this bears watching very closely.

Day 228: Sunday-June 21, 2009/Day 197-Iraq takes me back to a story concerning the increasing suicide rate in the military. Not surprisingly, a report has surfaced that alcohol abuse has soared in the Army as well! Since the war began in 2003, the number of soldiers being treated for alcohol dependency has nearly doubled **(Stars & Stripes, 20 JUN 09, More troops being treated for alcohol abuse, source: USA Today).** Now you already know we cannot have alcohol in a war zone as this would be a violation of General Order #1, so much of this must be manifesting itself in soldiers after they return home and find it difficult to transition back to normalcy. And we should also realize that alcohol is a depressant, which can lead to depression, which can lead to suicide. So this is a beast that the Army must find a way to control and they have their work cut out for them. This is like the proverbial onion that the more layers you peel back, the more associated issues are revealed, and the more you begin to cry!

Next we look at another story that I have left alone for a while which relates to recruiting. As I mentioned several days ago, the Army claims they are not having any issues recruiting people yet there have been claims that standards have been lowered to allow more people in. Of course the Army disputes this allegation, but a federal judge in San Francisco overruled two city ordinances that forbid Army recruiters from contacting people under the age of 18 **(Stars & Stripes, 20 JUN 09, *Judge strikes down California recruiting ban*, source: AP).** In fact, one city passed this referendum last November by an overwhelming 73% vote. The Justice Department immediately sued, which led to the ruling from the federal judge that the city ordinances were interfering with the government's ability to raise an army and protect the country. There is no doubt that the military gets an awful lot of young kids right out of high school and I still believe that those under 18 must have a parent sign any contractual agreement. I had to do this way back in the mid-70s! So philosophically I do not take issue with the Army recruiting in high schools. The military is a career choice like any other occupation. So with all due respect to Johnny Paycheck, if mothers don't want their babies to grow up to be soldiers, then they better start working on their college application while they are in elementary school!

Today is also Father's Day. I will call my dad later in the day to wish him well on this special day. Speaking of my dad, if I got my brains from my mother, then I got my common sense and practicality from my father. My dad is a Korean War vet who was in the US Navy aboard the USS *Oriskany*. He grew up in the Madison/Milwaukee area and he was a blue-collar worker who did the best he could to raise three children. My dad is also a realist and doesn't trust any politician or attorney. Who can blame him? Father is also who I got my love of sports from. I still remember the day when Bart Starr snuck into the end zone on the frozen tundra in Green Bay to win the NFL Championship in December 1966. My dad and his brother, my Uncle Jeff, were at a local watering hole watching the game and I got so nervous I hid in the bathroom until I heard all the cheering. Not sure why that was, but it only happens when I watch the Packers for some reason! And even though I think our country puts far too much emphasis upon professional sports and its star players, many of whom are overpaid for playing a children's game, I still love to watch sporting events. I hope to get out on the golf course with father when I am home on R&R, which is now right around the corner. My dad just goes golfing for the exercise and to drink beer. He really could care less about the golf game. In fact, we normally do not even keep score! But anytime out with dad is always a special day. So here's to you dad, wishing a wonderful father a wonderful Father's Day.

Happy Father's Day Dad

Day 229: Monday-June 22, 2009/Day 198-Iraq is brother Brad's 50th birthday! **Nerd-Bob**, as I affectionately call him, and in turn he calls me **Lizard Breath**, is taking the day in stride. Brad is a former Marine who went into the Corps a few years after I did. In fact, we were in at the same time back in 1977-78. Brad was stationed out at 29 Palms, California and really did not enjoy the desert heat. Brad has had some tough luck in his life, but is a good brother and watches out for the parents. Brad is normally not much of a talker and doesn't show emotion often, but he has sent over some very nice letters that have been from the heart. So here's to you, brother, on your 50th. Have a cold one for me.

Happy 50th Bro

Not much new news in the BUA today, but the situation in Iran remains very interesting. It does sound like a real power struggle between those seeking freedom and a brighter future versus those seeking to repress such an attitude. Looks like the old Rockin-rolla Ayatollah has got his hands full on this one! As I have often told students in my leadership lectures, the day of autocracy has long since passed. This form of leadership had its day, but in this country this leadership style has evolved into a more participatory/transformational type of leadership. Autocracy doesn't even play well in the all-volunteer military anymore. Some individuals will still try and use this antiquated style of leadership, but it just simply is not very effective anymore given today's society. So those individuals that have not adapted their leadership style with society have their work cut out for them in trying to achieve sustained success.

Most people want to be free, especially young people, and so what I see going on in Iran is a clash of generations. One generation which has been brought up under autocratic rule and knows no other way, and the youth of Iran, which is a very young country, who want more freedom and a different mode of leadership. We have witnessed part of this interesting dichotomy here in Iraq. The people who have most embraced the newly-found freedoms here are the youth and female populations. It is the old hard-liners that grew up under Saddam's regime that know of no other way that are struggling with the new concept. Sounds like they need some leadership courses in Iraq and Iran to teach these old dogs some new tricks! And believe me when I say that we have not yet mastered this transformation in our own country, so it will take decades to change in others.

Well, as the day is fairly slow I will talk of my physical fitness routines this week. I am sure you are all so interested, but this is just another little nugget to inform you of my own unique experience over here. As I have told you previously, because there is little else to do here, developing my mind and body fill the gaps in time. So if I am not in learning mode, then I am in physical fitness mode. As such, I have learned a great deal about the situation here and the history behind it. Part of this professional development has been a result of my studies in Naval War College and some has been through my reading of books about this war. On the physical fitness side of things, I work on a cardio routine three days a week, do my perfect pushups and exercise bands three days a week, and then do sit-ups five days a week. Then I take Sundays off. If it was good enough for the real big guy (give me a G, give me an O, give me a D – what does that spell – the real big guy!), then it is good enough for me!

I will say because of this routine I am in the best shape since I went into the Marine Corps over 34 years ago. Back then I was running close to six-minute miles, and the Marine Corps physical fitness test run event was three miles. The Army runs two miles for their test, but years of pounding the pavement has taken its toll on my knees and back so I just cannot run as quickly as I used to. Losing 30 pounds, so I do not have to lug that around with each step, has also helped the cause.

Anyway, I vary my running routines so I do not become too bored. I begin each cardio workout with a brief warm-up for 2 minutes at 3.5 mph. Then I cool down after each routine for 2 minutes at 3.5 mph.

On Wednesdays I incorporate elevation into my workout and on Fridays it is speed day (which I will go over in more detail on Wednesday and Friday of this week). But Mondays is my distance day. I call it the **Fuzzy Jurgensen** cardio routine because it helps me remember what to plug into the old treadmill! Now see if you can follow my logic on this one!

Fuzzy Thurston was one of the Glory Years Packers who played for Coach Vince Lombardi. He was the guard that played opposite of Jerry Kramer and used to be one of the pulling guards that led the famous Packers sweep. Fuzzy was a two-time pro bowler and is in the Green Bay Packers Hall of Fame. Anyway, his number was **63**. Stay with me now.

Sonny Jurgensen used to play with the Washington Redskins and was a five-time pro bowler and was enshrined in the Pro Football Hall of Fame in 1983. Jurgensen also played for Vince Lombardi when he coached the Redskins in 1969. Sonny wore the number **9**. Anyway, I was fortunate to grow up during the Vince Lombardi years when the Green Bay Packers were dominating the National Football League for much of the 1960's. I was a huge fan then and still am today. As someone who was born in Wisconsin, I think it is written in the fine print of the state constitution somewhere that you have to be a Packer fan or you are immediately deported to Illinois or Minnesota!

Now with this backdrop, my Monday workout is at a **6.3** mph clip for **9** laps, or a little over two miles. Hence, the Fuzzy Jurgensen! Weird? Perhaps. But it works just fine for me and you haven't seen anything yet!

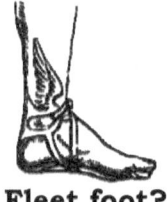

Fleet foot?

Day 230: Tuesday-June 23, 2009/Day 199-Iraq did not start off too well. To begin with, I overslept for the second time since I have been here so I missed Skyping Shari. However, I was not late getting into the palace. Once at work I observed a report that 8 more Iraqi citizens were killed in suicide attacks with another 40 injured. As General Odierno mentioned yesterday, AQI will kill anyone. They are not just trying to kill coalition forces—they are killing innocent civilians. All of this killing will only move the people further away from supporting their efforts so this strategy is incredibly flawed on their part. But if the goal is to incite sectarian violence and retaliatory attacks, this strategy is also failing miserably. If it is AQI, this just goes to show that if this type of extremist will resort to this type of violence here, they will do it anywhere. This is just another reason why we must keep the pressure on these people and not allow them to rest.

The situation in Iran is creating some interesting dynamics here. Let us see if you can follow the bouncing ball on this one. This is politics at its best – Middle Eastern style! The US Ambassador, Christopher Hill, said the US supports efforts to "restore healthy relations" between Iraq and countries such as Syria, Kuwait, and Iran **(MNFI, 23 JUN 09: *Iraq of Tomorrow*)**. Keep in mind the term "healthy relations." Several Iraqi lawmakers hope that the political unrest in Iran will make it more difficult for them to meddle in Iraqi affairs. Another lawmaker believes that Iran may fuel more conflict in Iraq to divert attention away from their internal strife **(MNFI, 23 JUN 09: *Washington Post*)**. So on one hand Iraq wants good relations with Iran, but covertly they keep meddling in Iraqi business! Keep in mind that Iraq and Iran do not like each other all that much. So it will be a delicate balancing act indeed to shake hands with one another while Iran is trying to stab Iraq in the back! Just not sure the trust is there with the current regime in Iran. After all, wasn't there just a sham of an election there or something? I have probably mentioned this once or twice thus far that I wish the Iraqi people the best, but I cannot wait to get back to America. One thing I can state without reservation about this place – it is never boring that is for certain!

Iraq is experiencing its own growing pains with democracy. Several Iraqi lawmakers have noted that the Council of Representatives is both ineffective and inefficient **(MNFI, 23 JUN 09: *NY Times*)**. A Kurdish representative said that the Iraqi people have a right to feel negatively about the council because they have not proven that they are working diligently to improve services in Iraq. Now, a couple of interesting side notes here. First, keep in mind that the Kurds are very suspicious of the Iraqi government, which has led to great tension between the two groups. The Kurds want to be a part of the new Iraq, but are unsure if they will be adequately represented by the government. If you recall, they basically control three provinces to the north and have a done a good job of carving out their own infrastructure. The Kurds have their own separate elections coming up next month. So they are a part of Iraq, kinda sorta! The other dynamic at play is that in order for the government to legitimize itself in the eyes of the people, they need to protect the populace and provide necessary services. This is a work in progress and very slow to develop. But if it does not develop, the people may revolt against the government, which then creates fertile ground for insurgents to take root, and we head right back into chaos.

Prime Minister Maliki is beginning to understand the ramifications of this delicate and fragile diplomacy as he stated:

> Parliament's system of democratic consensus is strange, conflicts with democracy and has many inherent problems. I think the presidential system is better than the parliamentary system if it will be accountable to the people.
> **-MNFI, 23 JUN 09:** *NY Times*

And to further validate what is at stake here, an Iraqi legal advisor to the government stated:

> If there was decent electricity and clean drinking water, there might be more patience from people. But on some of these laws, especially the ones that aren't controversial, the Council of Representatives has to do something or it risks losing legitimacy.
> **-MNFI, 23 JUN 09:** *NY Times*

So there are many dynamics to take of note here. The balancing act is a very fine line indeed and it will take effective leadership in order to move Iraq forward and stabilize the situation. The national elections early next year will be interesting to watch and see if the Iraqi people believe the current Prime Minister is the man to get the job done, or if they feel someone else is better suited for this major task.

Finally, today we go back to my workout. While I am certain this workout preview is much less interesting than watching Kiana Tom do her bikini workout, remember, this is my story! Anyway, Tuesdays, Thursdays, and Saturdays are perfect pushup days. I also do sit-ups, but I do these on cardio days as well. For whatever reason, I have always excelled at doing sit-ups. While many struggle with this exercise I never have. In fact, when I was in the Marine Corps stationed at Quantico back in the mid-70s, I was the unit sit-up champion as I cranked out 124 in two minutes! Now that is smoking! I only need a mere 66 in two minutes to max this event in the Army at my age. I have not really tried to see how many I could do past that point because it would all be wasted movement and tire me out more for the two-mile run which follows the pushup and sit-up events. Anyway, back to the perfect pushups. These devices are great and I have been using them for about two years now. A Navy Seal developed them so you know they must be good! But you get a good burn in your chest and arms and because of the rotating action of the pushup device, there is little to no strain on the wrist or shoulder joints. As I mentioned, since the Army APFT consists of pushups, sit-ups, and a two-mile run, this is what I have geared my workouts toward. I may have also mentioned that even at 52 I can still max my pushups and sit-ups, but I just cannot run as quickly as I used to and this event is where I need the most improvement. I hope to take a record APFT before I leave Iraq to see where I stand and how much I have improved, and before I get back into my bad habits upon my return (you know the barley and hops, chicken wings, etc.)!

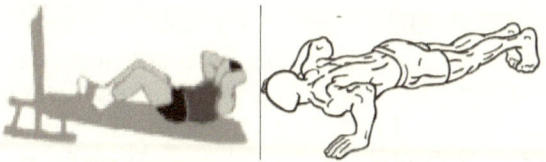

I am here to pump you up!

Day 231: Wednesday-June 24, 2009/Day 200-Iraq is a day I was going to take that furry little groundhog and stuff him right back into his little hole! I was bound and determined to break this cycle of one continuous day. One reason for this change was that my chariot was in need of routine service. It has been about four months already and slightly over 3000 miles. The vehicle has been running nicely and the AC works well, but I wanted to get service accomplished before I left on R&R. Then I should have one more servicing just before we get ready to leave Iraq this fall. So I dropped off my vehicle and hoofed it over to the Sports Oasis for breakfast before going to the palace. Remember, Wednesday mornings are a little more relaxed because there is no BUA on this day.

It was a 20-minute walk to the DFAC and it was already about 90 degrees F. at 0700! So by the time I arrived, needless to say, I was a little warm. My return trip this afternoon ought to be really nice as it is forecasted to be about 115 today! My breakfast was tasty, but once again my servers gave me too much food so I had to throw some away – again!

Today would continue on a slightly different course as my car was not expected to be finished until late afternoon. So my normal routine of going back to GRC for my workout and then lunch at the DeFleury was not going to happen today. I would either go back to the Sports Oasis, or perhaps I would walk to the Coalition Café, which is again open and see if the Pop's Burger had yet returned to the menu! Then I would hope to leave the palace a little early to retrieve my chariot and return to GRC for a later workout. I wanted to call Shari before I left the palace because the Internet was down this AM, so no Skype! Not sure what the deal is this time, but it is down again! WTF! My timing would need to be pretty good as well as tomorrow is GRD BUA day. So I would hope to hang our finalized slides and practice the script before I left to get my vehicle. But there are days when it takes forever to get the slides and the script. I like to go back to the peace and quiet of the barracks to rehearse a few times. Hopefully, SFC Maltes would not have any issues getting into the NEC tomorrow so she can brief, but if she cannot, much like the last brief I will be ready to spring into action like the jungle cat leaping upon its prey! So we shall see how this works out today.

Little in the way of news today, although that is not necessarily a bad thing! If you recall, yesterday I mentioned the recent AQI attacks. Well, General O was not convinced that these latest attacks were perpetuated by AQI even though suicide bombings are their signature. The other extremist groups know this as well. Today there was a report that many of these recent attacks are being carried out against Shi'a neighborhoods, possibly by violent Sunni extremists **(MNFI, 24 JUN 09: McClatchy)**. Well, guess what? That would be called sectarian violence. This is where the ISF must step up and crack down on such groups. While they may be operationally capable of accomplishing this, it is still unclear if they are politically and culturally ready to accomplish this major undertaking. As I mentioned several days ago, this type of violence is simply a part of the fabric of this culture. There should be no delusion that this will completely cease when we depart here for good. They key here is that the ISF contain this violence much like our police forces in the US and keep the mayhem at an acceptable level. And acceptable is defined differently dependent upon the culture. At this stage in the war, it is really up the Iraqis to determine the level of acceptability, not us.

To continue my little workout vignettes for you, today was again cardio day. My workout today is what I call my **Fuzzy Everest Blast Off** workout. You are probably asking yourself – what the hell is that? Well, let me enlighten you. As I mentioned a few days ago, I try and vary my workouts so I do not become bored. So my three cardio workouts during the week are all a little different. After more experimentation on myself in regard to higher intensity versus longer duration workouts, and the effects of both on my heart rate, I decided to go with the higher intensity workouts. They are a bit shorter and I can elevate my heart rate more effectively. So after my normal warm-up I begin running at **6.3** mph (hence **Fuzzy** #63), which is the speed I stay at for the entire workout. I then incorporate elevation into the routine (hence the **Everest** reference). At the eight-minute mark I move to 2% elevation for **3** minutes; at the 14-minute mark I move to 4% elevation for **2** minutes; and then the final **1**-minute, I move to 6% elevation. The **Blast Off** part comes in due to the 3-2-1 format similar to the countdown for the space shuttle launch. The goal of this workout is to reach the second of two benchmarks, both the 20-minute and the two-mile marks. With this routine at this speed, I hit both almost exactly at the same time! Yes, I realize this is a bit bizarre, but again, it makes this particular workout easy to recall. Just remember, the goal is not how fast you go, nor how far you go. The goal is to just get out and go!

The Fuzzy Everest Blast Off cardio routine

▮▮ ▮▮ **Day 232: Thursday-June 25, 2009/Day 201-Iraq** was once again GRD BUA day. This would be my last until R&R. Then upon my return, I would only have about 6-7 more of these briefings to go.

I have begun to read a series of books on counterinsurgency operations by the leading people in the field. This select group includes: Australian David Kilcullen, who appeared on the Stephen Colbert Show on 23 JUN 09; John Nagl, a former armor officer who participated in the early stages of this war; David Galula, who wrote of the Vietnam experience; and Thomas X. Hammes, who participated in the early stages of this war with the United States Marine Corps. As I have already mentioned, I have started off my professional development in this area with the Nagl book, ***Learning To Eat Soup With A Knife***. Remarkably, Mr. Nagl was mentioned in the BUA this AM. He is now the President for the Center of New American Security and believes that stability in Iraq directly correlates to US security presence **(MNFI, 25 JUN 09:** ***Christian Science Monitor*)**. Well, we are supposed to be reducing our footprint here in Iraq and allowing the ISF to take over. Even though US combat forces will be out of the cities by the end of this month, this reminds me of taking the training wheels off of a child's bicycle. The child will be on his/her own to navigate and steer the bike, but the parent is close by in case the child might need some assistance. Our forces will not be that far away from the cities and if the ISF does need assistance, we will be there for them very quickly. We will just be a little further away in order to allow them to develop further on their own and we will basically transition to a larger QRF (Quick Reaction Force). Mr. Nagl goes on to state that:

> This is a different, stronger Iraq today, and that fact, plus a still-substantial American commitment for the next several years, mean it's likely Iraq will not resume civil war and will not export instability to the region.
> **-MNFI, 25 JUN 09: *Christian Science Monitor***

I hope he is correct on his assessment, but all of this will depend upon the leadership in Washington and in Baghdad. We have already witnessed too many times in our short history what happens when the wrong leaders are in the right positions. Often, the inability of this leadership to overcome personality conflicts leads to poor decision-making and adversely affects troops in harm's way. After reading about the British experience in Malaya, our experience in Vietnam and in this war, it has become crystal clear that the military's organizational culture and our government's bureaucracy continue to lead us down a similar path. In other words, as great as America is, we have the innate tendency to continually get caught always looking forward, and by doing so, we always seem to lose sight of the lessons of the past! This is not conjecture, nor is this guesswork on my part. The facts are out there for all to see and the bottom-line is that we continually fail to learn from history. I have made this point in the fields of EMS and the fire service as well, so this is not a disease strictly quarantined to the military and government. This seems to be a part of the American fabric. This is almost counterintuitive to all of the museums and displays that we have in this country to capture past history, yet we simply do not learn from it nor apply its lessons to the present. Where is the logic, Mr. Einstein?

Today I will get you up to speed on my dietary habits over here, as I am certain you cannot wait to hear about it! Well, despite my eating habits and my own experiments upon myself, I am holding steady at between 78-80 kg (171-176 pounds). I eat what I want, but control the portions. My eating habits are pretty consistent during the week as I have a salad day; a cheeseburger day; a chicken sandwich day; a gyro day; a taco day; a turkey wrap day; a pizza day; and an occasional chocolate chip cookie or piece of cheesecake. I know I am not eating enough vegetables or fruit and the dietary people reading this will be shaking their collective heads, but what can I say. So my portion control theory is being validated. Now with this being stated, I am not losing any more weight, but neither am I gaining any. My 30-pound weight loss came with better eating habits and no beer! My theory will be seriously tested when I go home shortly and partake in some real barley and hops liquid refreshment. I will also have some Johnsonville brats, chicken wings and fries, and all kinds of other good stuff. Now I do plan on still working out while on R&R and watching my portions. So I will report on my weight prior to leaving and upon my return to Iraq.

The <u>Waite</u> control program!!!

Day 233: Friday-June 26, 2009/Day 202-Iraq was mail time once again. Today I was shipping a small box of books from my Naval War College program. I am sending these books home in waves from the sessions I have completed thus far. It took less than 15 minutes to complete this task on this day and it would be my last mailing before heading home on R&R.

The last few days have been pretty brutal here with several high profile attacks. Many innocent Iraqi citizens are being senselessly killed. There was also an incident that injured ten of our troops yesterday near Baghdad. While it may be difficult to identify exactly who is to blame for these attacks, they have an AQI signature. Many of these AQI insurgents are foreign fighters and do not care if they kill Iraqis, whereas many of the other insurgent groups from Iraq will draw the line on killing their own. It also appears that once AQI becomes degraded to a point of ineffectiveness, they choose to commit higher profile attacks and do not care who the target is. In fact, US intelligence officials are claiming that the number of AQI extremists are waning because the group is finding fewer foreign fighters willing to conduct suicide bombings **(MNFI, 26 JUN 09:** *AP***)**! Finally, some sanity! These same officials claim that the higher profile attacks are also not succeeding in renewing sectarian violence so the people of Iraq know what is going on.

> Pentagon Press Secretary Geoff Morrell stated:
> We think we have beaten back AQI to the point where they are now conducting attacks that are basically propaganda campaigns to make it look as though they are driving us out of Iraqi cities.
> **-MNFI, 26 JUN 09:** *AP*

> Middle Eastern expert Joost Hiltermann commented:
> There is no way to stop every bomber. But strong and wise leadership should be able to prevent the kind of lethal revenge in 2005 that culminated in civil war.
> **-MNFI, 26 JUN 09:** *Reuters*

🏀 So incremental progress continues and we shall monitor the violence as we head toward the 30th of June.

Today I also bid farewell to another member of **'power row'** – LTC Bob 'Mad Dog' Pritchard. Mad Dog is a member of the USMC and is going to Atlanta to visit with the family before heading back to Tampa and CENTCOM HQ. Bob is a reservist and a home builder by trade. Bob has done a great job while here for his four-month rotation. Of course, as has been my policy, I provided him with one of my Chief's coins for his collection. Great job Mad Dog.

Power Row Salute

Today was also time for my Green and Gold cardio routine. Why the Green and Gold routine you ask? Stay with me on this as the bouncing ball will be moving quickly! After my usual warm up I begin this workout running at **6.5** mph until the **5** minute mark when I begin my **1-3** cycle, which is one minute at **7.0** mph and then back to **6.5** mph for three minutes until I reach the half way point of the workout. The second half of the routine begins at the **10** minute mark where I jump up to **7.5** mph for one minute and then back to **6.5** mph for three minutes until the workout is completed, followed by my normal cool down routine. The routine ends when I reach either the two-mile mark or the 20-minute mark, whichever comes first. The **Lombardi Trophy** is to signify first as it is awarded to the world champion (1st Place) in the National Football League. Bizarre to you perhaps, but it makes perfect sense to me. See if you can figure out the logic behind the workout in correlation to the figures below. Good luck!

The Green and Gold cardio workout

Day 234: Saturday-June 27, 2009/Day 203-Iraq was relatively slow so I will take this opportunity to provide a short analysis on a book I recently finished reading that was written by LTG Ricardo Sanchez called ***Wiser In Battle***. I am glad I read the book and got his version of the events. LTG Sanchez has been chastised heavily for his actions in the early days of this war in every other book I have read thus far. So after getting both sides of the story so to speak, I will provide you with my opinion on the matter and again, it is only my opinion. It appears that LTG Sanchez is defending his actions throughout the entire book. It is almost as though he is on the defensive, but perhaps this is his way of responding to all of the criticism levied against him. This certainly does not make him guilty of anything, except being human. But a few big-ticket items I will comment upon are these.

It has been stated by several different sources that LTG Sanchez and Paul Bremer, who was the highest-ranking civilian in Iraq after we had reached Baghdad, never got along. While LTG Sanchez, who is a very religious man, does not come out and state such friction existed in his book, you can tell by the tone that there were significant philosophical differences between these two men. The problem with this in a counterinsurgent war, which the General states they recognized, is that this action was very counterproductive. The military and civilian instruments of power must work closely and cooperatively together in order to achieve success and this clearly was not the case.

Secondly, if General Sanchez recognized the war as one of counterinsurgency as he stated, the actions that were taken on the ground during his time as coalition commander do not validate this assessment. In fact, many of things that were going on in Iraq under his command were actually fueling the insurgency. One need only read the books and manuals on this type of warfare by its leading experts such as Galula, Kilcullen, Hammes, and Nagl to understand the actions executed were not in alignment with sound counterinsurgent principles. Past counterinsurgent war history also was not taken into account, which could have provided the template for success.

Perhaps the situation that really throws a cloud over LTG Sanchez's command was the Abu Ghraib situation. There are many conflicting reports as to whether he authorized more aggressive interrogation techniques or not. He states that he did not issue such direction, but others claim that he had signed documents to that effect. As the coalition commander, many people see him as being culpable for this unfortunate incident in this war, which effectively ended his military career.

Now with all of this being stated, I believe LTG Sanchez was a good commander. He cared about his troops and did the best job he could while he was in Iraq. He most certainly was not given the resources to be successful here and was not supported as he should have been by the Bush administration. His command group was understaffed, under resourced, and basically had to develop a national strategy while fighting a war because the government and CENTCOM had failed to create a coherent post-combat strategy once they had reached Baghdad. It was not until after General Sanchez was reassigned in mid-2004 that another command layer known as Multi-National Forces-Iraq was created. This is the same command group that I work with here now. This command staff basically oversees operations and deals with the more strategic political realm and nation building, while the Multi-National Corps-Iraq command staff deals with operations associated with fighting the war. This adjustment in structure obviously tells me that there was a need for such a command group, yet it was not provided to LTG Sanchez even though he requested such assistance.

Then, LTG Sanchez was to be given command of CENTCOM and his fourth star, but because of the high profile Abu Ghraib issue and the embarrassment of the incident to the Bush administration in an election year, these actions were never completed despite promises to LTG Sanchez by Secretary of Defense Donald Rumsfeld. So it appears as though LTG Sanchez became one of the fall guys for Abu Ghraib. And while I found it interesting that LTG Sanchez never took any responsibility or admitted making any mistakes while in Iraq, which in the atmosphere in which he operated, given a task never before given a commander, and being woefully understaffed, it would have been virtually impossible for him not to have made mistakes. After an investigation into Abu Ghraib, the incident that received such great visibility in our media turned out to be a case of a lack of leadership within the organization, lack of oversight by the command group, and just basically screwing around by a group of undisciplined and ill-trained soldiers. There were never any beatings, or other forms of torture by this group. But what this does suggest is that this was the proverbial tip of the iceberg and the highest ranking officials in our government needed the military to take the fall on this incident so deeper, more troubling issues associated with aggressive interrogation techniques were not uncovered in an election year. If this were to occur it may have cost George W. the election! This was a clear case of not wanting to open Pandora's Box. Any deep investigation would have led to more questions being asked of our officials at the highest levels in the government. As the situation became more transparent years later, this is exactly what has happened, although the full depth of the incident has yet to be discovered. So I would state that I believe LTG Sanchez was unfairly chastised in this entire affair, but because of the situation so close to the presidential election, sacrificial lambs had to be offered to the public. I guess I never really knew what the **DC** in Washington actually stood for until I read this book. Most people believe it is a shortened version for the District of Columbia. But perhaps it actually stands for **D**is**C**laimer, due to the fact that nothing is ever the fault of our elected officials so we should not expect any meaningful legislation while they are in office! It could stand for **D**uck & **C**over because whenever issues get a little dicey in old Washington many of the politicians are heading for Duck & Cover bunkers to avoid the shrapnel! Perhaps it stands for **D**isorderly **C**onduct! But I think what it really stands for is **D**amage **C**ontrol! It is clear that George W. Bush wanted re-election so badly that many people in his administration were willing to do anything to ensure this occurred. Having a scandal lead back to the White House perpetuated by the events at Abu Ghraib could not be allowed during an election period. This is the dirty side of politics that many people never see. As far as I am concerned it is a disgrace to our country. We seem to always prefer to look outside of the Washington fishbowl at countries like Iraq and Iran and talk of how messed up they are. All they have to do in Washington to see a messed up system and the people that perpetuate such a putrid atmosphere is to take a look in the mirror! Perhaps we need to fix our country, as great as it is, before we try and assist others. One media commentator stated something I found intriguing about the elections in Iran. Many of the people took to the streets in protest of obvious fraud in the election process. These people wanted to make a statement that their government was corrupt and had cheated them out of a chance at a brighter future. Yet, when we had election issues in past elections, and we realize our government glosses over or withholds information especially in election years all for the sake of getting elected, our people just say, "Oh well, I guess that is just the way it is here." Perhaps that is what separates a civilized society from those that resort to violence. Civilized countries prefer to vote the dead wood out of office and try and replace them with others that can represent them honestly. But from what I have seen recently, it doesn't appear that we have quite accomplished all of the surgery necessary to cure the patient as of yet!

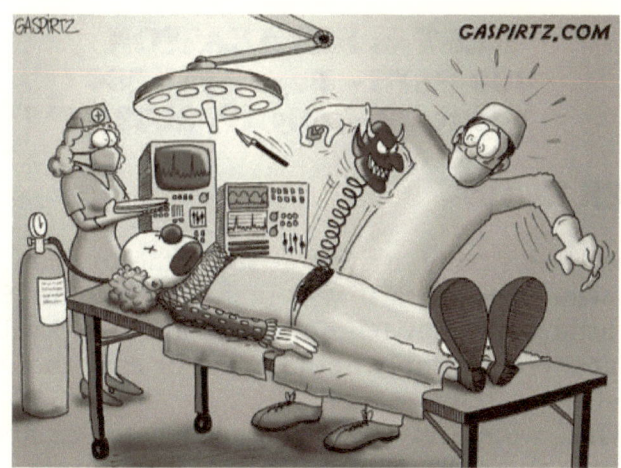

The patient is in trouble!

REST & RELAXATION
400 DAYS – Days 235 - 254
June 28 – July 17, 2009

THE TREK HOME

The day had finally arrived. It was time to head home for R&R and what a process it was just to get home. So let me start this little section by telling you all about it. I will start by stating that there is simply nothing fast or easy that occurs in a war zone!

I left GRC, where my barracks are located at about 1630 Baghdad time the afternoon of the 27th. There were four other people traveling with me today, including MAJ Fred Owens from the 416th TEC whom I had not seen since my unit split up upon arrival in Iraq last December. We were driven over to BIAP and checked in. Fred and I took our time waiting on our flight by having dinner at Sather Air Base, which will continue to remain my number one DFAC on VBC. Then we went back to BIAP with about 70 other troops and waited on the Air Force flight heading to Ali Al Saleem Air Base in Kuwait. This is the same place my unit had arrived when we came to war seven months ago.

The weather was about 125 today and we had to wear our battle rattle (body armor) on the flight. This is another one of those Spock moments where this process possesses little logic. You are not required to wear the 'battle rattle' on VBC which BIAP is a part of. You then turn in your battle rattle as soon get to Ali Al Saleem. So you basically only wear it on the flight from Baghdad to Kuwait. Now if the plane goes down, the battle rattle will not do much to save you! Fred thought that maybe we might need it if we went down and then had to battle some insurgents. Problem with that scenario is that you do not carry any weapons when traveling home on R&R! So unless it going to be hand-to-hand combat with the insurgents, we would be at a severe disadvantage! So not real sure about the logic in this one, but that is how the process works.

Once we arrived in Ali Al Saleem at about 2330 we were ushered to a few briefings, turned in our battle rattle, filled out our paperwork for our tickets home, and then got into our tents for about three hours of sleep. We were back up at 0530 to attend an accountability roll call. After this event, I had some breakfast, grabbed a cup of coffee at the Green Bean coffee shop, read the Sunday *Stars & Stripes*, and awaited our next time hack which was 1300. That is the time when we went into lockdown mode as we reported to a staging area, all 270 of us heading home on R&R! During our lockdown, we had several briefings, went through customs, received our flight itinerary, and then waited around until about 1830 when we conducted another roll call and then stood around in 129 degree heat until 1930 when we boarded the buses to head to Kuwait International Airport and get on board our freedom flight back to the US.

129/138

We packed into the Omni International Airway jet like sardines and began our 15-hour flight home. Along the journey we stopped in Shannon, Ireland for a few hours to refuel and stretch, and subsequently into Atlanta about 0800 Monday morning. Once there, we had yet another Spock moment. We unloaded the aircraft and were ushered to an area where we were checked in, our leave pass stamped, and we were given some paperwork explaining the process upon our return to Atlanta after R&R. This process actually went very smoothly despite the fact there were 270 of us!

Then we were guided to the ticket counter of the airline that would take us to our final destinations. While the line was long at the Delta counter, it was staffed appropriately and this too went very smoothly. Then from here it got a bit strange.

I had approximately four hours to wait until my flight out to Detroit. So I asked a young Captain how I could I get to the USO from where I was currently situated. He explained the route to the USO, which is an organization that caters to service members when they are in an airport for a prolonged period awaiting flights. They provide a sanctuary where there is food, television, fellow service members, and computer hookups. But in order to get there you had to go back through a security checkpoint with the always friendly TSA personnel. Common sense took another break on this process! Each soldier that was coming back from a war zone, after we had already been processed through a thorough customs inspection before we arrived in Atlanta, was now subjected to going through another inspection upon arrival in our own country! Some of the guys even had to remove their combat boots and empty all pockets in order to successfully pass through the metal detectors! Welcome home, troops! To me, if you are an elected official or a high- ranking officer in the military reading this excerpt, every service member felt this process was degrading and unnecessary. There was no reason for troops returning to their home soil to be greeted so rudely. And you need not take my word for this. Ask any soldier, marine, sailor, airman, or coast guardsman that went through this process so void of logic. As you might surmise, after a 15-hour flight with little sleep, there was lots of grumbling. One Navy Petty Officer got into a disagreement with a TSA Supervisor and mentioned to him that uniformed soldiers are not subject to having to disrobe or take their uniform off per Pentagon guidance. The supervisor stated that he knew of no such directive and they were simply following their guidelines. So either this guy was ill informed or the process needs to be changed. Otherwise, once again the federal government is wasting a lot of money and our time having us go through the long, tedious process before we return home. The process I have described is in Atlanta. The only other hub for troops returning home on R&R is Dallas and I am unsure if their process is as ridiculous, but I will ask that very question upon my return to Iraq.

Well, I did not have to disrobe at all, but once again I did have to empty my pockets to get through the security checkpoint. This may not sound all that bad to you, but a military uniform has lots of pockets and you typically put lots of stuff in those compartments! Despite this shameful greeting home, I then asked a guy at an information booth if I went to the USO from that point would I have to again go through this same security checkpoint? Believe it or not, he said yes! So to get to the USO you had to go through this security checkpoint where a lot of troops had hours to wait. Then to get to their gates, they would again have to go back through the same ridiculous process! It seems that very little logic went into this process. I will submit a letter to my elected officials about this insane process, but I suspect the letter will be absorbed into the beast known as our bureaucratic system and the process will never be changed. However, it is worth the effort for our troops.

Despite these trials and tribulations along this trek home, my flight to Detroit went without problems. So I called Shari in Detroit and told her the last leg home was on time and I would be in the Motor City for only a few hours. Well of course, nothing can be easy! One of the flight crew did not show up so there was a 45-minute delay in getting the replacement to the airport and our flight on its way.

Once at the airport in central Wisconsin the long awaited and anticipated homecoming finally occurred. I was greeted by my lovely bride Shari who immediately began to cry. We held on to each other for what seemed like an eternity. It was great to be back in familiar surroundings. We then traveled home and the homestead looked more beautiful than I remembered it and all of our neighbors had put yellow ribbons on their trees or mailboxes which lined my route down Pine Haven Court to our beautiful estate at the end of the cul-de-sac. What a beautiful site it was indeed.

Tie a yellow ribbon 'round the ol' Oak tree

THE WAITE R&R PLAN

The second biggest day in Iraq for me had finally arrived. Of course the biggest day will be when I see Iraq in my rearview mirror for the last time this October. The good news is that the longest stretch is over and I should have a little over three months remaining once I get back to Iraq. So anyway, before I begin my R&R I will provide my plan and will let you know how it turned out upon my return to Iraq.

First and foremost, this deployment has been difficult on both Shari and I in many ways as I have mentioned several times to you already. Our relationship has been strained and I hope we can recapture the magic that we had before I left. As neither of us are kids anymore, we know what it will take to get back on track. And we shall.

Then of course, I want to see my parents and spend some quality time with them. I pray that nothing happens in the three plus months I am gone once I return to Iraq. I want to see Shari's parents as well, and of course brother Brad, sister Allyson, my daughter Lexie, and my granddaughter Kayleigh. I hope to talk with Lex about her future plans as I have not spoken with her for a while now due to her limited communication capability. Then I should also see two of Shari's three boys, Matt and Jeremy. Josh, the chef-in-training, is back in Orlando completing his internship. I hope the three pooches, Buster, Bailey, and Sadie Mae, will remember my friendly face. We shall see how good a dog's memory really is! I shall miss good old Jenna, my parent's Sheltie, who passed away in April and left a large void in their lives. Then of course Drew turns Sweet 16 this year, and if all goes as planned I should be home for the big celebration on July 3rd.

I hope to spend some time in my studio and go through the items I sent home in boxes and footlockers. I may work a bit on war college, although I am far enough ahead at this point in time that I could probably just take a well-deserved hiatus until I get back to Iraq. I will read a bit and continue with my counterinsurgency education. I will work on refining this book and get ready to ship off my EMS leadership book to my editor. Then perhaps I shall see a few close friends, although I will resist the temptation to stop in at the fire station and city hall. I really wanted my R&R visit to remain very low profile, but Shari has blurted it out to many people so it is no longer a well-kept secret. But the reason I am going home is not to return to my normal life just yet because it is not yet time to do so. It is time for family, some cold beer, a few good boiled-in-beer Johnsonville brats, perhaps some golf, and some quality time around the house and in the pool. This is a very simple plan. It is not time for a homecoming just yet. That will come this fall.

THE EXECUTION OF THE WAITE PLAN

The two weeks went far too quickly as you might suspect. The wife and I got reacquainted. The house looked great and it was nice to see green again! Iraq is nothing but sand, at least the parts I have seen thus far, but Wisconsin in the summer is beautifully green. We also live in an area with 70-80 foot pine trees so it is indeed a stark contrast to Iraq.

The Waite Estate

The weather was also a real contrast. When I left Kuwait it was 129 degrees! When I landed in Wisconsin it was about 65 degrees. We also had some rain and a few cloudy days, neither of which I had seen for months in Baghdad. And there was no dust!

Shari and I stopped at few of our watering holes and despite the fact that I have not had any alcohol in almost eight months, the beer tasted good, but it only took a few to reach my stopping point. I am not a big drinker and I drink very slowly anyway, but I know when my stopping point is and it was reached quickly. But it was still nice to be back in friendly and familiar confines.

We celebrated Drew's 'Sweet 16' party while I was home, which was very nice. We had about 30-40 people over to the house and I cooked burgers and brats on our new Charbroil Infrared grill. It was a birthday she will never forget.

Happy Sweet 16, Drewski

Shari has done a great job of maintaining the Waite Estate and the place looked fantastic. She had done a lot with the inside décor and the yard work was very impressive.

The rest of my R&R was very relaxing as I was able to reconnect with my wife; visit with my daughter and granddaughter; see my parents, my brother and sister; visit with my aunt Ellen and her husband Bill; visit with the in-laws, Dick and Kathy Kertis; see two of the Shari's three boys, Jeremy and Matt; hit a few golf balls with my dad; drink some real beer; see some neighbors and friends; visit with some of the guys from the fire department; spend some time in my man-cave (my studio, aka garage) and relax.

My 'man-cave'

I also went four-wheeling in the great white north of Wisconsin with Shari's sister and husband; went to a baseball game in Appleton with Dave Kerkman, my second in command at the fire department and we watched the Brewers Class A farm club; had lunch with good friends Dave, Jason Joling, Scott Young, and Tim Desorcy; went out to eat several times and had some great steaks; and I even did some yard work.

We also celebrated Shari's oldest son Matt's 26th birthday. Matt will soon be getting married and also owns his own BMX team, which he has done a great job in managing. This is a hobby that he truly enjoys.

Happy 26th Matt

So my plan was executed almost as envisioned. I did not work on War College, but I did stop down at the fire station. I was glad I went home on R&R and it was great to see my family. But Shari made a comment one night at home that she believed I have changed since I left on my Iraq experience. She believes I am more quiet and angry. I guess after you go to war there can be no doubt that the experience will change you. I do not believe you can experience the savagery and chaos of a war zone and not be affected in some way no matter what job you perform. And my job is very tame compared to the brave men and women who must go out into the belly of the beast every day. These are the true heroes and the ones most affected by war who sacrifice more than anyone else.

After what I have observed and experienced I guess I am a bit angry. I am not angry at my family, but at being taken away from them and performing a job that I do not particularly care for. I am angry at our civilian and military leadership for their poor decision-making in the early stages of this war that led to the deaths of many brave men and women in uniform and cost taxpayers billions of dollars. I am angry with my unit because I feel I could have done so much more to assist in the war effort and better help the Iraqi people in a different position. I am angry at the Army bureaucracy and its inherent flaws in procedures and processes. I am angry that I had to go back to Iraq and again be taken away from my family and a life I truly enjoy.

But the support from my family and friends has been indescribable. Without it, this experience would be even less tolerable. The guys wearing red shirts upon my arrival at the fire department, which is a show of support for our troops and me, was truly an impressive moment in time. When I rounded the corner of the street leading to my home and noticed all of the yellow ribbons on the trees and mailboxes on each of my neighbor's homes, and the big pines at my home, I felt a lump form in my throat. My wife had also put up some red, white, and blue bunting on our fence and it was oh so beautiful.

The boys of WRFD

I told Shari that I guessed she was correct and that I had been changed by war and assured her that it will take some time to adjust back toward normalcy once I return for good. This is neither unexpected nor unusual and I believe I can make the adjustment without too much difficulty. At least I hope so. But I also told Shari that I sensed a change in her as well and this too should pass once this experience is well behind us. She has had a very difficult time with this deployment and we discussed my options in relation to both my fire service and military careers once I returned for good. Perhaps it will be time to move on and forward after 25 years in the fire service and over 34 years in the military. I could not envision leaving her again on another deployment. So I would have many major, life-changing decisions to make upon my return home this fall. I will need to make these decisions by the 1st of December in order to allow the fire department to make any necessary adjustments, so I will inform you of my decision before this book concludes.

While it was difficult to leave my family once again, the trip had been well worthwhile. The goodbyes this time were not as tearful nor as painful. The long part of this deployment is behind us and I only had about 100 days remaining once I returned to Iraq. Time will tell if this deployment had been worth the year stolen from me and my family. Perhaps it would lead to a more lucrative job opportunity or a chance to do something to assist our military to improve its processes and structure. If nothing positive occurs from this experience then I can unequivocally state that it was not worth it. But it is far too early to make any analysis on this as of yet. Time will tell the ultimate tale as it always does.

See you soon, honey

THE TREK BACK TO IRAQ

The time had come to return to Iraq for the stretch run. I told Shari that in order to get back, I had to go back. This was just another step in this experience that would eventually lead me home in the fall. So she took me to the central Wisconsin airport at zero dark thirty to begin this phase in the process. While there was nothing of note at the central Wisconsin airport, there was another inconsistency in TSA procedures and how they handle service members, which they need to get a handle on. There were no such issues in Minneapolis, which from there took me back to Atlanta. The biggest issue in

Minneapolis was whether or not Brett Favre was going to turn rogue and don the purple and gold of the Vi-Queens! I guess the Packers will just need to beat him and his new team twice this coming football season if he does sign with them!

Atlanta does a good job with those personnel returning from R&R. The only thing I can state about this experience is that it is far too long. We all had to report to Atlanta by 1300 and get our R&R paperwork stamped. Then we basically hung out at the airport for about 8 hours before we boarded the aircraft to take us back to Kuwait. I can understand some lead time to take into account late flights and missing baggage, but this seemed a little longer than was necessary. But as you may have heard a time or two, 'hurry up and wait' is truly an art form within the military community!

So after my last little slice of Americana as I enjoyed a McDonald's southwestern chicken meal, we boarded the big metal bird that took us back to Shannon, Ireland and into Kuwait City. I do not care how you cut it, this is just a long haul, although my accommodations were much better on this flight than when I came back on R&R. I was able to grab some quality winks on the flight and watch two movies, **Duplicity** with Julia Roberts and Clive Owens, which was entertaining, and then **Marley and Me** with Owen Wilson and Jennifer Aniston which was very funny and sad. It reminded me of the great effect pets have upon people and took me back to the recent loss of my parent's dog Jenna this past April. They still feel the loss and my dad gets choked up every time he talks about it. And yes, just like Old Yeller, I shed a tear when Marley died.

We arrived back in Kuwait City about 2030 hours and then we basically just did everything in reverse that we accomplished on the way over on R&R. As we deplaned, another group of troops were getting ready to board and head back on R&R. We boarded the shade-drawn buses and were taken to a staging area where we again waited for awhile as baggage was unloaded and we awaited our Kuwaiti police escort. The weather was very hot, and after some nice, cooler weather back in good old Wisconsin, the return to this oppressive heat was not very welcome.

We began our hour-long bus trip up to Ali Al Saleem Air Base, north of Kuwait City, and once there we were checked backed into the war and were told when to report for our flight update back to our final destinations. However, this time I was able to check in very quickly, gather my body armor without waiting in a long line, and then head over to a trailer that the Army Corps of Engineers had established on the base. I should have done this on the way over, but I am a quick study and learn from my mistakes. Once at the trailer I ran into two acquaintances from my unit that I had not seen since we arrived in Iraq last December – MAJ Mark Gregris, whom I introduced you to during my MRX at Fort Lewis last October, and SFC Donna Sendelbach. After we caught up on each other's R&R and Iraq experiences, we grabbed a cot in the trailer and caught some long overdue quality Z's until we had to report at 0730 for our flight information.

Unfortunately, because of some poor weather to the north, there was a backlog of troops heading out of Ali Al Saleem so my flight to BIAP would not happen until late evening on this day. So I basically hung out all day, watched some television, grabbed a few more winks, and visited with another unit compadre, MSG Harley Hansel, who is the GRD RSOI representative at Ali Al Saleem. MSG Hansel is a computer programmer and fellow Packer fan so we chatted of home, the upcoming season, Brett Favre's brain fart, and many other topics, including how we both were less than thrilled with our assignments. MSG Hansel told me he had a more difficult time saying goodbye the second time than he had leaving last fall. For me it was the exact opposite, but MSG Hansel has young children at home, which does make a difference. From my own experience and the many people I have talked to about the topic, I guess the jury is split as to whether it is easier or harder when you leave the second time. But MSG Hansel stated that you basically go home for the family and not yourself. This statement seemed very profound, but I believe R&R is also healing for the individual and an opportunity to get away from the madness of war and recall what normalcy feels like. So I think R&R is good for both the troops and their families.

Then, before heading to my departure briefing I ran into SPC Dale Barker, another fellow 416th TEC member, who was heading home on R&R. But once again, it was 'hurry up and wait' as we checked in and then waited about four hours before getting on the buses to take us to the Air Force flight north to BIAP. So with all of my battle rattle back on we stuffed into the uncomfortable and hot cargo jet that would take us north to our final destination and back to war. For some reason we had a long delay before we actually left Ali Al Saleem, which ate up much of the time we had gained by leaving an hour earlier! With a temp of about 100 at 2330 inside a hot plane it became very uncomfortable very quickly. As I sat there I thought this is just another thing that most people do not have to experience, including many Generals who should know what it is like so they can fix yet another process gone awry! If we had sat there much longer I suspect there may have been some heat casualties.

We finally took off and began the hour and half flight back to Baghdad. While uncomfortable, it did cool off as we gained altitude, which was very welcome. We landed without issue, got off the jet much as I had eight months earlier when we first arrived in Baghdad, and then I quickly found our GRC RSOI representative. There were five of us heading to GRC and this was done quickly and without incident. I got back to my little home away from home at about 0200, quickly unpacked, and then drifted off to sleep now safely back to continue with my Iraq experience.

GRD soldier delivering toys to children

CHAPTER 14
400 DAYS – Days 255-280
July 18 – August 12, 2009

Whether Iraq reaches its potential is of course ultimately the product of Iraqi decisions. But the involvement and support of the United States will be hugely important in shaping a positive outcome.

--US Ambassador to Iraq Ryan Crocker
2007-2009

Day 255: Saturday-July 18th, 2009/Day 224-Iraq was back to the grind. I did sleep in and would treat today as a Sunday due to the late night last evening in returning to GRC. The last task I needed to accomplish to be back 100% was to retrieve my weapon from the GRC vault. I talked with LTC Tony Jocius who seemed much more relaxed since I last saw him. This may in part be to the new commander at GRC who took command while I was on R&R. According to Tony, the new commander has a much different leadership style than his predecessor and one more conducive to a productive working environment. Tony also mentioned that he and LTC Mike Girone would be extending their tours by 124 days. This would put them back at Ft. McCoy sometime in late February.

Apparently there are another six to eight personnel from the 416th TEC doing the same thing. We have also had one or two personnel sent home due to discipline violations, so of our original 74 people who ventured to Iraq last December, we may be heading home with only a little over 60! So be it, but I would be one of those heading home in late October and would be very happy to be leaving this place.

There were a few new faces in the SOC since my departure on R&R. I spent the afternoon weeding through my many emails and getting back on track. The GRD G3 section had also established residence on VBC in Building 51 since I had left. Tomorrow I would visit them in their new digs. But it was back to the heat and the dust for another three and a half months.

SIGNIFICANT EVENTS IN IRAQ SINCE I LEFT ON R&R

Despite a few new personnel in the SOC since my departure on R&R not much had changed. Well, it had only been about three weeks! But a few things of note that occurred while I was gone included: the US met its deadline and honored the stipulations as set forth in the security agreement by moving out of all Iraqi cities by the end of June, which in turn became a National Day of Celebration in Iraq; the Iraqi Security Forces continued their evolvement and have been successful in quelling violence across the country; there has been a delay of the security referendum until the national elections early next year due to a price tag of about $100 million for just this one event and if this referendum is successful next year its effect will be to move US troops out of Iraq well before the end of the currently signed security agreement; the integration of the Sons of Iraq (approximately 90,000 people) into various jobs around the country continues to progress; several NFL coaches, both current and retired, visited with troops over the 4th of July; Vice President Biden, who has a son serving in Iraq, visited the country; sandstorms in Iraq are on the increase; the Iraqi Army continues to grow and evolve into a very professional force; the Iraqi Police have grown to over 315,000 personnel and are slowly gaining the respect of the Iraqi people; the US Army's advisory and assistance brigades continue to evolve from the brigade combat teams in order to address the specific needs in Iraq; Prime Minister Maliki has drafted a five year plan for the war on insurgents in his country; the democratic process continues to take shape; and Iraqis have not called for any assistance from US combat forces in the first 17 days since withdrawal of US forces from Iraqi cities **(18 JUL to 28 JUN 09)**. So all in all it is a lot of positive news and continued evolvement in Iraq. The bad news is that while I was away another 10 US troops were killed, thus reminding all of us that we are still involved in a war.

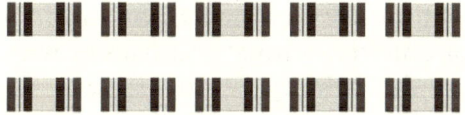

Day 256: Sunday-July 19th, 2009/Day 225-Iraq was a bit of a repeat of yesterday's schedule, but with a slight twist. I would fall back into my regular Sunday routine by working a little in my room until about noon, then head over to get gas for my vehicle, and then grab some lunch. My sleep patterns will take a few days to get back to normal as I slept little last evening and was up at 0400 working on my computer. But I got much accomplished getting up that early.

After lunch I would head over to the main PX to get a haircut as it was getting pretty shaggy after almost a month. According to my calculations, I would only have about 4 more cuts to go during my tour. But after my haircut I would go over and visit the new GRD HQ in Building 51 just down the road from the palace. LTC Chuck Samaris is getting ready to leave and head back to Fort Leonard Wood so I wanted to bid him farewell before he departed. With this accomplished I also ran into several other members of my unit and chatted with them for a while and welcomed them to VBC. The Command Group and the G1 would be heading over to VBC within the next week and then I believe all sections would be at their new home.

After my visit I returned to the palace to accomplish some work. To that end, it appears that the Iraqi Security Forces are getting the job done. Yesterday they arrested an Iranian-backed militant who had launched a rocket attack in southern Iraq which killed three US soldiers **(MNFI, 19 JUL 09: AP)**. The ISF is also working diligently to protect the thousands of pilgrims honoring Imam Musa Al-Kadhim **(MNFI, 19 JUL 09: Reuters)**. This is the first 100 percent Iraqi security plan. Thus far there has been little violence. So they are getting better and the people are gaining more confidence in their abilities as well.

Today concluded by calling Shari and the parents and a dinner meal of chicken wings and fries. I was hoping to watch Tom Watson make history at the British Open today before drifting off to a good night's sleep. I will let you know how that turns out tomorrow!

Day 257: Monday-July 20th, 2009/Day 226-Iraq was back to business as usual. We would start the day with the morning BUA and I would get back into my full routine. This day also marks my unit becoming double digit midgets as we move to 99 more days in Iraq if we use the published 26 OCT 09 as the target date. However, this date seems to be in some dispute as my unit compadres, MAJ Mark Gregris and SFC Donna Sendlebach have heard an earlier date, as has LTC Rick Ramsey. The good news is that the dates they have heard are anywhere from 8-9 days earlier than the date published. If this becomes reality, then we are at about 90 days remaining! I will continue to use the 26 OCT 09 as this appears to be the latest possible date we will be in Iraq. If we leave earlier that will just be a bonus. The date should also become more clearly defined within the next few months as preparations to go home will pick up in earnest.

Today I was again up early. My sleeping pattern is getting closer but just not quite there yet. Hopefully in the next few days it will be back to normal. I will resume my workout routine today as well and get back into the flow for the last 90-99 days. As I was once again up earlier than usual I took the time to be productive and began to post my resume on several different search engines to get a feel for what may be out there if I retire from the fire department and the military. I have several different passions and directions I could go in and will need to make some critical life-changing decisions over the course of the next several months and before 1 DEC 09.

Gees, you leave for a few weeks and everything changes on you! I went into the operations center a little early today and was awaiting the morning BUA at 0730, which is the time we have used since I have been here. Well, 0730 came and passed and I came to find out that the BUA time had changed to 0800 on all days except for Fridays, which shall remain at 0900. I guess I will need to go through a week of my battle rhythm to see if there have been any other changes since I left a mere three weeks ago!

I guess I underestimated the number of pilgrims the ISF had to protect as it was reported today that over five million Iraqis traveled to the shrine of Imam Musa Al-Kadhim in Khadimiya **(MNFI, 20 JUL 09: *Washington Post, NY Times, LA Times, McClatchy, BBC*)**. So it can be stated that the ISF, in their first real test on their own, did an excellent job and this is a very positive sign indeed.

And by the way, Tom Watson gave it a great try but missed a putt on the final hole, which sent him into a playoff where he eventually lost to champion Stewart Cink at the British Open. Still, that is not too shabby an effort for a 59 year old competing with a bunch of young bucks.

Day 258: Tuesday-July 21st, 2009/Day 227-Iraq was going to be a repeat of yesterday as I am settling back into my battle rhythm. My sleep patterns are getting back to normal although not quite there just yet. I am also easing back into my workout routine as even though I worked out while on R&R it was not six days a week. I weighed myself before I left Iraq at the end of June and I was 80 kg, which is about 179 pounds. My first day at home I weighed in at 179 on our home scale, but after many cold Leinenkugel's, limited workouts, and much good food I knew I must have gained some weight which I could afford to put back on my frame. Unfortunately, I forgot to weigh myself before I left home to come back to Iraq, but after a disruption in my schedule, no more beer, and less eating I knew I would drop a few pounds before I had a chance to weigh myself again. Well, that chance came yesterday after my run and I weighed in at 80 kg! So any weight I had gained was now gone. Oh well, such is life in the fast lane.

Yesterday, Secretary Gates announced a temporary increase in the size of the Army by about 22,000 soldiers **(MNFI, 21 JUL 09:** *AP* **& Pentagon Channel, 20 JUL 09: 2100 GMT+3 and Stars & Stripes, 21 JUL 09,** *Congress praises Army expansion,* **written by Leo Shane III)**. This is equal to about another 4-5 brigade combat teams. This will primarily provide some relief in regard to all of the deployments and allow for more training time for troops while at home station. I know the price tag is high on this but it really is necessary at this point. The Army is spread too thin right now and the strain on the troops is immense. As we drawdown in Iraq these numbers may also be reduced unless we get engaged in another conflict elsewhere. So we shall see just how 'temporary' this increase actually is.

The Iraqi Prime Minister is also in Washington to visit with the President. Prime Minister Maliki will visit Arlington National Cemetery to honor those who sacrificed so much to assist his country **(MNFI, 21 JUL 09:** *LA Times*). This display will no doubt help to solidify the relationship between both countries and it is nice to see someone else other than Americans pay tribute to our fine men and women in uniform who have sacrificed so much to assist in Iraq's development.

Day 259: Wednesday-July 22nd, 2009/Day 228-Iraq was a no BUA day. Well, my faith in the Army system continues to dwindle. The list for Colonel came out yesterday and moa was not on it. Now I will say that this is a very competitive process and there are only so many slots available, but perhaps it is time for this old dog to just retire. I am just not sure what else I could have done to position myself better in the process, but again, there are many qualified individuals out there. So I am not certain how the Army selects these positions as this appears to be as mysterious as the selection process for war college, but sometimes it is just being in the right place at the right time. In order to be fair and keep all Army functional branches in alignment with one another, the board must promote so many people in certain categories. For example, there may be more qualified candidates in one branch than another, but some simply cannot get promoted at this time. Some personnel are getting their last look and if not promoted this time they will have to retire, so they may get additional consideration. Then of course there is the equal opportunity aspects involved in the process as well. But I am sure the board tries to be as fair as they can be and no system is perfect. I am certain there are many happy people who are on this list and many more that are disappointed because they are not. This is just the way it is and I can live with that. If I continue to pursue and complete Army War College then I believe I have a good chance of promotion sometime down the road. The problem is that the Army's timeline, so that they can better manage their officer corps so it does not become too bottlenecked at certain ranks, and my timeline may simply not mesh. The timing for this promotion may simply not be there but dependent upon my fire service career, I may give the process one more year even though realistically speaking the outcome is likely to be the same. I also got a later start as an officer as I was an enlisted member for my first 10 years in the military. I have adjusted my military career goals several times and made it much further than I ever thought, but reaching the rank of Colonel and finishing the crown jewel of military education, war college, would have put the finishing touches on a great career. Perhaps the Army wants younger officers with more time remaining. At my age I only have about seven more years before they kick me out anyway. While disappointing it is not devastating. Shari will be pleased as this will make my decision to retire that much easier. So perhaps one more year and if war college and promotion do not occur by this time next year then I will simply hang up the desert tan boots and call it a career before I would have to worry about another deployment. Dependent upon my civilian career decision, this timeframe may be condensed. Retiring would finally leave all of my weekends open after 34 years. I would no longer have to miss any family functions or events due to reserve duty. I would not have to worry about another deployment and put Shari through this ordeal again. And I really would have no regrets. It has been a good, long, and in most cases, a rewarding career. It has often been stated that many people just do not know when the right time is to retire. I have seen this firsthand in the fire service on a few occasions. Just ask Brett Favre his viewpoint on this subject! Perhaps all of these events are simply indicators that a brighter person would take as signs to retire. Maybe the time has come.

There is also a lot of news in the media and consternation on the part of General Odierno in regard to the allegations that US and Iraqi forces are not getting along after we moved out of the cities **(Stars & Stripes, 21 JUL 09, *Iraqi curb US military operations in Baghdad*, written by Qassim Abdul-Zahra and Deb Riechmann)**. General Odierno is challenging commanders on the ground to make certain they get the message out that this is simply incorrect. To me this is a good sign that the Iraqi Security Forces are taking the lead and want to prove that they can handle the situation on their own. It has been reported that the US has been turned down a few times after requesting Iraqi approval to conduct some operations within certain cities. Well, this is their country and per the terms of the security agreement it is up to the ISF to determine if they need assistance or will allow such operations to take place. Outside of the cities the US is still able to move freely but many believe the US has simply become prisoners on the larger bases that still remain in the country **(JASG, 21 JUL 09: *AP*)**.

Day 260: Thursday-July 23rd, 2009/Day 229-Iraq I found out that my old friend Mike Girone had finally made the list for Colonel. I also believe Mike was the only one from our unit that in fact made this elusive list. While selection process does appear to be a crapshoot at best, Mike was getting close to his last look by the board and this was already his fifth attempt. This was my first look by the board, but it is highly unlikely that I will stay around long enough to have five looks! Mike had a strong board packet and did finish Army War College last summer so this may have been the tipping point for him. Again, the process of selection is quite mysterious and there appears to be little rhyme or reason associated with it but I was happy for Mike. It looks like he will now need to stick around for at least three more years in order to retire at that rank.

Today was also GRD BUA day and SFC Maltes came over to the SOC to brief as she is now working at the new GRD HQ on VBC. She will be taking her R&R next week so I will have the briefing duties for the next one or two briefs in August.

It appears as though the power struggle continues in Washington as politics is again knocking heads with common sense! As I mentioned last month, Secretary Gates has taken a pragmatic approach to the military budget this year and allocated more funding toward the needs of the present than the perceived needs of a long since dead cold war or conflicts that may never materialize. There is only so much money and the US already spends a fortune on defense spending in comparison to other countries around the globe. So better armored vehicles to protect our troops in Iraq and Afghanistan; better and lighter body armor; more flexible weapons systems; and lighter sea vessels have all taken precedence over the more expensive aircraft carriers and fighter jets. Now with this being stated, the politics gets involved because someone from an aircraft company knows they will be losing a lot of money if certain fighter jets are removed from the budget process. This is capitalism at its very best. The loss of these aircraft will also equate to lost jobs in these political districts. This is certainly a negative side effect for that particular area, but it may also lead to the creation of jobs by increasing the manufacturing of the other military equipment I have outlined. So the arguments are in some cases a bit weak, but still very understandable. However, it looks like Congress stepped up to the plate to handle the issue before it reached the President's veto pen and they cut spending of the F-22 fighter jets which cost $1.75 billion for an additional seven jets **(Stars & Stripes, 22 JUL 09, *F-22 funding cut out of Senate bill*, written by Leo Shane III)**. It appears as though common sense won out this time on Washington!

It appears that the Prime Minister's trip to Washington, DC was a success. Not only was political progress discussed, but also the topics of economy, industrial, and education cooperation between the two countries **(MNFI, 23 JUL 09: *TIME, LA Times, Washington Post, Newsweek, NPR, ABC,***

BBC, Fox, Reuters, AP, AFP). One US official commented on the historic nature of this trip:

> This visit is a sign of a comprehensive and long term partnership between Iraq and the United States; it goes beyond security cooperation, we are not just looking at the short term, this is the beginning of a long-lasting, normal bilateral relationship with the nation of Iraq.
> **-MNFI, 23 JUL 09:** *AFP*

These are all signs of stability and progress within Iraq and hopefully this trend will continue.

▌▐ ▌▐ **Day 261: Friday-July 24th, 2009/Day 230-Iraq** I continued my onslaught on the job market as I have applied for several different jobs, from associate professor at universities in Wisconsin and Minnesota, to teaching positions at West Point and Army War College, to a position with the RAND Corporation, to senior director at the National Defense University, to a fire chief position on a military base. While I may not land any of these positions, it will be interesting to see how marketable my skills are. The good news is that I still have a job to return to at the Wisconsin Rapids Fire Department, but if I am to put my Ph.D. to work and get out of the military I will need to land a different job. So it should be an interesting ride for the next few months.

I informed you a while ago about the tensions between the Kurds and the Arabs in the north. The Kurds, who have control of three provinces in the very northeast corner of Iraq, are very autonomous and have recently opened a new oil refinery **(JASG, 23 JUL 09:** *TIME*). This is significant because even though the Kurds sit on a very oil rich area of Iraq they have never been able to produce any oil before. Saddam Hussein kept the Kurds repressed for decades, but despite this repression they have managed to set up their own government and infrastructure. Many Kurds would much rather secede from Iraq and become their own nation, which would basically amount to a two-state solution here in Iraq. In fact, ultimately this solution may be discussed more seriously as the two groups are at odds with one another and great tension exists which may result in a civil war. Many commanders here are very worried about this situation, which could result in Iraq's destabilization process. The Kurdish politicians would like to become a part of the new Iraq but do not feel they would be represented fairly, and it still remains to be seen if old grudges and sectarian differences can ever be left in the past for the future of Iraq. Many experts do not believe this divisiveness will ever be overcome, so this situation bears close monitoring.

Prime Minister Maliki appears to have had a successful visit to Washington **(MNFI, 24 JUL 09: *Diplomatic LOO*)**. His leadership, or lack thereof, will be the key to the success in Iraq. While national elections will be held early next year, it is likely that Maliki will be re-elected as the Prime Minister. His detractors claim he does not possess the ability to overcome his bias against the Sunni and Kurdish factions within the country. If he cannot become a change catalyst and bring all factions together for the common good of Iraq, then this democratic experiment in the Middle East is likely to fail. But if he can overcome his personal bias and emerge as a strong leader in the political process, then the future of Iraq looks much brighter.

President Obama commented on the Prime Minister's visit:

> Overall, we have been very encouraged by the progress that's been made. What we've seen is that the violence levels have remained low, the cooperation between US forces and Iraqi forces has remained high, and we have every confidence that we will continue to work together cooperatively.
>
> **-MNFI, 24 JUL 09: *LA Times***

Finally today, Angelina Jolie was in Baghdad. No, she was not here to visit the troops, as I do not believe she sings or dances, but rather as a goodwill ambassador for the United Nations **(CNN, 24 JUL 09: *Internet story*)**. I think it is nice that she is involved in such humanitarian efforts. She is bringing attention to the plight of the millions of displaced Iraqis. I am not certain if she provides money for this effort or any other assistance other than her celebrity status. But still, not all entertainers are as engaged as she has been and I find that very noteworthy. For you ladies that may be reading this book, no, Brad did not accompany her on this trip!

Well done AJ

Day 262: Saturday-July 25th, 2009/Day 231-Iraq I was greeted by an email that informed me I had been approved as a candidate for a faculty position at West Point. This by no means indicates I have a job there and I would need much further investigation in order to determine if I would even accept the position if offered, but it is still a good first step and tells me that some of my skills and experience are marketable.

Iraqi Prime Minister Maliki is wrapping up his visit to the US and he and President Obama believe our withdrawal from his country is right on schedule **(Stars & Stripes, 24 JUL 09, Obama: Iraq withdrawal on schedule, written by Leo Shane III)**. Some observations I have mentioned throughout this book, further validated by a good book on counterinsurgency that I am currently reading written by expert David Kilcullen titled the ***Accidental Guerilla***, outlines key areas necessary to focus upon in order to maintain stability in this type of warfare. One of the most important aspects is the ability to protect the population from the insurgents, which helps to legitimize the government in the eyes of the people. Prime Minister Maliki and Secretary of Defense Robert Gates discussed just this element during his recent visit **(MNFI, 25 JUL 09: al-Iraqiya)**. Another key area of concern is to provide essential services to the people, which is also beginning to take shape in Iraq, and which greatly improves the quality of life for the people **(MNFI, 25 JUL 09: *Aswat al-Iraq*)**. I will mention more on this book by David Kilcullen, a former member of the Australian Army, in the days ahead. He clearly understands this type of warfare and makes some very blunt, but accurate statements concerning the poor decisions which were made by our leadership, which validate some of my observations earlier in this book. But learn and adapt we have during the course of this war, and I truly believe that this country is going to make it. Much better decisions are also being made in relation to Afghanistan after years of floundering in that country and so Iraq's example provides a glimmer of hope for that country and the war on terrorism.

Day 263: Sunday-July 26th, 2009/Day 232-Iraq the day began nicely as I slept in until 1000! That was close to 12 hours of sleep which tells me my body has finally figured out the time zone change and I believe I am back to normal. As normal as I can be that is! My morning constitutional was the best since I have been back as well. It is the little things in life!

So I got up, brewed a pot of coffee, and began work on finalizing my EMS leadership book, ***The EMS Leadership Challenge – A Call To Action***, which I would send out to my editor Rickey Pittman on or about August 1st. I also watched Alfred Hitchcock's ***The Birds (1963)*** with Rod Taylor and Tippi Hedren. I had not seen this movie in years and did not realize it had been nominated for an Oscar. Not sure if Hitchcock had any meaning behind this film or not, but it is a classic.

I want to revisit to a topic I have not mentioned in awhile - suicide prevention. A recent article in *Stars and Stripes* mentions that analysis is the key to success **(Stars & Stripes, 24 JUL 09, *Analysis key to anti-suicide efforts*: editorial in the Washington Post)**. I am happy to see that the leadership in Washington is still looking at avenues to improve in this area, because as I have stated, the Army has really been placing a band-aid on a gaping wound and more thoughtful and meaningful solutions are necessary in order to have any significant impact upon this major issue. Again, this is a cultural issue that will take great persistence and patience in order to impact the negative trend. Mental health issues still carry a stigma within many professions, which include the military and the fire service, two occupations I am intimately familiar with. But even as I state this, since about March there has been very little discussion on this topic since the Army mandated that all soldiers go through a course in relation to suicide prevention. This was the band-aid I mentioned. It is just hard to fathom that if our leadership is placing such great emphasis upon this subject, why there has not been a more persistent and continuous message. Because of this lull in messages on a very important topic within the military community, this tells me that the command emphasis is just not there yet. But the article does suggest that at least discussions at higher levels are continuing.

Day 264: Monday-July 27th, 2009/Day 233-Iraq I have reverted to an irregular sleeping pattern. I thought after a good night's sleep, finally, on Friday night and a 'Rip Van Winkle' on Saturday night that I was back to normal. I guess I had it wrong. Much like the situation here in Iraq, my sleeping pattern is 'fragile and reversible'! I got in a solid two hours last night before awaking and then took several hours to get back to sleep. I did not hear the alarm go off at 0500 but awoke at 0630 and was still out the door at 0700. My window to depart my barracks to head in to work now that the BUA is at 0800 is between 0630 and 0700, so despite the little REM faux pas I still made it into work on time.

What I have failed miserably to do over the past week or so is to inform you that there were big elections in Iraqi Kurdistan on Saturday. As I mentioned way back in January when Iraq had its provincial elections, the three provinces in the north that are a part of Iraqi Kurdistan have their elections separately. Saturday was that day and from all indications, voter turnout was very good (78%) and there were very few security violations **(MNFI, 27 JUL 09: al-Jazeera, al-Arabiya, al-Iraqiya, al-Iraq News, al-Shariqiya)**. In comparison to the fiasco we witnessed recently in Iran, these elections were overseen by international monitors and the Iraq Independent High Electoral Commission, better known by the acronym IHEC. In fact it was so low key and so well run that there was little buzz about it where I work and one reason I have failed to report this earlier to you. All in all, my bad reporting is a good sign that the situation is stable in the north. Iraqi President Jalal Talabani said the chances of fraud were quite low due to the oversight of international monitors **(MNFI, 27 JUL 09: *LA Times*)**. Iran did not allow international monitors during their election process if that tells you anything! It will now take a few days to sort out the results and determine the winners in the election, but no surprises are expected.

Finally, today there is still continued discussion about friction between US and Iraqi forces. General Odierno emphatically states that this situation has been blown out of proportion in the media, while other sources state that there have been a few isolated cases of misinterpretation of the security agreement. Some of this misinterpretation has been on both sides and it will take time to sort through. Some US forces still want to continue the fight we have started and can launch operations if in the scope of self-defense. Iraqi Security Forces want to take the lead and manage their own country, which is a point we need them to get to, and so we have a few minor disputes that we must work through. All in all, the transition out of the cities has been relatively smooth. Prime Minister Maliki stated:

> The US forces withdrawal from Iraqi cities gave a very positive image to the Iraqi people, it supported the credibility of the Americans, it vouched for their good intentions, it created an atmosphere which is conducive to long-term relations. It embarrassed all those who cast doubt on this relationship.
> **-MNFI, 27 JUL 09: *Washington Post***

Today I also applied for a research professor position at the Army War College in Carlisle, PA. I guess if you cannot beat them – join them! Since I have had no luck getting into Army War College the past few years, although I am trying again this year and the board meets in September, my only opportunity to get there may be as a faculty member! Very ironic, but it sounds like a good job, with a decent salary, meaningful work, at a very important military educational establishment. We shall see how this one goes.

I also sent in my organizational change article to Joint Forces Quarterly for possible consideration for publishing. Not sure how the message will be received, but they are always looking for articles and so nothing ventured, nothing gained.

Day 265: Tuesday-July 28th, 2009/Day 234-Iraq was off to a good start. I got a good night's sleep, had a cup of good coffee on board, and was headed into work a little early. Today took me to the 90-day mark left in Iraq, but realistically it appears as though we may be leaving a little earlier than that. That is fine by me!

Today is also my Aunt Ellen's birthday. I saw Ellen and her husband Bill while home on R&R. Ellen is a wonderful human being and takes after her mother, my late grandmother Grace Lagerquist. It was nice to see Ellen and she has stayed in touch while I have been over here. Because she is so conscientious I have also made her the executor of my will. Anyway, Happy Birthday, El!

Happy Birthday Ellen (with Kayleigh)

Not much in the way of news from the BUA today. However, yesterday in our post-BUA huddle it was mentioned that General Odierno wanted everyone to remain vigilant when it comes to force protection. There is some information floating about that some insurgent cells are looking to kidnap a member of the US military. Much like we have seen with the situation in Afghanistan, the insurgents will use this event for propaganda and leverage. So I sent this word forward to GRD who later that day issued some guidance that stated that all GRD members in Iraq are to ensure they are vigilant and move with at least one other individual even when on secured military bases! Yes, even here on VBC we are to move in groups or at least with a partner. Now for most people this is not an issue. For me, as I am the lone GRD member at the palace and drive to and from work, this is much more of a challenge. But I have always been very vigilant when it comes to personal security as I always thought that if insurgents really wanted to infiltrate our bases and wreak some havoc that they could find a way to do so. Just another one of those things you do not really worry about in the US, but another stressor over here in a war zone. This too I will never tell Shari until she reads it in this book. However, as this is already about 700 pages, not sure Shari will read all of this or get this far into the book until after I am long gone from Iraq and it is but a distant memory!

Day 266: Wednesday-July 29th, 2009/Day 235-Iraq Secretary of Defense Robert Gates made his first visit to Iraq this year **(MNFI, 29 JUL 09: *NY Times, CNN, Bloomberg, Reuters, AP, AFP* and JASG, 28 JUL 09: *NY Times* and Stars & Stripes, 29 JUL 09, *Gates checks on US advisory role*, written by Kevin Baron)**. He was over to discuss the current security situation with General Odierno and also review the landscape ahead as all US forces are to be out of Iraq in less than two and half years. He was also going to visit the first Advisory and Assistance Brigade (AAB), which is the Army's latest rendition of the Brigade Combat Team (BCT). An AAB is basically structured to provide more personnel of senior rank to train and assist the Iraqis. This new configuration has evolved because of the counterinsurgent mission we are faced with in both Iraq and Afghanistan. This type of unit configuration may not be as functional during a conventional style war, but certainly seems a logical metamorphosis given our current environment.

As I have mentioned to you on a few occasions already, Iraq is in the midst of a severe drought **(MNFI, 29 JUL 09: *unclassified*)**. Iraq has been negotiating with Syria and Turkey to get more water released as a result of the lack of precipitation. Few people outside of this region may understand that Iraq is not all desert wasteland. There are two major rivers that run through Iraq, the Tigris and Euphrates, and there are areas that are quite fertile, especially in southern Iraq. But without adequate water flow Iraq's grain production is down about 50%. Correspondingly, for effective irrigation the Euphrates requires about 500 cubic meters per second of water flow and currently it is about 250 cubic meters/second, or about 50% capacity. This circumstance leads to more importing of grain, which subsequently leads to less money for Iraq to use to rebuild its own country. So many factors are at play besides the insurgency. Decreasing oil prices, drought conditions, and corruption all make life more complex for Iraqis as they struggle to stabilize and progress.

Today was a ***shamal*** event. A ***shamal*** is a northwesterly wind that blows strongly across the gulf region. While it can happen anytime during the year, it typically occurs during the summer months. This sandstorm kicks up a lot of the fine dust particles and suspends them in the air for several thousand feet and it reminds me of fog back in Wisconsin. It is difficult to see and unless you protect yourself you will breathe this dust in and it gets over everything. There is typically very little movement by air during these periods as well. While there always seem to be dust in the air here, today was one of the two worst I have observed since I have been here.

So this is a shamal – Pictures taken at high noon!

Finally today there was a report in *Stars & Stripes* yesterday concerning a disturbing trend in the Army. Yes another one! This one discussed a Colorado-based Army unit that has been accused of some very serious crimes after returning home **(Stars & Stripes, 28 JUL 09, *First under fire, then under arrest*, source: Associated Press)**. This particular unit was deployed to Iraq in 2004 and since that time have had ten soldiers accused of murder, attempted murder, rape, DUI, drug deals, kidnapping, suicide, etc. This unit saw a lot of action and lost 64 soldiers during its deployment to Iraq. Many in the unit blame the lack of discipline within the unit itself for such circumstances. Other unit members stated that commanders ignored their pleas for assistance, while others turned to alcohol and drugs to cope. One of the soldiers mentioned that the Army trains you to kill and this is what they did, sometimes indiscriminately.

This is a very disturbing story indeed. And this is one unit of many, so as you begin to analyze the issue, some questions that beg to be asked are: Are there other units out there with similar issues? Was there a leadership vacuum in these cases or was it the system in which the units operated that failed? What type of culture has the Army created to fight and win this type of war? What type of intervention has the Army injected into its system to assist soldiers? And, has the system improved significantly since 2004? This story coupled with the alarming increase in suicide rates and alcohol and drug abuse all point to a system struggling to keep up. Many of these same circumstances arose during and after the Vietnam War, yet we were still unprepared to deal with them almost a half century later! Remember what I said about failing to learn from history? I will say that the Army is really trying to get a handle on this and have tried several programs to assist returning veterans, but clearly there is much work yet to be done. This appears to be the proverbial snowball rapidly moving down the hill, picking up speed, and the Army has thus far been unable to stop this free fall and get ahead of the situation. And if the Army does such a great job of gearing soldiers up for a war, why can they not exhibit the same effectiveness to gear soldiers down after war? The Marine

Corps is no different and while they have had instances of similar circumstances, it does not appear to be as widespread. The ostriches who read this will bury their heads in the sand by claiming that this differential between the services is because the USMC is much smaller than the Army. While this is a true statement, I am not buying it. Perhaps it does come back to leadership and the lack of strategic planning and vision. We plan incredibly well for tactical operations and 'hard power' events, but we continue to struggle with 'softer power' events such as mental health, suicide prevention, de-escalating stress, and preparing soldiers adequately for life after combat. We must do better, and our soldiers and their families deserve a much better effort from the Army.

The Ostrich Syndrome

Day 267: Thursday-July 30th, 2009/Day 236-Iraq I sent my ***ledocracy*** article into Fire Chief magazine as well after tweaking it a bit from the version I had submitted for possible publication in Joint Forces Quarterly. As a ***ledocracy*** is of my own creation, a hybrid version of an adhocracy, I have bantered this term about city hall and my fire department in Wisconsin Rapids for a few years now. This is a term you cannot Google, at least not yet. I am hoping to get it into print in a national periodical and show the boys at the fire department that this concept is now on a national stage! And after being on active duty for several months now, my resolve for wanting to change the bureaucratic system within government is even stronger than it had been. A ***ledocracy*** is basically an anti-bureaucracy. It is similar to an adhocracy but places more emphasis upon the aspect of leadership and therefore falls somewhere between a bureaucracy and a more free-flowing adhocracy on the organizational model scale. I hope to write a book on the topic within the next few years, but getting an article printed on the topic in a national periodical that many government officials will read is a step in the right direction. I will let you know how this turns out as I should hear something before my tour here ends.

There is much news today to report. I will try and do so in a rapid-fire manner.

- The Iraq Independent High Electoral Commission (IHEC) reported that incumbent KRG President Massoud Barzani easily was reelected by getting 70% of the vote. The Kurdistan List party won 57% of the vote while the upstart change party known as Goran won a respective 23% of the vote **(MNFI, 30 JUL 09:** *al-Iraqiya, al-Sharqiya, al-Hurra, Radio Dijla, PUKmedia***)**.
- As I have already reported to you, Secretary Gates has been here and left, but during his rapid visit here he met with the Iraqi Prime Minister to discuss the possible sale of US weaponry and equipment to Iraq. He also traveled north to meet with the KRG President to offer US assistance to mediate any Arab-Kurdish disputes **(MNFI, 30 JUL 09:** *Washington Post, NY Times, Financial Times, Reuters***)**.
- Secretary Gates also came over to see first-hand the state of the current relationship between Iraqi Security Forces and US combat forces since they moved out of all Iraqi cities. His observation, validated by General Odierno's assessment, is that the situation is stable and there were bound to be a few growing pains along this path toward progress, but he is confident that the ISF is up to the task to secure their urban areas and soon all of Iraq. He added that the US is standing by to render any assistance should it be necessary **(MNFI, 30 JUL 09:** *AP***)**.

- General Odierno has noted continued Iranian influence in Iraq, but stated that it is now more targeted as they attempt to influence the upcoming national elections **(MNFI, 30 JUL 09: AP)**.
- The Swine Flu is still in the news as 38 US cases and 10 Iraqi cases have been confirmed by the MNF-I Surgeon **(MNFI, 30 JUL 09: *unclassified CJ1/4/8*)**. On our way out of into Iraq on R&R we were all screened for the H1N1 virus (swine flu).
- The USMC definitely gets 'it' as I have mentioned on several occasions. In regard to their increasing suicide rate, their Commandant General James Conway stated that the answer to this problem revolves around the aspect of leadership **(Stars & Stripes, 29 JUL 09, *Marines train ranks to address suicide increase,* no author)**. He is right on and it will take effective leadership in all the services to have a significant impact upon this negative trend.
- The Inspector General blames military leaders and a major contractor (KBR) in the electrocution death of a Green Beret in 2008 **(Stars & Stripes, 29 JUL 09, *IG lays blame for electrocution,* source: AP)**. This is very damning testimony by an expert witness and will no doubt lead to a big payout of damages in a civil lawsuit. It may also lead to a criminal suit as well. This case may also lead to the flood gates being opened on this alarming issue as 17 others, primarily US troops, have been electrocuted under similar circumstances since the war began. As I have mentioned on several occasions previously, I hope justice is served in these cases.
- Finally, the Department of Defense is going to modify its two-war strategy after decades **(Stars & Stripes, 29 JUL 09, *DOD to modify its two-war strategy,* written by Leo Shane III)**. For years the US has structured the size of its

military to be able to fight a two front-war. The modification to this decades-old strategy is to maintain a force that is capable of fighting multiple smaller engagements without burning out troops, but will still be robust enough to fight in two major theaters of war simultaneously, such as Korea and Iran. If this is the case, perhaps the fundamental philosophy may have shifted, but it appears to be a case of smoke and mirrors if we are still sustaining a force structure to fight in two wars! I believe the intent to change is there but something must have been lost in translation!

Day 268: Friday-July 31st, 2009/Day 237-Iraq I will follow up on a few stories I have previously mentioned.
1. Back to the environment and the drought I mentioned recently. The Agriculture Minister stated that 90% of Iraq is either desert or suffering from severe desertification. One US Army advisor mentioned that Iraq today reminds him of early 20th century Oklahoma when it was known as the Dust Bowl of America **(MNFI, 31 JUL 09: *LA Times*)**. I guess this would explain all of this damn dust!
2. The Iraqi Minister of Defense recently visited the US and mentioned that he was surprised to find such great support from the new administration for his country. In fact, he got the distinct feeling that the US would provide broader support in the future in a variety of areas such as education and economic trade, and he also believes that the new administration has a belief in Iraq's future **(MNFI, 31 JUL 09: *Asharq al-Awsat*)**.
3. I have mentioned to you the number of troops here in Iraq and the corresponding number of contractors. Well, the numbers continue to dwindle and I can even get a *Stars & Stripes* at the DeFleury at lunchtime now, which tells me

there are less people here on VBC. Since I arrived in December 2008, approximately 23,000 troops have departed Iraq **(MNFI, 31 JUL 09: unclassified)**. We are also down about 16,000 contractors as well. Of course, some of these troops were reassigned to Afghanistan, which is now the main effort.

4. Finally, I must make a correction on Angelina's recent visit. Apparently she did visit with the troops over here on VBC. Why was I not informed of this! She was named a Goodwill Ambassador for the United Nations in 2001 and this was here third visit over to Iraq **(MNFI, 31 JUL 09: unclassified)**. She also had her picture taken with some of the troops, so even though I said she could not entertain by singing or dancing she did so with her pretty face. It is not every day you can get a picture taken with a famous movie star. So I stand corrected and apologize humbly to Angelina. I think her commitment to such humanitarian efforts is very commendable and noteworthy. I haven't seen Jennifer Aniston making any visits over here!

It also seems like we are experiencing a shortage of some of my favorite items over here recently with no rational explanation available! Orange Gatorade is nowhere to be found and has been MIA for the past few weeks. This is easily the most popular flavor but there must be a shortage of orange flavoring somewhere in the world! My Frosted Flakes are getting harder to come by as well as bananas. The last straw was that the DeFleury has been out of pudding the last few days as well and they have not had such a shortage since I have been here over eight months. Is this all part of the drawdown strategy?

A food shortage?

Day 269: Saturday-August 1st, 2009/Day 238-Iraq we began another month. My unit should only ring in two more months after this and hopefully we will be back home when we welcome November into 2009.

There was good news to bring in this new month and that is that July saw only 7 US troops die in Iraq, which is the lowest monthly total since the war began **(JASG, 31 JUL 09, source: AP)**. Of course, for the 7 brave troops that died and the families they left behind this factoid is meaningless, but it is a good trend that we hope to see continue into the upcoming months. I also do not want to get into the habit of reducing our losses to mere statistics. Everyone we lose here is someone's son, daughter, husband, wife, father, mother, or friend. Each military member that has lost his/her life has sacrificed more than most people will ever realize. So the cost of freedom should not ever be reduced to mere numbers because it is not statistics that keep us free, it is the brave men and women of our armed forces and each one has a name and a story. Lest us never forget that or take it for granted.

I am just about back to the point I was before I left on R&R in relation to my physical fitness training. I had a little lull while on R&R and so I decided to complete about two weeks of transitional workouts before I attempted to plug in my pre-R&R numbers on the old treadmill. I believe I am about right on schedule but we shall see next week. I also noticed a little soreness the first few workouts of doing my perfect pushup routine, but that too has subsided so next week I will attempt my pre-R&R numbers of reps and sets and see where I am at. And, I am still holding steady at 80 kg!

I have also replenished my in-room stock of goodies to bring this back to a pre-R&R inventory. I reduced the inventory before I left so I would have little food in my room and hopefully prevent any unwanted visits from little critters. And in case the power went out while I was away and the frig went down I didn't want any spoilage to return to.

I call my stockpiling 'pack-ratting', which my wife accuses me of often. I believe I inherited this trait from my father because he is the kind of guy that will 'dumpster dive' and retrieve items my mother has thrown out. Maybe it's just guy thing! But the path to curbside and garbage collection moves from my parent's house through the garage. My father's domain is in the basement and my mother's is upstairs. Unless doing laundry, my mother rarely ventures downstairs amongst all the clutter. However, a few years ago I had some remodeling done for my mom and dad and I had a garage built for them. After years and years of never having a garage, and having the cold Wisconsin winters take its toll on their vehicles, I thought this would be a nice addition for them. And it has been, but father has kind of turned it into his 'man cave' and it has simply become an extension of the basement! This is also the clearinghouse for any items leaving the homestead with curbside being the final destination. The garage is strategically located between the house and the curb. The problem is, much like the Bermuda Triangle, items that venture into the garage rarely find their way to the curb! Yes, my father simply keeps it there because you never know when there may be a use for the item in question. I am not as bad, although my lovely wife may disagree with that statement, but on a rare occasion I actually do go through all of my stuff, and I have a lot of stuff, and I do throw items out that I know have no personal or historical value. I have most of my personal items willed to family and friends, and if they have no need or desire for the item then it will go to an estate sale so they can get money for the item. Yes, it is all about planning in detail. President Bush should have had me on the team back in 2003 and this war would have been much shorter in duration that is for certain!

I will also try and reroute as many items to the Goodwill as possible, because one man's trash may be another man's treasure! But I do 'pack rat' a little over here to keep enough items in my room in case I get trapped there for some reason! After a careful and thoughtful analysis, I believe I have enough of a coffee supply to take me to redeployment back home. I will maintain my current stock of goodies until about October 1st and then slowly reduce the inventory to zero by the end of the month when we are scheduled to redeploy. I also believe one more care package from my sweetie pie and one from mother in-law Kathy Kertis should be all that I need with my remaining time here. In case you have not figured it out yet, yes I am a planner. I have even planned out my final wishes in my will down to the music to play at my after funeral celebration! Anal retentive? Perhaps, but the way I see the world is that we plan for vacations, holidays, retirement, our kids futures, and we all know that we are going to die eventually, yet so many people fail to plan for this actuality. Not me. As I have stated, fail to prepare, prepare to fail. And I want my death to be as easy as possible for those I leave behind, and that comes with proper pre-planning. Shari is the spontaneous one and that is why we complement one another so very well. Anyway, that is the inventory reduction plan and I believe it will be executed to perfection. I will let you know the result in the final chapter.

113/118

Day 270: Sunday-August 2nd, 2009/Day 239-Iraq I must share a good story concerning our wildlife around this place. I have already chronicled my battle with the dreaded desert mouse a few months ago where we fought to a draw. Well, SFC Holczer is a fellow member of my unit, the 416th TEC. He also bunks in the room next to mine in the barracks. He told me that a few days ago he had been paid a visit and asked me if I heard him in the middle of the night. I informed him that I had not as I sleep with earplugs in and often doing not hear a lot of peripheral noise. Well, apparently he was resting comfortably in bed, lying in the supine position with his hands crossed over his chest. I call this the casket pose. For some reason he simply awoke and felt something sitting on his chest the "size of a Coke can". Startled by the unannounced midnight rendezvous, SFC Holczer bolted out of bed and "screamed like a little girl"! He flipped on the lights and noticed a big, burly rat racing around the room. Much as I had done a few months earlier, SFC Holczer grabbed his trusty broom and went on the offensive. He too failed to initially capture his quarry as the 'rat bastard' was simply too elusive and cunning. As the trained warrior that he is, and following in my footsteps, SFC Holczer developed a strategy that would allow his unwelcome visitor a chance to escape and fight another day. So as soon as SFC Holzcer opened up this avenue of escape, the rat bolted for the door. While SFC Holzcer was right on his tail, literally, once the little varmint hit the hallway he was gone in a flash with no trace left behind. These furry little insurgents have quite a good escape plan I must say! SFC Holczer also believes this critter came in under his door where there was a good-sized gap. Now, he places a rolled up towel on the floor to plug this gap and impede the path of any more unwelcomed visitors! I tamp my rug under my door to seal the gap as well and at night I also place a towel down to further deter any more unannounced night stalkers! As of this date, 416th personnel have battled to two draws with the furry insurgents. Our enemy is very elusive and cunning. While there will be no Purple Hearts or Medals of Honor awarded for these battles, both SFC Holczer and I hope we do not have to fight such a battle again during our deployment!

Another epic battle in the desert

Day 271: Monday-August 3rd, 2009/Day 240-Iraq was some very interesting news concerning a Navy Captain that has been missing since the first days of the first Gulf War in 1991. Captain Scott Speicher's remains were located in western Iraq recently **(MNFI, 3 AUG 09: *AP, AFP, CNN*)**. While a sad ending indeed, it does bring some closure to the family after 18 years of searching. This may be an opportune time to also mention a group that we have here in the SOC called the PRD (Personnel Recovery Detachment). This group of dedicated professionals is comprised of military and civilian personnel who are always searching for those that are missing. We currently have about 23 individuals that are missing throughout Iraq whose disposition is unknown. But this group will continue to follow up leads and to search as long as our government allows them to do so. This group has a very difficult and often frustrating job but they do outstanding work.

I have been playing a trivia game on the Green Bay Packers Web page. This contest is part of their 90th anniversary. While founded in 1919, the Packers did not actually join the NFL until 1921 and are one of the oldest franchises in the league. The Steelers may have won the Super Bowl last year and their 5th World Championship, but they are still well behind the Packers tally of 12 championships, which is the most in the NFL. Even the Cowboys have only won 5 titles so how did they become America's Team? Anyway, each day there is a new question and thus far I am 13 for 13. I think the big prize is free tickets for a game in September. Yes, I know I will still be here but perhaps I can transfer the tickets to Shari, or give to a wounded veteran to enjoy the experience or a sick child. But, before I worry about that I must win and I am sure there are many others out there who are also perfect in their answers as they have not been hard to track down thus far. The contest ends August 7th.

Go Pack Go

Day 272: Tuesday-August 4th, 2009/Day 241-Iraq today marks the Coast Guard's 219th birthday. As *Cheers* character Cliff Clavin would tell you, "Well, Sammy, it's a little known fact" that the Coast Guard has troops serving in Iraq. While there are only about 43,000 personnel in the Coast Guard they perform some critical missions including port and river security. Unlike their sister service the US Navy, the USCG does have arrest powers and can board ships without permission under reasonable cause. So in essence they are like water police and their primary mission is to protect the extensive US coasts. They are a very versatile service and would be another good candidate for my ***ledocracy*** organizational model due to their adaptability and flexibility. Anyway, Happy 219th Birthday, Coastguardsmen!

Happy 216th USCG

After the Kurdish elections were very successful recently, Prime Minister Maliki visited KRG president Massoud Barzani to discuss disagreements between the KRG and the Government of Iraq **(MNFI, 4 AUG 09: *al-Sharqiya, al-Iraqiya, al-Hurra, al-Jazeera*)**. In turn, the KRG will send a delegation to Baghdad to further discuss the issues between these two groups, which has been a real source of tension over here. Perhaps recent events such as the Prime Minister's visit to Washington, Secretary Gates visiting Baghdad, and Vice President Biden's visit have all helped to put pressure on the Maliki government to seek fair and reasonable non-sectarian solutions to national issues.

Prime Minister Maliki, on the state of GoI and KRG relations:

> We are in the process of building a non-dictatorship state, we are in the process to build a state that is based on a central and developed system. There will be disagreements, but we will not be pessimistic and will not let these disagreements fail the democratic process and the achievements that were achieved with efforts and bloodshed...the disagreements that we have are considered small.
> **-MNFI, 4 AUG 09: *al-Iraqiya***

This critical situation must be monitored closely as it could easily destabilize all gains achieved if it deteriorates. Washington obviously realizes this and that is why they have placed so much emphasis upon it.

On the home front, I am becoming more familiar with **Facebook**. I now have about a dozen friends, most of which are some of my firefighters, but it is a start. I hope to expand this list and use this venue to publicize my books and expand my knowledge. My firefighters on the other hand are using it as a social network (which I think is what it is actually supposed to be used for) and are playing some game called Mafia Wars. In addition, I am now 14 out of 14 in the Packer trivia contest, as I am sure many, many others are as well. Only four more questions left I believe. And training camps in the National Football League have all commenced which means fall is around the corner. Fall is one of my favorite times of the year as that means cooler weather, a change of seasons (in some parts of the world!), and football. This year it is even more meaningful as it means I am getting closer to getting out of here!

113/120

❙❙❙❙ Day 273: Wednesday-August 5th, 2009/Day 242-Iraq as today is a no BUA day, and it has been relatively quiet here in Iraq (knock on wood), I will look a little further east as there are some interesting parallels with Iraq that are shaping up in Afghanistan.

General McChrystal has just about completed his assessment of the situation in Afghanistan. It appears as though he may request additional troops for the fight there **(Stars & Stripes, 1 AUG 09,** *General likely to seek more troops for Afghanistan,* **written by Anne Gearan and Lara Jakes)**. The exact number of troops he may request remains unclear, but it sounds as though he is looking at another surge in order to clear areas of the country and then in turn hold that same piece of ground. This is exactly the same strategy that proved to be effective in Iraq, but again there are also many differences between these two countries as well. It will remain to be seen if this same strategy will prove effective in Afghanistan as the dynamics are a bit different. And, this will be a real political hot potato anyway so unless some progress is being exhibited with the increase in troops already sent, it may be an uphill battle to get any more troops approved by Congress. This situation certainly bears watching.

As I have been very critical of the Army and many of its processes in this book, in all fairness I must now pay them a compliment. While talk of suicide prevention has been very quiet since about March, the Army is still researching many different avenues to reduce this alarming trend **(Stars & Stripes, 1 AUG 09, Casey:** *Anti-stress training could help stem surge in soldier suicides,* **written by Susanne M. Schafer).** It appears as though there will be awareness training woven into the training fabric of the Army, from basic training and into all of the professional development schools at every level. This is certainly a good step forward. There is much yet left to do in this area of concern, and I still say that the military must overcome a culture that is still very resistant to softer issues such as dealing effectively with the stigma of mental health issues. But they are trying diligently and for that I give them great credit. Much like the two wars we are currently engaged in, I just hope the Army continues to pursue this issue with the same patience and persistence they have exhibited in Iraq and Afghanistan because it will be a long road to success.

113/120

Day 274: Thursday-August 6th, 2009/Day 243-Iraq
Well, today is going to be a bit different than many I have had in the past eight months. After becoming entrenched in the Strategic Operations Center (SOC) as the Gulf Region Division (GRD) Liaison Officer (LNO), I will be taking on new duties today. LTC Chuck Samaris, who I have introduced you to previously, was our G3 (Operations) but has redeployed back to Ft. Leonard Wood, MO. His replacement is LTC Rich Pratt, who I went to the Engineer Officer Basic Course (EOBC) with at Ft. Leonard Wood back in July through November 1990. Ironically, it was during this military training that Shari met her husband Brett! Again, we were only acquaintances at that time! Anyway, Rich called me in the SOC yesterday and asked me how I felt about moving over to GRD HQ, which is now just down the road from the palace. GRD has and is still in the process of moving operations from the International Zone over to Victory Base in case I did not tell you that previously. So it is a short move and I believe a good one. The work in the SOC has really slowed down, and even though I occupied my time through professional development activities, I feel I can do more to assist GRD by taking on these new duties, which include taking over the GRD transformation planning as they draw down troops and offices just like everyone else as we plan our exist from Iraq at the end of 2011. So my battle rhythm will need to be amended for my final two months here, but I look forward to the new challenge.

As the Iraqi theater of operations attempts to remain relevant as the main effort shifts eastward toward Afghanistan, General Odierno stated that US forces are still going to be needed here in a training and advising capacity, even after the significant drawdown of troops by August 2010 (**MNFI, 6 AUG 09: *AP***). General Odierno was fielding some questions concerning the statements made by an Army Colonel in a memo that was mistakenly published which stated that Iraq was secure enough right now that US forces could leave the country sooner than planned (**Stars & Stripes, 5 AUG 09, *Odierno: Too soon for US to pull out*, written by Kim Gamel**). I do believe the General is being cautious with his assessments, and many people both inside and outside of the military community would like to see our timetable for withdrawal sped up, but since General Odierno has much more experience in this theater of operation than probably anyone, I will defer to his assessment and December 2011 is really not that far away.

Day 275: Friday-August 7th, 2009/Day 244-Iraq well my first day with my new battle rhythm was interesting. I believe things will work out very nicely, but there is going to be a bit of a transitional period as GRD HQ continues to get themselves established in their new home. Due to this, I may actually split time between the palace and GRD HQ for a while. This way we can maintain a presence in the operations center and I can assist GRD with their upcoming transition. So it may be that I only need to tweak my routine a bit over the next few months as opposed to a complete overhaul. Either way, this development will spice things up nicely I do believe.

There was an interesting article in the *Stars and Stripes* yesterday concerning the increasing rate of alcoholism in the military, which I alluded to back on Day 266. One opinion on this issue is that soldiers should be allowed to drink a few beers a day. This opinion stated that during the Vietnam War there was drinking allowed and there was never a widespread problem (**Stars & Stripes, 5 AUG 09, *2 beers a day, and keep the enemy at bay*, written by Charles A. Krohn**). There was a limit on the number of drinks and enforcement of the rules is a leadership responsibility. Of course that was a different time and generation, but some people believe that by soldiers having to go so long without any alcohol leads to binge drinking and longer-term problems. So this makes for an interesting debate. I enjoy beer when at home, but I really have not missed it since I have been in Iraq. But because it is another major issue for our overburdened troops, which leads them down associated paths such as depression, domestic violence, and suicide, this contributing factor will undoubtedly warrant more research, but it is very unlikely that General Order #1 (no alcohol in war zone) will change during this war!

Day 276: Saturday-August 8th, 2009/Day 245-Iraq it appears that the situation is shaping up in the north. The Kurdish Government is talking with the Government of Iraq and the two parties can hopefully "mark a new era" according to Iraq Prime Minister Maliki **(MNFI, 7 AUG 09: *Reuters*)**. This is one of those conditions that must be resolved in order for this country to prosper and remain stable. So hopefully this is a step in the right direction. Talks will continue, as the KRG will be traveling to Baghdad for further discussions.

The atmosphere over at GRD HQ is a bit different than my perch at the palace. The lighting is better, but there is no CNN or BBC playing constantly. I do have many of my fellow 416th TEC compatriots over there to talk with as it looks like I may be splitting time in both worlds over the course of the next few months. I attended a few meetings yesterday that discussed our redeployment home. This was all good, but there is a lot of work to do. You do not just get up one day, pack your duffel bag, and head to the airport. There is a lot of equipment that must travel back ahead of us, the mission must continue with less personnel as our unit is not being replaced, and our whole organization is drawing down as it continues to move from the IZ to VBC. It now appears as though we will be leaving around the 21st of October, which is right between the originally project 26th and the rumored 17th. That means only about 73 days remain in Iraq. Excellent! Then we are to be at Ft. McCoy for demobilization for about four days and then we will be released to go home. On this timeline, I will be home for my granddaughter's birthday and for Halloween. So it is getting very exciting right now around here as you can imagine as the days continue to dwindle down.

My job search continues, but I have not had to look for work for decades so it is all a little new to me. The good news is that I do not have to leave the fire department at this point as I love that job, but I am exploring my marketability even though we are in difficult economic times and the type of employment I am looking at is very competitive. So far my resume has been too much for an eastern university and a directors positions (I was reaching high on that one anyway), and a private sector research think tank (probably reaching a bit high on that one too). But unless you try and find out where your limits are you will never know. So if nothing else, I will be bracketing my next career level in this effort and see where it may take me. Otherwise I will simply go back to the fire department for a short period, continue to explore my options, and hopefully the economy will rebound and more opportunities will open up. The future positions I really covet I am still awaiting word on and this may not come until I get back home. But I will keep applying, as I will be selective with my next career choice. Even though I have applied for about 20 positions thus far, this does not mean I would have accepted certain positions even if offered. Location, money, and flexibility in schedule are all important variables to me, and many of the job descriptions I read lacked detailed information on two of these three criteria. The location was about the only constant. So time will tell what happens next, but given some more time, something really good may open up and I will continue to tweak my resume and search for that next golden opportunity. It is important to just never give up and to keep trying. This is easy for me to say with a job awaiting my return, but if I were looking for employment in this economy and had no job to make ends meet, I can see how people can get discouraged and depressed. But you still must never give up and I am a walking poster child for patience and persistence. Given time, very often these efforts will be rewarded. Otherwise, I have no issue with be a wing man (cook) at a Hooters restaurant as I collect my fire department and military pensions! There is always a bright side. And as I have told Shari, she could be the manager and keep all the girls under control while I work my magic as a grill master. It may not be the best plan, but it is a plan nevertheless!

My next place of employment? ☺

Day 277: Sunday-August 9th, 2009/Day 246-Iraq today of course is my favorite day of the week. But even though this is my half-day, there will be plenty of activity. GRD is working on reducing its footprint within Iraq over the next few months, all while still moving to VBC from the IZ. So GRD is right now a flurry of activity and one reason why they have pulled me from the SOC in order to assist them in their efforts. Yesterday I attended three meetings, and then I also spent a few hours in the SOC to catch up over there. Today, there will be three more meetings, all in the afternoon, including an update of the drawdown plan to MG Eyre. So no rest for the weary, but it is all good and challenging and will make the final 70 plus days go by quickly.

In other Iraq news, there will be around one million Iraqi pilgrims gathering at Karbala, which is just south of Baghdad. This time of year marks one of the biggest festivals in Shi'a Islam **(JASG, 8 AUG 09:** *BBC***).** These types of events seriously challenge security forces in trying protect this many people. There have already been several bombings and casualties. This is the first time the ISF have taken on such a challenge by themselves and without coalition assistance. Despite these acts of violence in this year's pilgrimage, the ISF has done a good job of limiting such events. One Shi'a pilgrim stated that security conditions were much improved and:

> I'm 65 years old and have felt free to practice my religious rights after the topple of the former regime. We

used to perform these rituals in the past, despite fear and oppression, but we fear nothing today.
-MNFI, 8 AUG 09: *Aswat al-Iraq*

These are more positive signs of progress you most probably will never know about!

In fact, normalcy is breaking out all over Iraq as they are even drafting anti-smoking laws to prohibit such activity in public buildings **(MNFI, 8 AUG 09: *Washington Post, Reuters, AP, AFP*)**. My hometown, Wisconsin Rapids (WI), has been dealing with this issue for several years now. In fact, the State of Wisconsin just passed legislation to prohibit smoking in public buildings, including taverns, throughout the state. And as Wisconsin has probably more taverns per capita than any other state, this has not gone over all that well with tavern owners! I believe this law takes effect next July. While I have never been a smoker, I do have an opinion on the matter, but that is really irrelevant. What I will say is that how the times have changed and it will be incredibly interesting to follow the incidents of lung cancer when this law takes effect versus the pre-law statistics. The anti-smoking advocates will sure look silly if the number of incidents do not decline significantly. And reading blogs concerning such issues is always entertaining, as an individual will complain of government controls and constraints on one topic, but then advocate the same use of such controls in incidents they believe in! Can you say hypocrite? But it makes for an interesting debate and should be interesting to follow over here as well.

Finally today, I am unsure if I have mentioned previously, but if not, the use of Predator drones has struck fear in the hearts of insurgents everywhere! Much like carbon monoxide, these harbingers of death are silent and lethal. The bad guys never know what hit them. While past strikes have unfortunately resulted in some civilian casualties, or what is called collateral damage, the use of such weapons is getting much more focused and more accurate. This is very bad news for the insurgents. These 'death from above' aircraft are flown remotely from a location far away from the actual site of the strike. This weapon has proven its value again and again, and whether it is gathering intelligence or sending missiles down range, it is very bad news for many insurgents.

Bad news for insurgents – Death from above

▮▮ ▮▮ **Day 278: Monday-August 10th, 2009/Day 247-Iraq** after yesterday's briefing to MG Eyre, it is clear that the train is moving down the tracks very quickly at GRD and we better buckle in snuggly for the ride! Many tasks will need to be accomplished over the course of the next few months as the drawdown continues at a frantic pace. But in just over two months, we (me and my fellow 416th members) will be gone and GRD will be transformed into a much smaller organization. So it will be some long days ahead, but it is all good!

To follow up on some stories I have mentioned previously, the smoking ban being discussed here in Iraq is going over like a lead balloon **(Stars & Stripes, 8 AUG 09, *Proposed ban on public smoking riles Iraqis*, written by Ernesto Londono).** Cigarettes are very cheap over here and many people smoke. In fact, one smoker went so far as to state "We want Saddam back...you could do anything during Saddam's time." I found this to be a very interesting comment and perhaps the Iraqis first of many exposures with government controls and mandates! Maybe they are more like us than we may believe! Many military members smoke as well. In fact, about 32% of military members smoke, with the Army having the highest rate. The civilian rate is supposedly at about 20%. If this becomes a military policy, it will no doubt hurt recruiting efforts for an all-volunteer service that is already stretched thin, and the global environment would suggest there is no end in sight for the foreseeable future in relation to the type of conflicts we are currently mired in. So this story will be very interesting to follow in the months ahead as it has created great debate and inspired much passion.

Something new that the Iraqi government is doing that I do not believe the US has ever attempted is to offer money for mixed sect marriages **(Stars and Stripes, 8 AUG 09,** *Iraq offers $2,000 for mixed-sect marriages,* **written by Bushra Juhi and Deb Riechmann)**. This development seems like a very good idea in order to develop a better relationship among the two major sects in Iraq. And, this trend is apparently a rapidly growing one. While sectarian violence will always be a part of Iraqi culture, ideas such as this exhibit Iraq's attempt to overcome centuries of bad blood and bring the two major sects (Shi'a and Sunni) together, as cooperation between these two sects is very important to the success of this country. The effects of such a program will not be known until long after this book is published, but it certainly bears watching.

There have been several cases of the swine flu diagnosed in Iraq and a 21-year-old Iraqi female just died from the disease **(MNFI, 10 AUG 09: *AP, Reuters*)**. This number includes 51 US troops, with another 71 who are isolated and suspected of having contracted the disease **(Stars & Stripes, 10 AUG 09,** *51 US troops in Iraq diagnosed with swine flu,* **written by Chelsea J. Carter)**. Hopefully this disease will not become a widespread health issue here in Iraq, but it is being monitored closely.

Finally today, another sign of progress that is unlikely to make Fox News or CNN and that is the removal of the many concrete barriers that lie within the city limits of Baghdad **(MNFI, 10 AUG 09: *al-Hurra, al-Arabiya*)**. These massive concrete slabs were placed in many areas around the country to better protect the populace from mortars, rockets, and other explosive devices. Taking them down to open up neighborhoods and streets is a sure sign of progression and the path back toward normalcy. But because you are unlikely to hear this news from the American media, I am reporting this 'good news' story to you now.

Day 279: Tuesday-August 11th, 2009/Day 248-Iraq are a few stories on our companion war to the east concerning expectations. As you are aware, General Stanley McChrystal was selected to replace General David McKiernan as the top commander in Afghanistan. He was given a few months to assess the situation there and then deliver an assessment to the President (JASG, 10 AUG 09). McChrystal is likely to request additional troops so that they can push the Taliban out of areas in the country and in turn hold those cleared areas. In other words, this could be considered 'Surge II' as this strategy proved very successful in Iraq in 2007. But whatever strategy is ultimately decided upon, it is likely to take several years and billions in funding before any significant success may be realized **(Stars & Stripes, 10 AUG 09,** *Experts: Afghan effort to last a decade, cost more than Iraq,* **written by Walter Pincus)**. And even after all of this, success may be limited or fleeting as this area of the world is much different than Iraq. In addition, Pakistan is intimately related to any success in Afghanistan and this is where Osama bin Laden is suspected to have hidden or still be hiding, and plotted attacks against the West. So even though the American public and correspondingly our politicians, may not have the stomach for years of more irregular warfare, the military understands that this war on terrorism is not over and will not be for years, perhaps decades to come.

Senator Lindsey Graham (R-SC), who is a member of the Senate Armed Services Committee and a member of the US military, will be traveling over here soon with Senator John McCain. He stated that we should not "pull another Rumsfeld" **(Stars & Stripes, 10 AUG 09,** *Senator warns against pulling a 'Rumsfeld',* **source: AP)**. This is in reference to former Secretary of Defense Donald Rumsfeld who grossly underestimated the amount of troops needed to achieve success in this type of warfare. Senator Graham hopes we have learned from history and will not place General McChrystal behind the power curve in the war in Afghanistan. But politics always has and always will play a major role in such efforts and we shall see what Washington will do in response to the situation in Afghanistan after General McChrystal presents his assessment.

Finally, you may recall all the bad press swirling around our friends at KBR. Well, the family of a Green Beret who was electrocuted back in January 2008 sued the company in criminal court. Not so surprisingly, the charges were dropped against the company because no malice aforethought could be proven **(Stars & Stripes, 10 AUG 09,** *No charges in KBR electrocution case,* **written by Kimberly Hefling).** In other words, the contractors who did this sloppy work did not conspire to kill troops, but this was the effect of accomplishing such poor work, coupled with poor leadership at KBR, and little oversight. So it is obvious that this particular court feels that stupidity and incompetence is not a criminal act! I guess in reality, if this were the case we would need a lot more prison space! So I feel very sad for the family of this soldier and the other 17 that were killed because of this incompetence, but civil action is still likely and while this company may not be criminally liable, it is likely that the families of these heroes will be awarded millions in damages. Certainly small consolation for the loss of a loved one! We shall watch how the legal system works in this one, and if KBR gets away clean on civil charges then people need to start asking some serious and uncomfortable questions. All I can tell you is KBR is very lucky I am not sitting on the jury!

Day 280: Wednesday-August 12th, 2009/Day 249-Iraq was haircut day again! After this one, I am calculating only about two more, maybe three before we are home for good.

Last evening was one of my later ones in awhile. COL Ted Fultz, who is a member of the 416th TEC and is now the SPO (Security, Projects, and Operations), was working on a slide presentation for tomorrow's presentation to MG Eyre. COL Fultz is very meticulous in his preparation and as such I was assisting him in getting the presentation ready and we did not start the process until after dinner. So, I ended up staying until 2130, and by the time I returned to my room it was pushing 2200. But it felt good to do something worthwhile and long days do not bother me in the least. Once I was back to my room, it was off to bed where I got a good, uninterrupted night of sleep.

So as we begin to wind things up in Iraq, there is still plenty of work to accomplish. Al-Qaeda is also trying to stir things up, but again they are killing innocent civilians in an attempt to stir up sectarian violence. In the last few days they have killed about 48 and injured another 250 **(MNFI, 12 AUG 09:** *LA Times* **and Stars & Stripes, 11 AUG 09,** *Iraq: Bombs*

***targeting Shiites kill at least 48,* written by Hamid Ahmed).** So they are trying to prove that the ISF cannot protect the people and destabilize the government. I am unsure as to the strategy behind this because it is more likely the government would request assistance from US forces, which they have not yet done since we vacated the cities at the end of June, than let AQI run amuck. This may only prolong our stay here as opposed to waiting another 12 months, when we draw down to 50,000 troops and another 16 months after that we are out of here. Then they really could have the upper hand. So the rationale is puzzling, but operations continue to target theses insurgents and the sanctuaries they seek.

We will finish up this chapter with a peek east toward Afghanistan. The next several months are shaping up similar to the first few months of 2007 here in Iraq **(Stars & Stripes, 11 AUG 09, *McChrystal: Casualties in Afghanistan will stay high,* written by Anne Gearan)**. There was a large influx of troops, more pressure put on the enemy, a change in leadership, and a shift in strategy. As such, US casualties mounted quickly. This is what is occurring right now in Afghanistan. The reason it is being allowed to occur is because the situation in Iraq turned around as the new strategy took shape and has proven to be a recipe for sustained success. The same hope is equated to the situation in Afghanistan. We shall see if this proves to be correct, but over the course of the next few months there will be some anxious moments for many US and coalition troops, their families, military leaders, and in the White House. But these types of campaigns, much like many counterinsurgent experts have proclaimed, are long drawn-out affairs. A former NATO commander who looks to become Britain's next army chief, states that it could take 40 years to achieve success in Afghanistan! While the US and the coalition may have a presence in that country for decades to come, it will not be at the levels of commitment we currently see **(Stars & Stripes, 11 AUG 09, *Britain's army chief says Afghan mission could take 40 years,* written by John Vandiver)**. So the lesson here is that these types of wars, which are expected to dominate the next half century or so, are long and take great commitment. Let us hope that our government and military leadership is paying attention to these lessons as we move deeper into the 21st century.

GRD soldier training Iraqi maintenance workers

CHAPTER 15
400 DAYS - Days 281-300
August 13 - September 1, 2009

In the end, Iraqis will decide the outcome of this struggle. Our task is to help them gain the time they need to save their country. To do that, many of us will live and fight alongside them. Together, we will face down the terrorists, insurgents and criminals who slaughter the innocent. Success will require discipline, fortitude and initiative - qualities that you have in abundance.

--CENTCOM Commander/2008-present (also served as MNF-I Commander/2007-2008 and Commanding General US Army Combined Arms center 2005-2007 and MNSTC-I Commander/2004-2005 and Commander 101st Airborne Division/2003-2004)

Day 281: Thursday-August 13th, 2009/Day 250-Iraq started with what may be my last GRD BUA update to General Odierno. Because of my new duties, and the fact that SFC Maltes is still on R&R, I may not brief after this day. She will be back for the next update in two weeks, and we really only have four more briefings to do until we depart Iraq. It is also likely that the one briefing we do in October will be the last one for GRD period. I am trying to get MAJ Sean Begley, another member of the 416th, to do the final brief. He is anxious to brief General Odierno and he would do a fine job. We shall see, but as GRD draws down and will be considerably smaller when we leave the country, it is unlikely that that the smaller engineer organization will continue to have the staff to put a briefing together for the General. All of this is a sign of progress and that we have accomplished our mission.

As I have mentioned to you previously, we have seen an uptick in violence against civilians here in Iraq this past week. It appears as though al-Qaeda is behind many of these attacks. Their goal is to incite sectarian violence and throw Iraq back into chaos. Prime Minister Maliki urges Iraqi citizens to "stand against any foreign agendas that serve the interest of foreign countries who want to interfere with the Iraqi future." **(MNFI, 13 AUG 09: *Baghdad Times*)**. Certainly, the Prime Minister believes that other countries continue to fund AQI efforts and are still attempting to interfere with the direction Iraq is taking.

An Iraqi professor, regarding attacks against Shi'a pilgrims, states that:

> These bombings are an attempt to return Iraq to square one. I expect these attacks to rise the closer we get to the elections. The coming months will be a very critical time for Iraq.
> **-MNFI, 13 AUG 09: *Reuters***

Other Shi'a leaders are urging Iraqis to resist the temptation to retaliate. Grand Ayatollah Ali Al-Sistani urged Iraqis to shun violent retaliation against innocent civilians **(MNFI, 12 AUG 09: *New York Times*)**.

> Another Iraqi cleric, Majid Al-Aasadi stated:
> We will not react against these efforts to ignite sectarian violence because this is exactly what our enemies want and not what our Iraqi people want.
> **-MNFI, 12 AUG 09: *New York Times***

Finally, an ISF spokesman stated that violent extremists are using new methods to target Iraqi civilians. This includes attaching explosives to bicycles, toys, and motorcycles **(MNFI,

11 AUG 09: *AK News)*. Despite these attacks, the ISF is confident it can adequately protect Iraqi citizens.

So while the situation continues to improve and stabilize, there are still forces at work that seek to destroy all of this progress. So the next several months will be very critical in the development of Iraq. They must overcome decades of a cultural mindset that called for "an eye for an eye" and retaliation against those that struck against you. But I must say that the Iraqi population is showing great restraint against this cultural urge and this tells me that the majority of Iraqis want an end to this needless violence and a better future.

Today was also my father-in-law's 75th birthday. Dick Kertis is a retired member of the Wisconsin Army National Guard and a handyman extraordinaire. Dick has about every tool imaginable and can fix just about anything. Dick is also very laid back and takes everything in stride. Shari has learned a great deal about lawn care and machinery from her father. However, unlike Dick, Shari can't fix things after she breaks them! She normally will call on her father when that happens. So happy 75th, Dick! Have a cold one for me.

Happy Birthday, Dick, and no I am not trying to steal your cake!

Day 282: Friday-August 14th, 2009/Day 251-Iraq was the day after what may have been my last briefing to General Odierno. If this is the case, I was pleased with my briefing and I hit all of the Iraqi pronunciations and had a smooth, steady flow to the presentation. It may have been my finest brief! And that's the way it is – August 13th, 2009. Rest in peace, Walter!

Yes, I did have my haircut yesterday right after the briefing and the young man did give me a George Custer! Thank goodness I only have about two to three more haircuts before we leave.

There were more suicide bombings in Iraq yesterday that killed more innocent Iraqis. By all appearances, this is the work of AQI and they are targeting a specific sect in northern Iraq. By the same token, it does appear that the Iraqis understand who is responsible for these attacks and are not taking the bait on retaliating against one another as AQI wants them to do. So democracy and stability may be taking root here. In fact Iraqi President Jalal Talibani commented on the country's evolving democratic infrastructure that:

> The forthcoming stage shall be the foundation for the creation of an Iraq that can accommodate the broad spectrum of the Iraqi populace with all their sectarian leanings, to live in peace and security and distance itself from authoritarian, partisan practices or the one-party system
> **-MNFI, 14 AUG 09: *al-Iraqiya***

It is likely that this unnecessary violence will continue until the national elections early next year, but if those go well and there is greater sectarian representation within the Government of Iraq, then Iraq is well on its way to taking care of itself and this will pave the road out of Iraq for US troops in 2011.

On the home front, my stepson Matt and his bride Gina were married today. While I wish I could have been there, I wish them both much happiness. Just another event sacrificed for good old Uncle Sam. And Shari and Drew are leaving for Florida today. They will be gone for about two weeks. Drew likes the water parks better down there and Shari will visit with some old friends and see her youngest son Josh who is doing very well with his internship as a Chef at an upscale Orlando restaurant. Once Shari returns from this trip, it will very close to the beginning of school for Drew who heads to high school this year. And I should be within two months of departing this place for good. Let the old hourglass keep on draining!

Time is running out my pretty!

Day 283: Saturday-August 15th, 2009/Day 252-Iraq would be my debut by my lonesome in setting up the working group meeting in relation to the transition of the Gulf Region Division. This entails preparing a PowerPoint briefing, setting up the room and the computer for the presentation, and then basically running the show. CPT Reed Goodell had been doing this mission, but this is one of the duties I assumed when GRD pulled me back over from the palace and pushed Reed out to one of the GRD districts to assist them as our transition continues. Reed was what we in the military call a 'Battle Captain,' which is an action officer who takes care of missions such as this. There are also Battle Majors over here that run a lot of different types of operations. These are their actual position titles and I am not making this up! But I am not aware of any Battle Colonels over here, but I could be pioneering a new way ahead! While this may be a truism in fact, even though my official title may not be reflective of it, I prefer to think of myself as a **JOAT** officer. Much as I have challenged many of my work mates at my new home here in Iraq, I will buy you a beer if you can tell what this acronym stands for! In the magic land of acronyms (the Army), I have created a new one! **JOAT** is my acronym for what I do and have basically done over here on my tour. I am a **J**ack-**O**f-**A**ll-**T**rades and master of none! But flexibility and adaptability are my hallmarks and I will take all of this in stride and do the best I can with whatever mission I am given. If I screw up, what are they going to do? Send me home? Hmm, something to ponder! But if they don't, only about 66 more days to go anyway, so I will be **Gumby** on steroids and remain as flexible as possible over the course of what will be a busy next two months.

Just call me JOAT!

Day 284: Sunday-August 16th, 2009/Day 253-Iraq I finished the David Kilcullen book the ***Accidental Guerilla***. This was a very good book and Mr. Kilcullen truly understands the complexities of this type of warfare. Now, that I have finished reading Nagl, Galula, Kilcullen, all considered counterinsurgent experts, it is clear that not only do many military members not understand this type of warfare very well, neither do our politicians or the American public. Clearly, if one pays any attention to history whatsoever, this type of warfare is and always has been a long drawn-out affair with a very high price tag. But once again, we get back to my wheelhouse and the topic of leadership. If leaders at the highest levels of government and within the military community do not comprehend this level of commitment by our troops and our country's resources, then these same people do our soldiers, marines, airmen, sailors, and coastguardsmen a grave disservice. It is also likely that the type of fourth generation warfare we are now engaged in will be the norm for the course of the next quarter to half century! Is our country prepared for this? Are our leaders paying attention? I hope mothers and fathers of young men and women who will be joining the military over this course of time are listening because you may want to ask these difficult questions of your elected officials before they are put in office.

Our men and women in uniform deserve this level of comprehension before they are placed into harm's way. So let's wake up out there people and pay attention to history as we look ahead to the future!

I have just begun to read the last of my counterinsurgent books called **The Sling and the Stone** by Thomas X. Hammes, a former Marine commander in the Iraq War. This will complete my counterinsurgent professional development during this tour and then I have only two more books to go and between our long flight home and time spend at Ft. McCoy during the demobilization process, I am confident I will finish all the books I have in my current stock which will bring my total to 15 books read during my tour.

Continuing with this theme, a recent article in Stars & Stripes outlined just how difficult a fight we are engaged in that the American public really has little visibility on **(Stars & Stripes, 14 AUG 09, *Flushing out the enemy*, written by Ann Scott Tyson)**. In order to engage the enemy, our troops must utilize some cat and mouse strategy to lure them into the open. This will ensure that innocent civilians that the insurgents hide behind and within are protected as much as possible. Because just following insurgents into a home or a crowd and then firing upon them may kill some insurgents, but it is also likely to kill some innocent civilians in the process. In turn, this will push the populace away from the influence of our troops and the government, and in essence create more insurgents. It is the old one step forward, two steps back. So the real goal in this type of war is to protect the people from insurgent influence and kill them when the time and opportunity is right. So the mission is population-centric and not simply killing insurgents. This is a very foreign idea to many military commanders and troops, but this is how fourth generation must be fought to achieve success. This is not as sexy or as clean as the portrayal in Hollywood movies in the minds of many Americans. The kind of war we are fighting now is dirty, it is hard, and it is difficult. The enemy knows that facing us head on would be suicide. We have too much firepower and technology at our disposable. So the enemy has adapted and become very smart. They are using tactics to negate our large advantage in technology and firepower. It is us that now must adapt and we are having great difficulty in doing so. But we are doing it, slowly but surely. But this type of war also comes with a great human cost involved. The flag-draped transfer cases will continue to arrive at Dover Air Base as long as we are engaged in this type of war. But as the host country security forces become larger and more adept at their craft, then and only then can coalition forces begin to drawdown. As we are seeing here in Iraq, as this transition occurs, then we see less coalition casualties. But there are still acts of violence that kill civilians and host country security forces. And even though this is still sad and tragic, the process is evolving and this is a point we must reach in order to be successful and in order to set up the host government to be successful. I hope this assists you in understanding this type of warfare a little bit better even though I may not have articulated it very well. But I just feel that many Americans really do not understand the complexity of this type of war and just how dirty and hard it is for our troops. The sacrifice of our troops has been incredibly large and I guess this is the point I hope the American public never loses sight of.

||| ||| Day 285: Monday-August 17th, 2009/Day 254-Iraq

I was beginning to get comfortable with my new duties at GRD HQ. I was beginning to get a clue as to how things worked and I knew that I needed to make some adjustments to some of the inefficient processes that were currently in place. The current process to disseminate information to those that need it, and in turn, these same individuals need to provide me with the most current data to present to the GRD Commanding General, is in need of revision. And I do not mean minor tweaking; I mean a major overhaul! I am determined to make the process easier for all involved, including me!

One of my duties at the 416th TEC before we were mobilized was to develop a catastrophic response plan for our units in case of a natural or man-made disaster somewhere in our area of operation. Now, just so you are clear on this, the National Guard and Army Reserve are not one in the same. The Guard has a dual mission, which means they can be mobilized to respond to incidents within their state, or a neighboring state for that matter. They can also be used in the event of a national crisis, such as we are experiencing right now in Iraq and Afghanistan. Guard units also receive funding from both state and federal sources. When mobilized by their respective Governor, he/she has control over how they are utilized in the crisis. The Army Reserve is strictly federally funded and currently has no role within state matters. But because of some antiquated policies that Army Reserve units operate under, they cannot be used to assist states in the event of a crisis. This seemed very odd to me, and we have a firefighting unit in California that was called upon by the local fire chief to see if they could assist him in relation to a major land wildfire. Because of this antiquated policy, and the dreaded bureaucracy that is our governmental and military structure, the local reserve commander could not get permission to assist the fire chief! So all of the equipment and personnel that could have assisted, and that have been paid for by taxpayers, just sat there while Rome burned! This is a figure of speech of course, but I think you get the idea. So our unit (the 416th TEC) embarked upon trying to bypass this antiquated policy by providing commanders some options should this type of situation occur again. Well, this well-meaning mission was again met by the bureaucratic monster and was quickly absorbed into it without any decision being made. How unfortunate for taxpayers and the local emergency management directors! There was still no good disposition on this issue when I was mobilized, and I do not believe there has been much discussion on it since that time. But an article in Stars & Stripes rekindled this fire within me and led to this excerpt **(Stars & Stripes, 14 AUG 09, *Pentagon, governors face off over military reserve*, written by Lolita C. Baldor)**. As you may surmise, the federal government and the state's governors are at odds on who should control this resource if they were to be mobilized to assist a state in a time of crisis. I am certain that governors are not against having Army Reserve units, or any of the other service (Navy, Marine Corps, Air Force, Coast Guard) reserve units for that matter, assist with any disaster within their respective states. But who pays the bill for such assistance, and who controls these valuable assets during the crisis, is an entirely different matter altogether. But I do hope this matter can be resolved equitably for all parties concerned because reserve unit personnel want to assist and it sure seems unfair and almost criminal if personnel and equipment that taxpayers have paid for just sit and collect dust while a disaster unfolds a few miles away from the reserve center. Let us please apply some common sense to this matter, although in the case of the government on many occasions, this truly is an oxymoron!

Day 286: Tuesday-August 18th, 2009/Day 255-Iraq was full of news. Shari, Drew, and Cole made it to Florida with only minor issues. I guess the tailpipe from my trusty Vibe fell off somewhere in the mountains of Tennessee, but this did little to slow Shari's journey south. The tailpipe on the car is the original and I have had the vehicle for almost seven years now, so for being from a cold weather state and all the salt that goes on the roadways during the winter months, it has held up pretty well.

Two members of an American think tank commented that Iraq has ably managed their affairs since the US transferred security responsibilities to them **(MNFI, 17 AUG 09: *Forbes*)**. These individuals also praised Iraqi's for their restraint from engaging in sectarian violence. These comments point toward sustained stability, progress, and hope for the future.

Iraqi Prime Minister Nouri Al-Maliki is warning of more violence as the national elections approach **(Stars & Stripes, 15 AUG 09, *Al-Maliki warns of more attacks*, written by Hamid Ahmed)**. The Prime Minister believes insurgents will do their utmost to disrupt these elections and attempt to re-ignite sectarian tensions. He urges all Iraqis to remain vigilant and not get suckered into such a trap.

There are also many Iraqi's that are becoming very uneasy with the thought of a faster US withdrawal **(Stars & Stripes, 15 AUG 09, *Growing number of Iraqis uneasy with idea of faster US withdrawal*, written by Kim Gamel)**. Because of the recent violence against innocent civilians, many Iraqis want to be certain that the ISF can adequately protect them. While much of this uneasiness may be related to the perception of insecurity created by some high profile attacks, the number of attacks has not significantly increased. But because the insurgents are getting desperate and running low on weapons, explosives, and funding they are resorting to higher profile attacks in order to gain more attention and remain relevant. It is obviously working!

Finally this busy day, there was an interesting article in *Stars & Stripes* in relation to a recent survey conducted in Afghanistan **(Stars & Stripes, 15 AUG 09, *Poll: Pakistanis revile Taliban more than US*, written by Kay Johnson)**. According to the article, from the Pakistani populace perspective, 70% of those surveyed view the Taliban unfavorably. Al Qaeda weighs in at 61% and the US at 68% in relation to being viewed unfavorably. So I guess we are gaining ground on the Taliban; however, we still have a little work to do to catch al Qaeda! As the new US strategy takes shape, hopefully we will be viewed more favorably by the population because this is what this type of warfare is centered upon and is an absolutely critical factor in order to achieve and claim success.

Day 287: Wednesday-August 19th, 2009/Day 256-Iraq was full of interesting information. To begin with, you may recall the referendum I mentioned long ago where the Iraqi people can decide when they want US troops to leave their country. This referendum was supposed to have occurred this past summer, but some shrewd political maneuvering moved it back to tie into the national elections next January **(JASG, 18 AUG 09, *Iraq may hold vote on US withdrawal*: Washington Post and MNF-I Early Bird News, 18 AUG 09, *Iraq may hold vote on US withdrawal*, Washington Post, written by Ernesto Londono and Stars & Stripes, 18 AUG 09, *Iraq to let voters decide on US withdrawal timetable*, written by Ernesto Londono)**. What this means is that if the Iraqi people vote for us to leave, then we have one year from that time until we need to leave the country. This would accelerate the timetable that was agreed upon under the terms of the signed security agreement by about 11 months. So much like the Brett Favre situation, this story will not go away either!

Next we shall head east and talk of Afghanistan. As I also mentioned a while ago, there is now more news surfacing about General McKiernan's removal as the top commander in Afghanistan. As surmised, it appears as though he was removed because of his inability to adapt to the situation he was faced with. This is no revelation to me or many others. It does not make us psychic, only observant. This also does not mean he was a bad general officer. It simply means he was from the old conventional-style school and was uncomfortable dealing with fourth generation warfare **(Stars & Stripes, 18 AUG 09, *McKiernan's ouster reflected new realities*, written by Rajiv Chandrasekaran).** Because of his failure to adapt, he was basically removed by Secretary of Defense Robert Gates, a decision supported by Chairman of the Joint Chiefs of Staff Admiral Mike Mullen. Both of these men did not want to continue on the same course in Afghanistan because it was costing American lives. They wanted a fresh perspective and their own hand-picked 'A Team.' General Stanley McChrystal was selected as the new commander based upon his experience and background in this type of warfare. Many people believe he is from the same mold as CENTCOM Commander General David Petraeus, with an innate ability to forge important relationships with leaders in Afghanistan and also adept at juggling all the balls coming out of Washington. And at this point in time, President Obama is getting some heat on his new strategy **(MNFI Early Bird News, 18 AUG 09, *Obama defends strategy in Afghanistan*, Washington Post, written by Sheryl Gay Stolberg).** General McChrystal is expected to provide an analysis of the situation in Afghanistan to the President within the next month. So while it will be quite interesting to see if the situation turns around in Afghanistan with this change in leadership and strategy, like Lieutenant General Sanchez early in the Iraq War, General McKiernan was not given all of the resources or advantages that General McChrystal is being supported with. These factors make any type of comparison simply apples to oranges. But the bottom-line is this. If all of this turns the situation around in Afghanistan, and we can get to a point of sustained stability that we are facing now in Iraq, and most importantly, save American lives in the process, then it will have achieved a level of success that everyone is hoping for.

Then we come back to Iraq where the situation in the north is still very tenuous. This is the last bastion of resistance for AQI and they are creating a lot of havoc. General Odierno is working with ISF leaders to see if he can convince them to allow US combat troops to assist them more aggressively in this area of operation **(Stars & Stripes, 18 AUG 09, *Odierno seeks to put US troops into northern Iraq*, written by Kim Gamel)**. Thus far, since the US vacated all cities within Iraq on June 30th, the Iraqis have not requested any US assistance. But with the situation deteriorating in the north, and the national elections quickly approaching, this situation may change. And it must be weighing heavily on General Odierno's mind because he was particularly ornery during the BUA on Monday. In fact, I have never heard him that upset since I have been here, so the mood here is changing yet again as resources continue to be diverted east toward Afghanistan. A very fine balancing act it is, and for those government and military people who really believe we are set up to fight a two-front war – these same people need to analyze our current situation in greater depth and pull their collective heads out of their asses! Clearly we do not have enough resources to fight on two fronts and be effective without jeopardizing force protection. To believe otherwise suggests that these people are detached from reality and what is really occurring on the ground, or they are sitting far away from the battlefield and simply have no clue!

Then we get back to the defense budget and all of its political trappings. The President condemned wasteful spending within this portion of the nation's budget and the mentality of appropriating money for weapons best used against the Soviet Union must be changed **(Stars & Stripes, 18 AUG 09, *Obama condemns waste in defense spending*, written by Liz Sidoti and MNFI Early Bird News, 18 AUG 09, *Obama: Spending on defense rife with waste, 'pork'*, written by Liz Sidoti)**. The problem with this old school mindset is that the Soviet Union fell back in 1989! Anybody in Washington read about that? We need more funding for MRAPs, body armor, surveillance equipment, care of wounded warriors, etc. Yes, we need to always look ahead, but we also need to take care of the present situation in which we find ourselves. But it seems that when we need to apply the lessons of the past to create sound war strategy, we simply ignore it. Conversely, when it comes to appropriating billions in defense funding, we like to look back and prepare for the enemies of the past. Again, can anyone in Washington spell C-O-M-M-O-N S-E-N-S-E?

Day 288: Thursday-August 20th, 2009/Day 257-Iraq was a sad day around Iraq. Yesterday there were two large explosions in Baghdad that killed approximately 95 Iraqi civilians and wounded another 400 **(Stars & Stripes, 20 AUG 09, *Iraqi capital shaken by bombings*, written by Sinan Saleheddin).** This is the worst such act of violence in quite awhile and the incidents have AQI written all over them. So the Prime Minister is calling for an investigation of the incidents and security is being questioned **(JASG, 20 AUG 09, *Iraqi PM orders security review*, BBC)**. The insurgents are attempting to discredit the ISF and the GoI to create chaos and incite sectarian violence. Thus far, their attempts over the past several months to ignite such tensions have failed, but very high profile attacks such as we experienced yesterday will certainly test Iraqi resolve.

I was in the GRD HQ building on VBC yesterday when the largest of the two explosions occurred. We felt the tremor in our building and are about 7-8 miles away from central Baghdad! GRD still has some people in the International Zone, although the movement of GRD HQ is about 85% complete at this point. There were some windows blown out by the concussive effects of the larger blast, and a few ceiling tiles were displaced, but really nothing of significance.

I called Shari in Florida and she asked me about the incident. The reason she inquired was because she had talked with my mother who saw the incident on CNN or Fox News. Shari has consciously made a point not to watch the news while I am over here, but mother has not taken this same approach! So when my mother saw the news of the bombings she just wanted to know if Shari had talked to me about it. I did not mention this incident to Shari yesterday after the fact, and had not planned on doing so. But since my hand was forced on this one, I explained the situation to her. This would have been another one of the situations that I would not tell her about until safely home, but I also understand the angst that many family members go through by watching the news and hearing of such tragedies in areas close to where they know their son, daughter, father, mother, or friend is stationed over here. This is just human nature to worry about such things. This is also another one of those situations that many Americans will never have to worry about, and thankfully so.

118/125

Day 289: Friday-August 21st, 2009/Day 258-Iraq witnessed another turn of events here in Iraq. To set the stage for you we must go back in time. As you recall, I have been stationed in Al Faw Palace the past eight months as the GRD LNO to MNF-I (did you remember all of these acronyms or did you need to refer to the acronym/symbol legend?). Less than two weeks ago I was basically pulled from these duties, although I still need to go to the palace occasionally to conduct business, and was plugged right into the main effort at GRD which is related to the inactivation of the GRD HQ within the next few months. This is all part of the entire theater-wide transformation process. Still with me?

My duties at GRD are very busy and my job is basically 'herding cats'! There are about 20 different sections within GRD that must now report the disposition of their personnel and equipment to me in the G3 section to track its status and to keep our CG informed. In other words, in a few more months, GRD will be a much smaller organization and after a buildup of personnel and equipment over the past five years, we must now jettison/find a new home/turn-in/redistribute/reallocate a lot of stuff! It is a busy time at old GRD to be sure. It is now my job to gather all of this information from all of the sections and put it together for a weekly briefing to our CG, MG Michael Eyre. So my days are full, as I still need to go over to the palace every few days to accomplish a few tasks there, but the days go by even more quickly now and that is all good as my tour winds down. Still with me?

Then, last night LTC Rich Pratt, who I have introduced you to and who was a member of my Engineer Officer Basic Course (EOBC) at Fort Leonard Wood (MO) back in 1990 (July through November), and who replaced GRD G3 (Operations Director) LTC Chuck Samaris who went back to Ft. Leonard Wood, was just summoned to appear at a court martial trial back at Fort Leonard Wood! He leaves tonight! Apparently a deposition is not good enough, or testimony by video teleconference (VTC), and I guess military attorneys trump a two-star General and a couple of full-bird Colonels, as MG Eyre, COL Goetz and COL Fultz all tried to get Rich out of this trip. But I guess the fact that we are at war and a critical stage in our mission here did not sway the powers in authority! All of this means Rich will be gone for about 7-10 days depending upon the travel situation, which I highlighted for you in painful detail in the last chapter. Still with me?

So after our normal SPO (Security, Plans, and Operations) huddle last evening at 1700, COL Fultz wanted the LTCs in the room to remain behind after the meeting. This select group included me, LTC Mike Ryan and LTC Carlos Rodriguez (another fellow EOBC class mate). Mike and Carlos were just promoted since we arrived here in Iraq. I was promoted in December of 2004, which means I have date of rank, which means I will be the acting GRD G3! Talk about moving up quickly! Well, I am not one to shy away from leadership opportunities, and I will not pass this challenge up either. But as I just arrived over here, and do not have the depth of knowledge on many issues GRD is faced with, it will be a team effort to hold the line the next 7-10 days until Rich returns. Unfortunately, there is no extra pay involved with this transaction and no promotion to the next rank. There is just more responsibility and more work. I might add that moving into this acting position does not relieve me of any of my other duties either, so I will do the best I can with the cards I have dealt in this very unique and unforeseen circumstance. Much like King Leonidas and the 300 brave Spartans at

Thermopylae (remember the movie ***300*?**), the rest of the G3 section will step into the gap created by Rich's departure and continue on with the mission.

**The G3 section shall hold the line
"Spartans, prepare to work!"**

Day 290: Saturday-August 22nd, 2009/Day 259-Iraq brought condemnation from world leaders on the senseless attacks on innocent civilians in Iraq **(MNFI, 21 AUG 09: KUNA, *al-Sharqiya, al-Sumariya*)**. Leaders from Turkey, Syria, Egypt, Japan, and the United States all issued statements concerning these acts of violence and reiterated their support for Iraq's security, safety, and stability.

Prime Minister Maliki commented on the attacks by stating that:

> The criminal operations that happened today no doubt call for a re-evaluation of our plans and our security methods to face the terrorist challenges. These heinous crimes are a desperate attempt to derail the political process and affect the parliamentary process.
> **-MNFI, 21 AUG 09: *AFP***

In light of the investigation into the security of Iraq by Prime Minister Maliki, Iraqi officials confirmed that the ISF must shoulder most of the blame **(MNFI, 21 AUG 09: *Reuters*)**. A spokesman for the Defense Ministry stated that the "ISF must admit our mistakes, just as we celebrate our victories." So an internal review is under way, and most probably an external evaluation as well, to shore up these breaches in security and provide some sense of hope for the Iraqi people.

LTG Helmick, who is in charge of the command that is training Iraqi Security Forces expressed his frustration with the recent bombings and indicated that the process takes time to develop and it is not developing very quickly **(Stars & Stripes, 21 AUG 09, *Lt. Gen. Helmick 'frustrated' in Iraq*, written by Pauline Jelineka and Lolita Baldor)**. More sophisticated elements of such training include forensics, intelligence gathering, and logistical capability which are lagging behind the basic training of such forces. LTG Helmick does not anticipate this type of advanced training will be completed by the time the current security agreement ends in December 2011.

Finally, we go back to the Army efforts on curtailing the increasing rate of suicides among its soldiers **(Stars & Stripes, 21 AUG 09, *Resiliency survey aims to prevent soldier suicides*, written by John Milburn)**. Again, I give them credit for continuing to pursue avenues to improve awareness and treatment of this serious challenge within the Army. Later this year, the Army will use an online tool that all soldiers will need to take to assess their resiliency level. This tool will provide a baseline for all soldiers. Then, in two years there will be a reassessment to determine if the soldier's score has improved. The problem with this is that it takes time, and this is something that should have been foreseen based upon historical references and instituted long ago. But despite this delay, these kinds of programs tell me that the Army's leadership is committed and serious about finding long-term solutions and they deserve great credit for this effort.

▌▌ ▌▌ Day 291: Sunday-August 23rd, 2009/Day 260-Iraq was my well-deserved half-day off. Our transformation working group meeting went well, as did the day in general. My first day as the acting G3 yielded many good results as I stepped up to the plate and swung for the fences. If I can continue this pace until Rich returns, then it would have been a successful week, with little drop-off in performance, and it would look good on my resume.

We capped a good day off with a cook out at GRD HQ with the supposedly famous Bubba Burgers and some brats. If you recall, when we first arrived here in Iraq it was one of my duties as the JOAT (Jack-Of-All-Trades) officer to buy some of the Bubba Burgers and have them transported over to GRD HQ when it was located in the IZ. The only place you could get these burgers was at the main PX right here on VBC. Well, last night I finally had the opportunity to partake of one, and while it was OK it wasn't in the same class as a Pop's Burger, which I miss dearly. Now that GRD HQ is located on VBC and most of the personnel go to the Coalition Café, previous home of the Pop's Burger, they no longer serve this tasty delight! Figures!

After our little cookout, I took LTC Carlos Rodriguez and LTC Mike Ryan with me as guests to the Victory Cigar Club meeting. Both Mike and Carlos are members of the Cigar Aficionado Club at GRD HQ, but the Victory chapter is much larger. Mike and Carlos enjoyed the experience and may become members even though our time here is coming to an end. In fact, Carlos won two of the five raffle prizes given out at this meeting so he was thrilled. In two weeks there will be a special meeting of this chapter at the Joint Visitors Bureau better known as the JVB hotel where many distinguished visitors stay when in Iraq. I have yet to visit this place on VBC so Mike, Carlos, and I will attend this special meeting in two weeks and check the place out. Before that however, we will attend the Cigar Aficionado Club meeting at GRD HQ, which should kick off an evening of cigar filled fun!

To close the loop on the recent violence we have experienced here in Iraq, a major crackdown and investigation are occurring. Apparently, Iraqi lawmakers called for a special meeting in the wake of this violence and intensely interrogated the heads of security on how this could have happened **(JASG, 22 AUG 09,** *Baghdad security shakeup possible in wake of deadly bombings,* **source: McClatchy and MNFI, 22 AUG 09:** *Aswat al-Iraq, al-Sharqiya, PUKmedia, al-Iraqiya, al-Iraq News).* One of the issues to this story is associated with the fact that there are a lot of security agencies in Baghdad, but not one single agency that oversees all of these organizations. Therefore, there is no common strategy or unity of effort and this creates gaps for insurgents to infiltrate through. In fact, eleven ISF commanders were detained and questioned as part of the government's investigation **(MNFI,**

22 AUG 09: *NY Times, Washington Post, AP, AFP, BBC, Christian Science Monitor, Bloomberg, McClatchy*). There has also been rumor of some collusion by some ISF personnel which may have led to the insurgents being able to smuggle this amount of explosive material into the IZ **(Stars & Stripes, 23 AUG 09, *Inside job?*, written by Sinan Salaheddin)**. This will be a constant struggle for the ISF and the GoI and it is all a part of their evolution. The problem is that the longer it takes for them to fix the problem, the more innocent civilians will pay the price for this evolvement.

But the Iraqi resolve is still strong and one storeowner put it all into perspective by stating:

> The security situation in Baghdad is good, except for what happened yesterday. The business in the market is strong, while the crowd is buying everything they need for Ramadan. Despite the bombings, Iraqi people stay united. We are a strong people who've been through a lot. Explosive attacks will not prevent us from enjoying our normal daily lives.
> **-MNFI, 22 AUG 09: *al-Sharqiya***

And yesterday was the beginning of Ramadan, which I will highlight for you in more detail tomorrow.

Day 292: Monday-August 24th, 2009/Day 261-Iraq was the third day of Ramadan, a Muslim holiday, and the ninth month of the Islamic calendar. It is believed to be the month which the Koran began to be revealed and is also considered the most venerated and blessed month. The most prominent event in Ramadan is fasting during the daytime. This month of worship is culminated with a three-day festival of fast-breaking **(JASG, 23 AUG 09, August significant events)**. But while this month will include a pilgrimage and celebration, it has been tempered by the recent violence.

Normally during this time of year in Iraq you see people out in the market places shopping and getting ready to celebrate the end of Ramadan festival. But because of the high profile attacks on civilians last week, the atmosphere is much more subdued and many people are staying home **(JASG, 23 AUG 09, *Fear of violence makes for a quiet beginning to***

***Ramadan*, LA Times).** All of this has left the Government of Iraq searching for answers in the wake of this violence, and I will say that they are taking swift action to rectify this gap in security. It remains to be seen if this action will also be decisive. For the sake of the Iraqi people and their future, I hope they succeed in this endeavor.

Well, yesterday was our weekly update brief to our Commanding General. While I put the slide deck together for this briefing, I have less of a prominent role than I do during the Saturday working group meeting, which leads into this briefing. But for just arriving at GRD HQ and being thrust into a key position due to Rich Pratt's situation, I thought things went pretty well. I am a quick study and have picked things up in short order, but I am not so delusional that I believe I am ready to assume the role of the G3 full-throttle! I still consider myself a JOAT (Jack-Of-All-Trades) officer and a gap-filler for LTC Rich Pratt until he returns. At this point, he should be back in the states and at Ft. Leonard Wood as the court martial trial he is testifying at is scheduled to begin today and run into Wednesday. But so far, it is steady as she goes with another busy week getting ready to greet me.

Day 293: Tuesday-August 25th, 2009/Day 262-Iraq was another busy day. With Rich Pratt still away at Ft. Leonard Wood, I was forced to attend a lot of the meetings he would normally attend. While much of it was interesting, it seemed as if we would have meetings to determine when we could have more meetings! Today was no different except for the fact that I had to attend a meeting for COL Goetz, our Chief of Staff, with MG Guy Swan, who is currently the MNF-I Chief of Staff. MG Swan was the CJ3 (Operations) for MNF-I when I arrived here in December and I would see and brief him often in that capacity.

The plot thickens on the security breakdowns that have occurred here recently. Apparently, the insurgents paid people at the various checkpoints about $10,000 in order to get their explosive-laden trucks into central Baghdad **(JASG, 24 AUG 09, *Iraqi suspect: It cost $10,000 to pass checkpoints:***

AP). So many people will be fired and arrested as this investigation continues to reveal the gaps in the ISF.

The Iraqi Foreign Minister commented, "Sometimes you can't fight these people with checkpoints. You should be mobile. You should go after them; you disrupt and penetrate their network to get human information. This is the key." **(MNFI, 24 AUG 09: *AFP*)**.

In another twist that led up this corruption-filled situation, the director of Iraq's National Intelligence Service resigned last week. Apparently the Cabinet of Ministers had requested his resignation as he had reached retirement age. It just so happens that the director also warned the Iraqi government of attacks that could occur in August! As we continue to move forward, it is beginning to take shape that this country may have a lot further to go than most Americans believe. It appears to be in a constant cycle of one step forward and two steps back, and it often seems as if they are on the fast track to nowhere.

And I will end today with more bad news as I feel it my duty to tell you the truth as I interpret it and not simply paint you a rosy picture. Admiral Mullen was recently on television **(CNN - State of the Union with John King: 23 AUG 09)** and described the situation in Afghanistan as worsening despite the recent influx of resources, including 17,000 additional troops **(Stars & Stripes, 24 AUG 09, *Mullen concerned over eroding war support as Afghan fight ramps up*, written by Richard Lardner)**. Richard Holbrooke, special representative to this region from the Obama administration, recently met with the commanders in Afghanistan and they all informed him that they do not have enough troops on the ground to achieve success **(JASG, 24 AUG 09, *US military says its force in Afghanistan is insufficient:* NY Times)**. As I mentioned recently, General McChrystal, who just took over command of this theater of operation, is expected to present his analysis of the current strategy to the Obama administration within the next few weeks. In counterinsurgency doctrine I believe the ratio of counterinsurgents to population is around 1:50 and we are well short of this figure, and the standing up of security forces in Afghanistan is very slow to evolve, as was the case in Iraq. So the next month or so may be a defining moment within the

Obama administration since the President will need to decide what he will do in Afghanistan. Senator John McCain wants commanders to request what they need and not what they believe will pacify Congress. Yet, the American people have grown very weary of both wars and this will greatly affect how politicians vote on the recommendations pushed forward by the Obama administration. So the next month should be very historic and interesting to watch. Let us hope and pray, that unlike some of the decisions made within this level of government and the military since the war on terrorism began, that these new commanders and politicians make good, sound decisions based on a well-defined strategy and a clear end-state. If they do not, once again, our brave men and women of the US military will pay dearly for these lapses in decision-making.

Day 294: Wednesday-August 26th, 2009/Day 263-Iraq was another busy day. Five more meetings and much more work as the countdown continues.

Despite the recent violence here in Iraq, there are still signs that the country is moving toward stabilization. In the western province of al Anbar lies a large lake known as Habbaniya. For the first time in awhile, Iraqis enjoyed a beach season because of the increase in security conditions **(MNFI, 25 AUG 09:** *NY Times*). One beachgoer stated, "I'm here to get away – from the bombs in Baghdad, from the sound of generators. We're here to have a good time. There's no difference between Shi'a and Sunni. We are all Iraqis" **(MNFI, 25 AUG 09:** *NY Times*).

Another sign of normalcy is related to the re-emergence of hair salons. Yes, I said hair salons! Because of the increased security and reduced presence of extremist militias, many Iraqi women have felt safe enough to venture to hair salons **(MNFI, 25 AUG 09:** *McClatchy*).

Now we can take these two stories for what they are. There are areas around Iraq that are simply safer than are others. This is not unlike the situation in the United States. Some cities are just safer than are others. Yet, in other areas of the country there is still a lot of work to do. Baghdad is a cauldron of emotion because it is so large (approximately six million people), it has a large mixture of sects, and it is home to most governmental organizations that attract extremist attacks. So this area will be volatile for a while. Then in the north lies Mosul, which is an area of ethnic tension and the last stronghold of al Qaeda in Iraq. This area too will take some time to become properly and adequately secured. But in a lot of other areas of the country life is returning to some level of normalcy, and despite the good, the bad, and the ugly stories we see every day in the news, these small indicators of progress tell me that this country is in transition, it is evolving, and it is progressing. The dichotomy of this struggle is palpable, but each day there seems to be another reason for hope.

Day 295: Thursday-August 27th, 2009/Day 264-Iraq was GRD BUA day and for the first time, except for when I was away on R&R, I was not involved in its production or broadcast. SFC Maltes has returned from R&R and flew solo in the SOC. Due to my new duties I watched the BUA from the comfort of GRD HQ.

Yesterday I attended five more meetings and I as I have mentioned, this organization has far too many meetings. When you are tied up in meetings all day, then there is precious little time to actually get any work accomplished. The military, GRD included, has also become so entranced by PowerPoint that we have a lot of highly paid and skilled people spending a lot of valuable time putting together slide presentations for briefings. We have become slaves to PowerPoint and I do not know what the military would do if they lost this capability! Perhaps communicate more effectively? Do not get me wrong, I think PowerPoint is a great tool to supplement communication, but the military has morphed this tool into a primary means of communication and I believe we have really lost some of the human interaction we had before the creation of such software. And then you have anal-retentive supervisors so caught up in the creation of a presentation that it becomes their primary focus, as opposed to the actual message you are trying to relay. In my humble opinion, we have simply strayed too far away from human interaction and become too deeply involved with technology. In the process, we have taken a step backward and perhaps this is why our military has struggled in this type of warfare that heavily relies upon human interaction and face-to-face dealings that we used to have well before the creation of PowerPoint and the computer age. And no, I am not a technology-hater, or just an old dog whose time has passed him by. I think we should use technology to assist our mission, but it should not become our primary focus. And in many cases, this is exactly what has happened.

Well, my care package from Shari finally arrived. This is likely to be my final care package as our time draws down here. Inside were a plethora of goodies and my retirement packet from the fire department. I discussed our future with Shari when I called her, as she is still enjoying a well-deserved vacation in Florida. The next few months will be very interesting, and while I am now leaning on going back to the fire department upon my return, I now have enough information to base a decision upon in regard to the numbers associated with my pension. It now becomes a matter of a better offer and the right situation that would draw me away from the fire department to a different location and a new adventure. Time will tell.

I also received my new exercise bands today, as I have broken my old set during this deployment. I am anxious to use them and they look very sturdy and should work out nicely.

Finally today I would like to honor my grandmother, Grace Lagerquist, who would have been 101 today. I remember her every August 27th. She was a wonderful woman and I miss her very much. Gram basically raised me, and my sister and brother, so my mom and dad could both work and provide for a better life for the Waite family. I remember these days very fondly. And like most grandmothers, every time you saw her she was trying to feed you! God bless you, Gram. I love you and miss you. Happy birthday!

Happy 101st Gram. Miss you.

▌▐ ▌▐ ▌▐ ▌▐ **Day 296: Friday-August 28th, 2009/Day 265-Iraq** was filled with news not from Iraq, but from the US in regard to the death of Senator Edward Kennedy. 'Teddy' outlived his two more famous brothers, John and Bobby, and served in public administration for several decades. So even though death by assassination has a tendency to make an individual more famous than they may have otherwise become, Ted Kennedy served in the public eye longer than both of his brothers combined. As the legend goes, there is a Kennedy curse and this family has been besieged with incredibly bad luck. Ted himself did not go through life unscathed as the tragedy at Chappaquiddick attests. But as the last of the three most famous siblings of Joseph and Rose Kennedy, an era has ended, and if there was a curse, we do not know that it ends with the death of Teddy. The Kennedy clan is very large and there are still off-spring involved in politics who will serve for years to come. And if there never was a curse, perhaps the luck of the Kennedy's will improve as the decades go by.

There was an interesting story in *Stars & Stripes* yesterday as there appears to be an issue in getting troops in Afghanistan to definitive trauma care in a timely manner **(Stars & Stripes, 27 AUG 09, *Rethinking the 'Golden Hour'*, written by Lara Jakes).** Because of the rugged terrain and difficulty in gaining access to certain areas, using a helicopter to evacuate casualties is more of a challenge than it has been in Iraq. In the world of EMS there is something we refer to as the 'Golden Hour.' Basically and simply defined, this means that from the time a patient is seriously injured, if initial medical care can be rendered within minutes, and the patient transported to definitive care within 60 minutes, which is typically surgical intervention, they have a very high probability of survival. The longer the patient moves past this 'survival window,' the chances of a positive outcome decrease exponentially. So the debate in Afghanistan is now is centered around this 'Golden Hour' and some medical professionals believe the troops would be better served by taking a little extra time and moving them to a higher level of care versus getting them to a medical facility that can provide some level of intervention, but not the definitive care the patient actually needs. In other words, if a soldier has a serious head injury, he/she most probably needs a neurosurgeon. Many medical facilities do not possess this level of expertise, so time is basically wasted by transporting them to such a limited facility. Then once there, the decision is most likely made to transport the troop further to a facility that has a neurosurgeon, but much time has now been wasted and the clock is running out. This is exactly the same debate we have in the US in the world of EMS. My newest book, **The EMS Challenge – A Call To Action**, highlights this deficiency in detail. There are many EMS organizations around our country that believe by transporting a seriously injured patient to the closest medical facility they are serving the patient's best interest. This is in fact a false assumption. Seriously injured patients need to be transported by ground or air to a Level I or Level II trauma center to receive the definitive care they truly require. Anything else is typically just a waste of time. But EMS directors and many medical directors still believe these patients must be transported to the closest facility. This backward thinking is not supported by any research. In fact, the research supports just the opposite, but it is just another cultural quirk that is difficult to change without effective leadership. This is exactly what I see going on in Afghanistan as well. The only loser in this debate are the patients who do not receive the specialized care they so badly need because of politics, personalities, and the lack of leadership necessary to rectify such a challenge.

The Golden Hour versus the best care available

Day 297: Saturday-August 29th, 2009/Day 266-Iraq was a sad day as yesterday we lost two more soldiers. The device used was an EFP (Explosively Formed Penetrator). The tell-tale sign of this device is a copper lining which heat up into a large metallic slug that can penetrate our thickest armored MRAPs. It has been surmised on many occasions that these devices are coming into Iraq from Iran. It has also been stated that because the ISF is still trying to chart its course without US assistance, the national elections looming on the horizon, and American equipment and personnel leaving the country in bunches, it will be a vulnerable time for US troops. In fact, all of these variables may make us more vulnerable now than ever before because there are many extremist groups that still want to claim they kicked us out of Iraq and will continue to kill troops until we leave to prove this point. There is a tendency to become complacent as you near that day you get to go home. I certainly want to go home and have since about Day 1! But I prefer to go home in one piece as opposed to missing a limb or in a flag-draped transfer case. So remaining ever vigilant over the next 54 days is going to be important. Yet another little nugget I will not share with Shari.

It appears that LTC Rich Pratt's visit to Ft. Leonard Wood is taking longer than anticipated. The court martial trial was supposed to end on Wednesday, and as of Thursday, they were still holding Rich as a corroborating witness that may be need to be recalled. So we are not sure when he will return!

But the beat goes on and the work is getting accomplished. While I enjoy the leadership opportunity Rich's absence has presented, I still have all of my other staff duties to accomplish as well, so the days are quite busy. Did I mention that we only have 55 days remaining? Thank goodness. As I work in a union shop at my fire department, was a union officer for several years and am now in management, the dichotomy of this organizational equivalent to Sybil and a split personality within the US Army of Engineers (USACE) is quite stark. USACE is primarily composed of civilians, as I may have informed you of previously. There are a handful of military personnel providing the connection to that side of the organization, but often the two sides do not communicate overly well with one another and are often just two different entities under one roof! The military personnel are often the first to arrive at work and the last to leave. Many of the civilians work banker hours and are paid more handsomely than their military counterpart. Trying to gather information often is often like herding cats and a real challenge. I will say despite these shortcomings that this organization does a lot of very good things and possesses a lot of good and talented people, but like the Army, USACE needs a structural overhaul. In the few short weeks I have had the opportunity to see this up close and personal, I can tell you that I am not impressed with what I have seen thus far. And I knew little to nothing about USACE and GRD before I came to Iraq, and when I go home, I will return to the 416th TEC, which is not a part of USACE or GRD. So the military has basically become slave labor compared to the civilian side of the house and this creates some hard feelings. The whole civilian/contractor piece of this war has really gotten out of control. So there is nothing but challenge for leaders over here, but I have been very disappointed in this respect as well. But march on we will despite these challenges. Only 53 days to go. Did I mention this already?

Herding cats at GRD!

Day 298: Sunday-August 30th, 2009/Day 267-Iraq was a day I found myself reflecting back upon the past few weeks. I have highlighted the trials and tribulation of the Gulf Region Division, but as I read my newest book, ***The Sling and The Stone*** by former Marine Colonel Thomas X. Hammes, his commentary upon why the US struggles so much in the type of warfare we are currently involved with in Iraq and Afghanistan, aligns very closely with my perceptions. First of all, the heart of the matter begins with the bureaucratic quagmire in which our military and government reside within. I have outlined *ad nauseum* for you throughout this book why a bureaucratic organization has great difficulty in being responsive in a dynamic environment. My creation, a **ledocracy**, is a much better fit for this atmosphere as we have discovered within my fire department despite a few growing pains, but no one out there is listening! Hammes has basically the same observation that our governmental and military organizational structure is not conducive to the type of warfare we are currently involved within.

Hammes also discusses the American affinity for technology. The same people that have ignored the organizational structure argument must be the same ones not understanding that for all of our technological advances, this factor has been greatly neutralized by fourth generation warfare. And from what I have seen here thus far, we have become too hypnotized by technology and spend far too much time using it. For example, almost every staff officer, every unit, every commander loves PowerPoint. It is almost beyond love, it is an obsession! I cannot tell you how much time and energy are spent by very intelligent people making sure the font on a slide is just right, or the color is pleasing to the eye, or the graphics are way cool! The problem is that we are also getting away from human interaction by doing so and relying upon software to send our message. Now don't get me wrong, I like PowerPoint but I believe it has gotten out of control within the military community. If these same people would use all of that time and energy on human interaction and analyzing information, as opposed to creating PowerPoint briefings for hours on end, we would be much more effective in this type of warfare. We have become nothing but **'PowerPoint Warriors'**, and I have been forced to follow this path much to my chagrin. The past few weeks have been nothing but PowerPoint slides and manipulating several different slide decks (presentations). Much of my best work here has come the old-fashioned way by actually talking with a person about certain information as opposed to cramming it all on a slide! Whatever happened to the methodology of actually talking to people? It seems like it is becoming somewhat of a lost art, yet every counterinsurgent expert will tell you that basic human interaction and face-to-face conversations are the keys to success within fourth generation warfare. So despite my protests, I have become a **'PowerPoint Warrior'** here in Iraq and I think this is a waste of my skill sets. I deal with elected officials all the time as a fire chief, as well as the general public, and do not need PowerPoint briefs to do so. I could have been more effectively utilized on a Provincial Reconstruction Team (PRT) or as an advisor to Task Force SAFE. In my opinion, this would have served the Army and the United States effort here much more effectively. But the process of trying to cram names onto a Joint Manning Document (JMD) to make them fit, the proverbial square peg in a round hole concept, is simply another issue within the larger organizational structural problem. But who am I to complain? I am just a citizen-warrior that has been ripped away from my family and forced into PowerPoint slavery. Sounds like a made-for-television movie! After all of this nonsense, all I can say is – 52 more days to go before I can escape all of this insanity!

PowerPoint Warrior Extraordinaire

Day 299: Monday-August 31st, 2009/Day 268-Iraq was one day closer to going home. As I have mentioned throughout this book, the list of my duties here in Iraq is endless. I have become a JOAT (Jack-Of-All-Trades) and have enjoyed the diversity. Unfortunately I have not been able to gain any depth in any one topic as a result of this diversity, but the breadth of the experience has certainly been an interesting ride. Despite the fact that we only have 51 days to go, it is unlikely that this diversification will end before we depart. Semper Gumby!

Today I think we will do a little rapid-fire exercise to catch you up on some backlogged news. We will begin in Iraq.

> General George Casey, the top general in the Army, stated that it is unclear how the recent violence in Iraq will impact the drawdown of US **forces (Stars & Stripes, 29 AUG 09, *Casey: Impact of Iraq violence on pullout* unclear, written by Diana Elias).** Even with this uptick in violence, it is very unlikely that the Government of Iraq will request US forces to remain past the end of 2011. In fact, if the

national referendum is passed in January, our departure will be even sooner.
- In Afghanistan, NATO's top commander believes the situation to be serious but hopeful **(Stars & Stripes, 29 AUG 09, *Admiral's blog details 'serious' but hopeful Afghan-war outlook*, written by John Vandiver).** Admiral James Stavridis believes there are four key variables that must occur in order for the mission to succeed: 1) making the Afghan people the center of gravity by winning over their support and not simply just killing the Taliban; 2) achieving an effective balance between military operations and diplomatic efforts; 3) broadcasting a clear and consistent message through strategic communication to counter Taliban propaganda; and 4) training Afghan security forces to protect their own country. All of these are classic counterinsurgent principles that have proven to be effective in Iraq. It remains to be seen if they will be as effective in Afghanistan.
- While the situation in Iraq continues to improve, the situation in Afghanistan is getting bloodier as August became the deadliest month for US troops in its eight-year history. August follows on and surpasses July 2009 as the deadliest of this war **(Stars & Stripes, 29 AUG 09, *August becomes the deadliest month for US in Afghanistan*, written by Heidi Vogt and Amir Shah).** This trend was not unexpected as more US troops have moved into Afghanistan and are fighting under difficult conditions and reacquiring territory that the Taliban regained once we turned our focus toward Iraq several years ago. So we now have young men and women dying because our government dropped the ball and now these same troops are asked to regain territory we once had gained from the Taliban! Isn't war grand? Yet another blunder on

behalf of our government that was not well defined or strategically thought out. In fact, we had absolutely no strategy, and the fact that we will see more casualties and conditions worsen before they improve because of incompetent leadership at the highest levels makes me very ill indeed.
- Finally today, we will talk of a good piece of technology that the US has effectively employed in Iraq and Afghanistan. Unmanned Aerial Vehicles (UAVs) will be used to combat piracy **(Stars & Stripes, 29 AUG 09, *US plans land-based UAV patrols to combat piracy*, written by Mark Abramson)**. The Reaper drone will be used to gather intelligence in the Indian Ocean area but it will not be armed. At least not yet! This intelligence can then be fed to naval vessels that can move into position to combat pirates more effectively. If this issue of piracy continues to progress, you are likely to see more assets used to protect cargo vessels as this becomes the navy's counterinsurgent war on the high seas. Good luck with that one, Captain Jack!

Day 300: Tuesday-September 1st, 2009/Day 269-Iraq

was somewhat of a red-letter day. First of all, it was Day 300 in this little magical mystery tour! It also marked 50 days to go before we left Iraq. Then a few days in Kuwait; a few days at Ft. McCoy; and then this little experience will be history. Again, just due to our tour and its unpredictable nature, taking 15 days of R&R, and having the remainder of our leave run off while we are at home (called terminal leave), we will come up a little short of the actual 400 day order, but the book title shall remain the same!

Five-O left to go! Book 'em Danno.

LTC Rich Pratt has returned. His little detour to Ft. Leonard Wood via Iraq took about 11 days. So while I enjoyed the leadership opportunity Rich's rapid departure created, it was simply just more work for me. So, welcome back, Rich! But although it may be just a rumor, I have heard that I may be receiving a letter from the academy on my acting role as the GRD G3! While I may not win the award, it will be an honor to be considered among the greats such as Anthony Hopkins, Sean Connery, and Hugh Jackman. I would like to thank my mother and father and....!

For award winning role as an operations officer in the Middle East, the winner is....

How about some real news? It appears as though the White House has created about 50 measurements to gauge progress in Afghanistan **(JASG, 30 AUG 09, *US sets metrics*** *to assess war,* source: **Washington Post).** Some of these metrics include assessing the effectiveness of Afghan army recruits, Pakistani counterinsurgency missions, and delivery of promised US resources. While this is all fine and dandy, it comes eight years into this war! Anyone else think this is just a little slow?

The national elections in Iraq are about four months away, and there is little doubt that this will result in a surge of violence as we get closer to this important day **(MNFI, 31 AUG 09: NPR)**. Many sources believe current Iraqi Prime Minister Nouri Al-Maliki will be re-elected. Some believe he has become consumed with power and cannot overcome his bias toward other sects within Iraq to effectively form a representative government of the people. This has created an opportunity for his challengers and while I would see him losing the election as an upset, it is not beyond the realm of possibility. Then we hope we do not see what happened in Iran occur here in Iraq. Handling defeat is never easy, but once the people here can do so without resorting to violence, then they may have turned a very large corner.

The Iraqi economy received a large boost recently when they were awarded a $1.8 billion grant from the International Monetary Fund to expand different sectors of their economy **(MNFI, 31 AUG 09:** *al-Sabah*). Certainly with the drought affecting crop production and the cost of oil exports down, this is very good news and it will provide a badly needed stimulus for this country still trying to find its way onto the global stage.

We are definitely on our way out of Iraq. We have a good start on moving about 1.5 million pieces of equipment out of the country, including 130,000 US troops **(JASG, 31 AUG 09, *American commander: US on road out of Iraq,* source: AP)**. All of this must be completed by the end of 2011, unless the infamous referendum kicks us out earlier. If that were to occur it will give military planners a serious headache! But there are about 130,000 contractors and civilians to get out as well and over 300 bases to close. So while a massive effort, and one sure to be very expensive, it is well underway. The better news is that I will be long gone from here by the time all of this effort wraps up and getting me and my equipment out of Iraq is a fairly large priority for me!

GRD employee handing out toys donated by the Gulf Region Division

CHAPTER 16
400 DAYS – Days 301-320
September 2 – September 21, 2009

> *We can call it quits and withdraw from Iraq. I think that would be a gigantic mistake. Or we can set a deadline for pulling out, which I fear will only encourage our enemies to wait us out – equally a mistake.*
>
> --Vice President Joe Biden/2009-present
> (also served as Chairman, Senate Foreign Relations Committee 2001-2003 and 2007-2009)

Day 301: Wednesday-September 2nd, 2009/Day 270-Iraq began with good news and bad news. The good news was that August was the least deadly for US troops in the history of this war. We lost seven soldiers last month, and while this is hardly any consolation to the seven families of these heroes, it does signify progress and that their sacrifice has not been in vain. August follows on the heels of July, which had the previous record low. So there is a downward trend in Iraq. However, in Afghanistan this trend is just the opposite. August overtook July as the deadliest month for US troops in that war and that trend is likely to continue for the next several months as we begin to apply more pressure to the insurgents in that country.

General McChrystal is creating a new strategy in Afghanistan, which will be outlined to the President very soon **(Stars & Stripes, 1 SEP 09, *McChrystal calls for new strategy in Afghanistan*, written by Jason Straziuso)**. It is also likely he will request more troops, which in turn, will cause the number of casualties to rise just as it did during the surge of 2007 in Iraq. Surge II is likely necessary in this new strategic model because Afghanistan is larger in land mass and has more people than Iraq, therefore, more troops will be needed to hold territory once it is cleared of insurgents. This strategy will also assist in protecting the people of Afghanistan until Afghan security forces are capable enough to protect them. This is the same strategy that has been used effectively in Iraq. The problem is that at the height of Surge I, we had approximately 250,000 troops in Iraq to effectively execute the new strategy. This came at a very high cost, but it did help to stabilize the situation. In Afghanistan right now there are approximately 62,000 troops, so they will need a large influx of soldiers in order to turn that situation around. And this will be a real hot potato for politicians to handle as the Afghan War has become very unpopular back in the US. This is certainly through no fault of the troops, but rather, our policy makers and military strategists, or lack thereof! Eight years into this war and we are just know developing a coherent strategy? Are you friggin' kiddin' me? We are almost three years post Surge I, and we are just now looking at replicating this success in Afghanistan? Are you friggin' kiddin' me? At this point I am not certain if we are just that slow, that stupid, or that arrogant. Perhaps a concoction of all of these elements could create a new drink called the Slow-to-catch-on! I need to stop now before I get too carried away!

Can I interest you in a Slow-to-catch-on my dear?

▌▍ ▐▎ ▐▌ ▐▌ Day 302: Thursday-September 3rd, 2009/ Day 271-Iraq was going to be back to a normal routine.

Yesterday was a bit atypical, but a good day overall. Instead of leaving at lunchtime to go get my workout in, we had an impromptu birthday party for COL Joe Goetz our Chief of Staff. We had pizza and birthday cake and it was all very nicely done. Come to find out that COL Goetz' family owns a supper club and pizza place in Rome, Wisconsin which is only about 10 miles down Highway 13 from my hometown. Small world indeed!

Lots of news concerning our drawdown in the media recently as General Odierno returned yesterday from R&R. Despite all of the conjecture about our future status in Iraq, it is highly improbable that President Obama will push any more troops or funding toward Iraq, or adjust the withdrawal timeline. The time has come for them to determine their future course. One Iraqi believes the Iraq of today is halfway between being on the verge of collapse or on the verge of salvation **(JASG, 2 SEP 09,** *Can Iraqis move past sectarian divides?,* **source: BBC).** This current of state of being has come at a very high cost, American persistence, and Iraqi resolve. Until we do leave at the end of 2011 as it stands right now, we will continue to assist the Iraqi government to rid themselves of extremists and further stabilize, but then it will be their show from that point forward.

The main effort has certainly shifted to Afghanistan as the pendulum has swung from the lowest US casualty rates in Iraq in its history to the highest casualty rates in Afghanistan in its history for the same period! But with General McChrystal's analysis of the current strategy in Afghanistan on the desk of President Obama, it will remain to be seen if more troops are requested and what commitment to this war the current administration will provide with a struggling economy at home and a controversial national healthcare package being debated **(Stars & Stripes, 2 SEP 09,** *Pentagon worried about Obama's commitment to Afghan mission,* **written by Nancy A. Youssef).** Here is what I see as the potential issue with all of this rhetoric. Clearly, we do not have enough troops in Afghanistan to finish the mission. Every counterinsurgent expert will tell you this much. How many are needed to achieve success and in what timeframe is the real question at hand. This is what General McChrystal's 'crystal ball', no pun intended, needs to tell him. Senator John

McCain told him to ask for what he needs and not what he believes will pacify Congress. Some retired General's have advised him to tread lightly and present a separate troop surge package from the strategy analysis itself. This war has also been and will continue to be costly. With a struggling economy at home and an expensive healthcare package being debated within the halls of Congress, the balancing act will be very fine indeed. But my advice to President Obama is not to shortchange the troops as was done in Vietnam and the earlier stages of the Iraq War. Either you are committed to the effort or you are not. You can't be sort of pregnant! So either make the commitment and provide the civilian and military assets necessary to achieve success in Afghanistan or pull the plug. The brave men and women in our armed forces deserve this type of leadership and their families will be forever grateful for such a thoughtful decision and effective leadership that we have not seen in awhile.

What is our future in Afghanistan?

Day 303: Friday-September 4th, 2009/Day 272-Iraq was the beginning of college football. Over here of course, Thursday night games are on television early Friday morning when I get up. Some of the early Saturday morning games I will see Saturday evening so that will be pretty cool. Bucky Badger begins its season on Saturday against Northern Illinois and there has not been much talk of Wisconsin in the Big Ten this year with the likes of Ohio State, Penn State, and Iowa, so we shall see how they will fare this year beginning Saturday.

Go Bucky!

There is a push on here to emphasize the improvements in security during Ramadan **(MNFI, 3 SEP 09: *McClatchy*)**. This type of strategic communication is very important in this type of warfare and since we had two large explosions a few weeks ago amidst allegations of corruption, it has been relatively quiet. Let us hope it stays that way.

Last night I watched Secretary of Defense Gates and Chairman of the Joint Chiefs of Staff Admiral Mullen give their weekly press conference on the Pentagon Channel **(Pentagon Channel, 3 SEP 09, 2100 GMT+3).** They mentioned that the new strategy for Afghanistan, as detailed by General McChrystal, was on the President's desk. This was about as much detail as they provided on the topic. There was no talk of additional troops even though some reporters pushed the subject, but this request is likely to come later. So the die on this war will soon be cast and we shall see in which direction the new administration is going to go.

My job continues to diversify as I grow more comfortable at GRD HQ. I have assisted the G1 (Personnel) with developing some new slides for our weekly briefing, and yesterday I assisted the Secretary to the General Staff (SGS) develop some new charts to brief the GRD Command Group. Yes a JOAT (Jack-Of-All-Trades) officer I am, but all of this activity is making the days fly by.

A JOAT I am, I am!

Day 304: Saturday-September 5th, 2009/Day 273-Iraq I had finished *The Sling and The Stone* by Thomas X. Hammes. I found his book to be a very good read. Mr. Hammes and I think a lot alike in regard to the antiquated systems within the military that are in great need of change. I have lamented over many of them over the course of the first several hundred pages of this book. Mr. Hammes not only lays out the issues as he sees them, but he also makes some solid recommendations to correct such deficiencies. However, it is unlikely, given the bureaucratic nature of the military and one of its major underlying problems, that such change will occur quickly if at all. There are many bad officers who have benefited from the current system and many other officers who will resist any change to the antiquated processes we have in place. In fact, one of the Colonels I mentioned to you earlier in this book, someone I would never hire because of his poor personnel management skills, has just been rewarded with his general's star! This is all he ever wanted and he did not care how it attained it. In fact, the very organization he used to bad-mouth he now works for! He sold his soul to the devil to take a position with this organization after he departed Iraq so he could get his star! This is just another example of the poor system that currently exists within the military that both Mr. Hammes and I have spoken of.

Because of all these variables, unless we begin to see some change, some sanity amidst all of this chaos, the ability for someone to see the subtle shades of grey through the black and white fog, then I fear this may be the beginning of the decline of the United States as a superpower, an empire if you will. Our resistance to adapt and our arrogance will eventually lead to our decline. It is already beginning, and bear in mind that every great empire of the past has fallen at some point in its history. These collapses have been attributed to misappropriated power, greed, over extension of resources, and simple arrogance. All of these empires had great militaries, all had thriving economies, and all were resistant to change and adapt. Sound familiar? Now I am no doomsayer, just someone who is pretty observant. A future study of mine will highlight the similarities and differences between the Roman Empire and the United States of America. While no expert in this matter, at least not yet, I have done some preliminary research and the comparison and contrast of both cultures is quite striking. But as the Roman military overextended its reach, left its own borders vulnerable to attack from invading armies, and greed and corruption seeped out of the walls of its government, so began the decline of the Roman Empire. Let us hope our arrogance does not blind us to the facts that lie before us. We have the ability to make necessary changes and adapt to the environment in which we now exist. But I have already highlighted that we, Americans, continually fail to learn from history, whether it is our own or someone else's. This failure may lead us down the path to collapse as a superpower. If this does occur, then of course our government and military leadership will claim that they never saw it coming. That is only because they would have had kept their blinders looking toward the future and forgotten all of the rich lessons of the past, yet again!

In this case, let us not do as the Romans did!

Day 305: Sunday-September 6th, 2009/Day 274-Iraq was another dry one in Iraq. Iraq is still negotiating with Syria and Turkey to release more water from their reservoirs into the Euphrates and Tigris rivers that flow into this country. But the other countries are experiencing a shortage of precipitation as well and releasing a significant amount of water may not be feasible **(MNFI, 4 SEP 09: *AP, AFP, NPR*)**. Turkey states that they cannot release anymore than they already are so crop production will continue to be severely hampered.

There is still much political infighting within the government of Iraq. Many people believe that the current Prime Minister is influenced by Iran because of his Shi'a roots. I have outlined the obvious Iranian influence in this struggle within Iraq throughout this book for you. And as violence continues to escalate, with US troops primarily relegated to the sidelines per the terms of the security agreement and awaiting final withdrawal, Iranian influence continues to take root. An intelligence source in Iraq was asked if the Americans should step back into the fight to restore order. His response was a bit surprising, but reflective of the current mood here when he stated that it is probably wiser to "stay out of it and be safe" **(JASG, 04 SEP 09, *Behind the carnage in Baghdad*, media report)**. This source was further pressed about he thought Iraq would look like in five years. His response – "Iraq will be a colony of Iran." Not a very pretty picture unless our relationship with Iran improves during the course of this time.

Yesterday evening was a series of firsts for me, not only here in Iraq but in my entire life. LTC Carlos Rodriguez, LTC Mike Ryan, and I left GRD HQ around 1830 to begin our double-header of sorts for the evening. We began at GRD HQ with a small gathering of the bi-monthly meeting of the Cigar Aficionado Club. There were probably about ten men gathered in the wooden gazebo outside the HQ as darkness fell across Iraq. I enjoyed a cigar, bought a nice coin from the club to add to my expanding collection, and engaged in some good conversation. So we were off to a good start this evening.

Mike Ryan and yours truly at the Cigar Aficionado Club

We then traveled down the road to the bi-weekly meeting of the Victory Cigar Club, which this night was meeting at the Joint Visitors Bureau (JVB). The JVB is a one-story hotel that sits on a parcel surrounding Victory Lake and overlooks Al Faw Palace. The night was warm and the moon was full which provided an excellent backdrop for such a nice and unique experience.

The JVB is the place where many distinguished visitors stay when they visit Iraq. Secretary of Defense Robert Gates has stayed here, as well as Stephen Colbert when he was here in June. This was the first time I had been in this building and I wanted to make it a point to see it before I left Iraq. So the was the first of my three first-ever experiences this evening.

While we did not see much of the hotel as it was closed off to the public and is composed primarily of guest rooms anyway, the patio on the backside of the building is massive. At this particular meeting there were about 50-60 people, including a few young female soldiers. There were hors d'oeuvres, which included steak kabobs and bacon wrapped shrimp. Scrumptious! So Carlos, Mike, and I grabbed a cigar, some food, and a near beer and sat down and enjoyed the evening to the tunes of the 1st Corps band.

After we finished eating we lit up our cigars. I had picked up a cigar lighter at the daily bizarre over by the main PX when I first joined this cigar club. It is a dandy and looks like the afterburners on an F-16 at night! I think you could actually do some minor welding with this thing! I used this torch to light into what I believe was a true Cuban cigar. Not sure how you get them here because I know you cannot send them back to the US or get them in the US, so this may have been a knockoff, although Mike and Carlos thought this was the real thing and that was good enough for me. While hardly a cigar connoisseur, I could not tell if this was a Cuban by taste, but it was a good cigar with a very smooth flavor. So this was the second of my first-ever experiences and perhaps my first and last Cuban cigar.

Then, on the patio floor were some golf mats that were placed to hit some golf balls into Victory Lake. There is also a sign posted on the concrete pillars by the golf club rack that states **"DO NOT HIT GOLF BALLS AT AL FAW PALACE"**. Well, that would be no issue for me and it would take one mighty swing to hit the palace from that location anyway. Even Tiger might have be hard pressed to hit the palace as it is a good distance away, but there must be a reason the signs were posted in the first place!

Now, I am a duffer and keep groundskeepers pretty happy because most of my time is spent in the woods on a golf course, and I had not played in well over a year. I did hit some golf balls with my dad while on R&R but that has been about it. But I thought this is probably the one and only time I will ever get to do this so I went for it. The mat I was hitting off of was flat with no plastic tee. The club I grabbed was a flat-faced driver. So this would be a challenge to get some height on the ball with this setup, but I just wanted the experience and was not auditioning for the PGA. My first attempt harmlessly dribbled into the murky depths of Victory Lake. But my second shot was a might swing that began to slice through the thick night air and took a bead on the palace. While we lost sight of the ball soon after it left the patio floor we heard a crack a few moments later and it sounded like glass breaking. Oh, by the way, General Odierno's office sits on the side of the palace facing the JVB! Someone thought they saw the office window shatter. Not good!

OK, I got ya on that one. If you believe this tall tale then I do have some swampland for sale! Seriously, I hit a few balls into the lake that came nowhere near the palace, but it was still a unique experience that will not soon be forgotten.

So all-in-all, with the camaraderie, the full moon over Baghdad on this warm summer evening, and a series of three first-ever experiences I gave Carlos and Mike a ride back to their CHUs before heading back to my barracks to conclude one of my more memorable days in Iraq.

Ready for the PGA?

Day 306: Monday-September 7th, 2009/Day 275-Iraq was the day after we celebrated Labor Day over here. The civilians got the entire day off while military personnel worked a reduced battle rhythm, with many working half days. As this was my normal half-day off, there was no advantage to me at all! Because of this, I stretched my time off and went in a little later than normal as there was no briefing to the CG today. I also left a little earlier than normal so I could feel like it was a holiday. It is the little things that get you by!

As GRD is a construction manager and has assisted in the completion of over 4700 projects over here since the war began, reconstruction is major deal in Iraq. There is much work yet to do, but it will need to be accomplished by the Iraqis themselves from this point forward. Prime Minister Maliki understands this reality and commented that:

> Iraq is in need for its brothers and friends in the entire world. Iraq is in need of expertise and partnership to improve the country. Therefore, after the improved security situation in the country, we must continue our

efforts in rebuilding and rehabilitating the infrastructure. We must renew it in a way to suit our people.
-**MNFI, 5 SEP 09:** *al-Iraqiya*

While billions of dollars have been used to repair and reconstruct critical infrastructure within Iraq over the course of this war, it is still not close to being completed. And with more exports necessary due to the drought and plummeting oil prices, it will take many more years to get Iraq to where they want to be. Hopefully the Prime Minister's plea is heard and other countries will invest in Iraq. This process has already started, but greater security and stability may be needed before many other corporations decide to invest in Iraq's future.

Former US Ambassador Ryan Crocker made the news today. He and General David Petraeus were the two individuals who adeptly combined the civilian and military efforts to a point where we finally began to achieve sustainable success here in Iraq. Had they not done so, this would have been a war lost and one in which many troops had died in vain. Mr. Crocker stated:

When I arrived in Baghdad, Sunni insurgents and Shi'a militias, as well as their backers in Syria and Iran, believed they were on the verge of driving the US out of Iraq. But instead of stepping back, we stepped forward, and not just with troops.
-**MNFI, 7 SEP 09:** *Newsweek*

Today came another story on suicides within the Army. There were 143 suicides last year in the Army alone and another 113 among the other armed services. This year there have been already been over 100 with four months remaining in the year. These are the highest numbers since records began to be tracked over 30 years ago **(JASG, 7 SEP 09: BBC)**. This time period would be post-Vietnam when we had another high incidence of suicides within the military community. And let's see, we are currently involved in two dirty, difficult wars, one going on eight years and the other over six; many troops have been deployed on multiple occasions; troops have stripped away from their families for far too long; there has been no anticipation of such issues by the military; forces are strained to the point that would lead to such high suicide rate; and the military is still searching for answers in the wake of rates that are not subsiding. You need not be a psychologist to figure this one out. I have rambled on about this very topic on several occasions throughout this book and told you that a mandatory, online suicide prevention program we were forced to suffer through was a band-aid on a gaping wound. All this told me is that the one-size-fits-all mentality that the military normally employs under such circumstances was a knee-jerk reaction to the bad press they were getting through the media. It also told me they really had no clue as to how to find long-term solutions to this ever-increasing problem. Once again, much like the bureaucracy in which the military exists, leaders tend to look outward for solutions as opposed to focusing inward at the organizational structure. False expectations by government and military leadership in this type of war, inadequate training for troops before being deployed, ignoring the lessons of the past, inadequate force structure for this type of fourth generation warfare, lack of anticipation by leaders, the cultural obstacles created with the military community concerning healthcare issues, throwing resources toward Cold War enemies at the expense of the fight we are currently engaged in, all point toward an organizational structure that is antiquated and unable to create change rapidly. And as this cumbersome, slow-moving, resistant-to- change operation plods forward, more troops are being killed or killing themselves. So this new research that the Army is going to be using is not a bad a thing, even though it comes pretty late in this battle, but to me it is not the real issue that will lead to long-term solutions. It really is not an attempt to make Humpty Dumpty healthy and wise, it is simply an attempt to put Humpty Dumpty back together to fight another day.

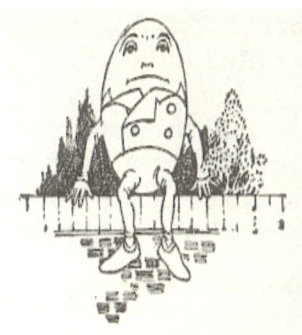

Can we really fix Humpty Dumpty?

|||||||||||||||||| Day 307: Tuesday-September 8th, 2009/Day 276-Iraq was a day to remember that we are still involved in a war and people are trying to do us harm **(JASG, 8 SEP 09: *Washington Post*)**. Yesterday we lost four more soldiers and the security detachment that was transporting our Command Sergeant Major Mitch Prater was struck by an IED! Fortunately, no one was injured in that attack, but a vehicle was destroyed and our three soldiers and the Aegis security team were lucky. The situation is much different since June 30th when all US troops vacated Iraqi cities. In fact, it is more dangerous now than when we arrived last December. So it is not a time for complacency and we must be even more vigilant over the course of the next few months. This is another story that will remain untold to Shari until after I am out of here for good.

Secretary Gates is using a little strategic communication to prop up the performance of the ISF thus far since US troops left all Iraqi cities at the end of June **(MNFI, 8 SEP 09: *al-Jazeera*)**. Gates stated that the ISF is held in high regard by US generals. While this may in fact be true and not simply information operations, there have clearly been some gaps in security. Today was another example. But no one expected them to be perfect right out of the gate and the extremists are trying to prove that the ISF is not fully capable of protecting the populace. As I said, it is a very dangerous time right now.

Anthony Cordesman from the Center for Strategic and International Studies stated that:

> It is often more prudent to keep forces at a high level than to rush to withdraw them, and then have to bring them back again. We're not talking about years, the election comes in January, and showing a few months of patience may be justified.
>
> **-MNFI, 8 SEP 09:** *LA Times*

I do not disagree with Mr. Cordesman, but it is unlikely that we will be returning troops to Iraq once we leave. At least not for a while, and not while the Afghanistan situation remains tenuous. But as a fire chief I can relate to this statement as it is always embarrassing when you leave the scene of a fire too quickly because of impatience and then must return a short time later to re-extinguish a rekindle of the same fire! Not a good thing at all.

In other news, Iraqis are paying more for meat and produce. The drought has hurt the ability to produce crops as I have mentioned, but strained relations with Syria have also apparently caused a reduction in imports, which as we know, then causes an increase in price due to high demand and not enough supply **(MNFI, 8 SEP 09, *NPR*)**.

The Army will now apparently embark upon another course to mentally strengthen soldiers. It is supposedly mind-strengthening training modeled after the Japanese Samurai culture **(JASG, 8 SEP 09, *Samurai mind training for modern warrior*, source: TIME)**. Yes, the plan is to provide this training to all 1.1 million soldiers to mentally toughen them up. Excuse my skepticism on this one! I think this type of thing is great and wonderful when the timing is right. But as the Army has great struggles with the alarming increase in suicide rates, overcoming a staunch culture that still attaches a stigma to those requesting assistance for mental health-related issues, an overwhelming conventional-style warfare mindset, a force structure not compatible with fourth generation warfare, etc. etc. etc. Does this seem like the right time to introduce yet another new initiative? I am certain that the practices of the ancient Samurai was a cultural mindset ingrained into the very fabric of their society. To expect similar results based upon a few classes provided to an already weary force, when training time to overcome greater obstacles is at a premium as it stands, seems like a real shot in the dark at this point in time. Perhaps I am wrong and have just grown cynical with what I have observed over the course of this past year. Then again, perhaps not!

Day 308: Wednesday-September 9th, 2009/Day 277-Iraq began with reinforcement of my theory that we have still not yet overcome our conventional-style warfare mindset. There is a story circulating in Afghanistan about tactics used by US forces to secure and clear a hospital. The tactics used were supposedly very heavy-handed **(Stars & Stripes, 8 SEP 09, *Charity:* US military invaded Afghan hospital, tied up staff, written by Kay Johnson).** These types of tactics fuel an insurgency and undermine all that the US is attempting to accomplish. Now keep in mind that this could be another case of the Taliban stretching the truth for their own benefit by using cheap but effective propaganda to turn the people against the US. They are masters at this technique. But the incident is under investigation and if US forces were at fault, the commander of this unit should be fired. The troops should be counseled and remediated, but it is the fault of the leadership if after eight difficult years of war, the guidance issued by General McChrystal, and the lessons of the past, there are little excuses for this type of incompetence. It is one step forward, and two steps back with us in many instances and we are often our own worst enemy. How does the old adage go? I have seen the enemy – and the enemy is us!

The White House is getting ready to debate the strategy provided by General McChrystal **(Stars & Stripes, 9 SEP 09, *White House to debate Afghan effort*, written by Karen DeYoung).** It will take a few weeks to determine any changes in course. But as we all know, as the American people go, so go the fickle politicians. The problem with this ridiculous pattern of response is that most Americans really have no idea what is going on in Iraq or Afghanistan. This is not a condemnation of any sort, as I am an American myself, but it is a reality. I always wonder about these polls that surface that Americans are not happy with the war in Afghanistan. Who are the people that are being surveyed? Are they counterinsurgent experts? Are they former military members? Or are they average Joe who knows most about either war through what they see or read in the various media outlets? So how is this a good source to gauge how the war is going is my real question? Yet politicians, who should know much more about the situation, will bend to the polls in order to get re-elected. I realize these people are elected to represent their constituents, and US troops comprise only 1% of the overall voting public, but this process just seems messed up to me. So I say again, I certainly hope the President either provides the resources necessary to achieve success in Afghanistan, or pulls the plug. Let us not do yet another half-ass effort and therefore jeopardize the wellbeing of the troops in harm's way. Let our leadership, both within the government and the military, stand up and shine and do what is right, not what is convenient or popular. Our brave troops deserve this from our leadership.

Day 309: Thursday-September 10th, 2009/Day 278-Iraq we will begin this day with another one of those stories that most Americans will never hear about and never have to deal with. This particular one deals with the topic of custody battles while service members are deployed **(Stars & Stripes, 6 SEP 09, *Custody battles can become a rude 'welcome home' for military parents,* written by Leo Shane III)**. It just seems despicable that a judge would rule against a parent while they are deployed and serving their country. Judges may claim they must simply follow the law and I say horse hockey to that excuse! Can judges not exercise a little common sense on occasion? If they cannot in such circumstances then our laws are really messed up! Congress also needs to get their collective head out of the collective ass and start addressing such inequities. We have been involved in war now for almost eight years and Congress does not hesitate to send troops into harm's way, but when it comes to taking care of them when they return, then it becomes a real 'slog'. So taking children away from parents when they are deployed is just wrong when temporary custody can be awarded until the parent returns home and can deal with the issue. Again, how about a little common sense here people!

Secretary Gates has stated that the US will not abandon Afghanistan and that we are in this thing for the long haul **(JASG, 10 SEP 09: *Washington Post*)**. Hopefully the President will see it this way as well. But if this is truly the case, I hope they provide the necessary resources to the military in order to achieve success and not just enough to keep the fight going.

Violence in Iraq continues but a police commander stated that "the attacks will continue. But the number of victims is limited because the Iraqi Police are capable of foiling the attackers and preventing them from reaching their destination" **(MNFI, 10 SEP 09: *NY Times*)**. In some cases this is the correct, while in others it is obviously not. But the effort to get better and more proficient at reducing violence is an ongoing, dynamic process for Iraqi Security Forces. This again is something that is going to take time to evolve and mature.

Finally, there was some interesting news on life expectancy in the Middle East. Israel and Jordan lead the way in life expectancy at approximately 81 and 79 years of age. Yemen is ranked last with approximately 63 years of age, and Iraq is just slightly below the mean average in Iran at about 70 years of age **(MNFI, 10 SEP 09: open source documents from the MNF-I surgeon's office).** So despite years of war and disease, this average is much higher than most people would have guessed. Hopefully with greater stability and increased healthcare opportunities throughout Iraq this trend will continue upward. This would be another success story here as the Gulf Region Division has managed many healthcare projects, such as hospitals and clinics, and all of these variables have resulted in greater longevity for the Iraqi people. Since the war began in 2003, the average life expectancy has steadily risen from just under 68 years of age to now just about 70 years of age. An indicator of success perhaps?

(-10) Day 310: Friday-September 11th, 2009/Day 279-Iraq brings us eight years to the day when we were attacked on US soil and we witnessed in horror the twin towers of New York crumbling to the ground and taking the lives of almost 3000 people in the process. This is the reason why we are here right now and in Afghanistan. It seems that the American public has an amnesic tendency to forget such high-profile events. This may be related to always-looking-forward syndrome I have mentioned to you previously. Whatever the reason, we must not fail in degrading al-Qaeda to a point that they cannot have the freedom to plot such terrorist attacks again. So simply leaving Iraq and Afghanistan before the job is done is inviting another attack and this is what the American people should stay focused upon.

There was another nicely done ceremony at Al Faw Palace today as we commemorated the 9/11 attacks. As do most Americans old enough to recall the event, I remember I was at Fire Station #1 in Wisconsin Rapids in my office when one of my firefighters informed me that another firefighter had just called and told us to turn on the television. At first the scene seemed surreal and simply a terrible accident, but when the second jet hit the second tower I knew we would be going to war. The next several weeks would be filled with heart-wrenching stories emanating from the rubble the lie in the streets of Manhattan, including the deaths of 343 New York City firefighters.

The perpetrator of this event, Osama bin Laden, was also remembered today and the fact that after 8 years he is still unaccounted for. We almost had him a few months after the attacks in Afghanistan, but he slipped away and we have not come close to catching him since **(JASG, 11 SEP 09, *Bin Laden still on the loose 8 years later,* source: Freep.com).** Again, as much as we would like to bring this criminal to justice it still would not end the war on terrorism. I do not believe the American people really understand this fact. Al Qaeda is a very loose organization based upon a guiding principle. It is not controlled by one man or even one group. So capturing or killing one individual really does little to deter such a misguided movement. We must continue to pressure them where they live and train, and give them little freedom to move or to plan more attacks. Americans must understand that this is a marathon and not a sprint. We need to be committed for the long haul or we can expect more attacks upon US citizens, perhaps again on US soil. Let us not forget!

Today was simply one of those days. It started off with a physical fitness test early in the morning. I would now get to see if working out six days a week and dropping 25 pounds would result in a higher PT score. My goal for this test was anything above a 270. My last two scores, when I was heavier in worse physical condition, were a 268 and a 269 respectively. If you score a 270 or above you are awarded a physical fitness patch that you can wear on your Army PT uniform. Today this was my goal.

It started off on an excellent note as I maxed out my pushups and my sit-ups thus validating my perfect pushup and sit up routines over the last eight months. So I had a perfect score of 200 heading into the run, my worst event. As I may have mentioned previously, as a young Marine I could run six minute miles, and the PT test back then in the Marine Corps consisted of pull ups, sit-ups, and a 3-mile run. I knew from all of my workouts that I needed my left foot to hit the pavement 80 times per minute to stay on pace for a 17:08 two-mile time. This would give me about 80 points and the achievement of my goal. The weather today was about perfect and the running course took us around Lost Lake. This route has four corners and one tricky one near the end. I had never before run this course, but it looked flat and fast. So off we went and when I hit the mile marker I was 5 seconds ahead of pace. I also hit the mile and a half marker right on time. At one and three-quarter miles I was beginning to feel the 'burn' but was still on schedule, perhaps just slightly behind. Then, with success within my grasp, the bizarre occurred. Remember the tricky little corner near the end I mentioned? Well, I had never run this course before and there was no one in sight in front of me to follow. So with no signs to guide my way, I had two choices – veer right or veer left. Veering right was much more natural and so I set my course. As I checked my watch and thought to myself "this last quarter-mile sure seems much longer than all the rest"! Indeed it was as I kept looking for the finish line that was never to come today. I knew then that I had run well over two miles and went the wrong way! I am a big believer in taking the path less traveled as a leader, but this gave the adage a new meaning! Just call me Wrong Way Waite! How embarrassing it was as success slipped from my grasp in that last quarter mile and that fateful turn right. I most probably would have come in at around 17:15 or so and finished the PT test around 280. But what are you to do? I got a good workout in this day, took a little good natured and well-deserved ribbing, and got a good gauge on my level of physical fitness. 1SG Scott McWilliams told me I was not the first to do what I did and no doubt I would not be the last, but in two weeks there would be another opportunity for a PT test and a chance at redemption. This gives me two more weeks to tweak my fitness level, and I have no doubt I will successfully complete that test around 280 or so. Such is life, and if this is the worst that happens to me over here in Iraq, I am just fine with that.

Sometimes you feel like a nut, sometimes you don't!

Day 311: Saturday-September 12th, 2009/Day 280-Iraq we will begin with more news generated by another electrocution death in Iraq **(Stars & Stripes, 10 SEP 09, *US contractor in Iraq reportedly electrocuted while taking a shower*, written by Kimberly Hefling).** You may recall the other 18 electrocutions throughout Iraq and our friends at KBR that have been linked to several of them. While I must state that KBR was not mentioned in this article, this organization is the one that has been assigned this particular task in Iraq so they may be linked to this incident as well. If so, the question must be asked, what is taking our lawmakers so long to get a contractor that is getting paid millions of dollars from US taxpayers under control, especially after all this time? To me, this is another case of our government not being very responsive or quick to take action. It also signals to me that they still have little oversight or accountability over the many contractors they have paid millions to and this is a very sad state of affairs indeed.

Today we shall also get to another poignant story from the war on terrorism. This particular story comes to you from Afghanistan **(Stars & Stripes, 4 SEP 09,** *Death of a Marine*, **written by Alfred De Montesquiou and Julie Jacobson)**. The situation in Afghanistan as I have reported to you is extremely dangerous right now. It seems the recent change in tactics has presented a new set of challenges for troops on the ground. Now I do agree a change in tactics is necessary, but when troops are in harm's way, and it is relatively clear that innocent civilians will not be in the line of fire, then the US should be authorized to unleash hell on the Taliban. The Taliban is using this change in tactics to become more aggressive in their attacks upon coalition forces and then retreating back into the villages and towns to blend in with the rest of the populace. As a result of this newfound aggressiveness by the Taliban they are claiming more US lives. Marine Lance Corporal Joshua Bernard was one of the heroes that fell that fateful day in Afghanistan. His father, a former Marine himself, was very upset with this change in tactics and believes it places troops at greater risk, which in fact is correct. Lance Corporal Bernard's story is one of many that could be told, as changes in tactics after years of war, and not properly resourcing military commanders for this type of warfare, will no doubt result in more fatalities. Again, in my opinion this is a failure in leadership from the highest levels that is now attempting to right a wrong in the eleventh hour. Soon we shall see how much support our troops will get from the current administration if we are truly going to be staying in Afghanistan for a while. But for Lance Corporal Joshua Bernard any assistance will come too late. Joshua Bernard was 21 years old.

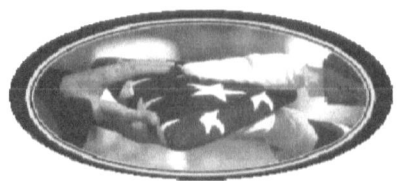

Rest in peace Marine.

▮▎ ▮▎ Day 312: Sunday-September 13th, 2009/Day 281-Iraq began with John Maxwell and his book ***The 360 Degree Leader***. I have read many of John Maxwell's books on leadership and enjoyed them all. In fact, I have used a few as supplemental reading for some of my fire officer courses I have taught. I am about a third of the way through this particular book and while it is good, it is not one of his best in my opinion. I also disagree with Mr. Maxwell on a few issues, including his use of the term chameleon leadership. He, like most people, see the chameleon as someone who is wishy-washy and moves to whatever side is to his/her advantage. In my opinion, the chameleon has gotten a bad rap because of this stereotype, so I want to set the record straight.

I believe a chameleon is one of the most adaptable creatures on this earth. This creature has the innate ability to blend into the environment in which it finds itself. I view this as a case study in flexibility and the ability to adapt to any environment. In fact, the military uses the new buzz term of 'adaptable' leadership. I believe a good leader must recognize his/her environment and adapt as is necessary. Each situation is unique and must be handled as such. A one-size-fits-all mentality, whereby you use one leadership style no matter the individual or the situation, tells me this person does not understand the art of leadership. So like the chameleon, leaders should vary their leadership styles and tailor it to the given situation. This is hardly being wishy-washy. It is being an adept leader that understands the nuances of leadership. So rock on, dear chameleon! I for one stand in your corner.

The often misunderstood chameleon!

More news from Afghanistan as the political machine begins to heat up. These are the types of instances when politics and military operations do not mesh together well. There are many 'grandstanders' in politics today. Many politicians run to get elected, and then once placed into office, they are constantly in campaign mode! Rather than do what is right and necessary, they make decisions based upon what they feel is the best methodology for re-election. As far as I am concerned, elected officials should have a shelf life and from what I have seen at this level during my tour, one term is enough for most of them! Then, if they are elected, they need not worry about getting re-elected! This would take that variable right out of the equation. It would also reduce corruption and prevent someone from occupying a Congressional seat for decades. Anyway, I digress. The political soft shoe will begin in earnest soon in regard to the Afghanistan situation **(Stars & Stripes, 12 SEP 09,** *Doubt raised on Afghanistan troop boost***, written by Julian E. Barnes).** I would hope that the politicians that do not believe we need more troops in order to be successful in Afghanistan are also the same ones that will vote to pull our troops right out of Afghanistan as well. However, my bet is that many of these politicians are angling for votes at the expense of this unpopular war, and by not voting for more troops and then not voting for immediate removal of troops, they are in essence placing our brave troops in greater jeopardy. It is the old whack-a-mole game in Afghanistan as it was in Iraq, and our politicians are not providing enough mallets to our military to allow them to be successful. This also signals to me that these politicians were not paying attention to what occurred in Iraq either. Because of this complete failure in leadership, as much as I love the military, I would never recommend to a parent to encourage his/her son or daughter to join the armed forces. Incompetence from the highest levels of government translates to dead soldiers on the battlefield. I hope I am wrong on this one, but it has happened before and it is shaping up this way yet again. God bless our troops.

And yes, the American public that really knows nothing of the wars in lands far away from their comfort zone have very little knowledge of the strategy involved. They also must have very short memories. It has only been eight short years since the terror of 9/11, yet people tend to forget this **(Stars & Stripes, 12 SEP 09,** *Connection between 9/11, Afghan war fades,* **written by Liz Sidoti).** As the American military only comprises about 1% of the overall population, the media reports nothing but bad news from the war zones, and our economy is struggling while billions are funneled for the war effort, it is not surprising that people have negative feelings toward the war. But overall, the American public is ill-informed on this topic and they are not reliable sources of information when it comes to polling numbers. Who are the people that are surveyed in these polls anyway? What does it really mean when the media reports that 60% of Americans are against the war effort? Who are these 60%? Are they well-informed counterinsurgent experts? Are they politicians with military experience who have talked to commanders on the ground? I would venture to guess they are not. I know I have never been asked my opinion on the subject! Yet politicians seem drawn to this smoke and mirror show like a moth to a flame. Perhaps I just do not understand the entire process. Perhaps the politician is severely misunderstood much like the chameleon. Perhaps the American public is better informed on the wars than I give them credit for. But one thing I do know is that half-ass resourcing of military operations will place many troops at greater risk. So when people equate the Afghan war to Vietnam, it is because this is becoming a self-fulfilling prophecy. It appears to be another case of the old Pygmalion effect, whereby if you expect something to happen, it is likely to become reality. Many people have mentioned these two wars in the same breath, and many of the politicians and military leaders of today have followed the same course as their predecessors during the Vietnam era. Is it any wonder that history is again repeating itself?

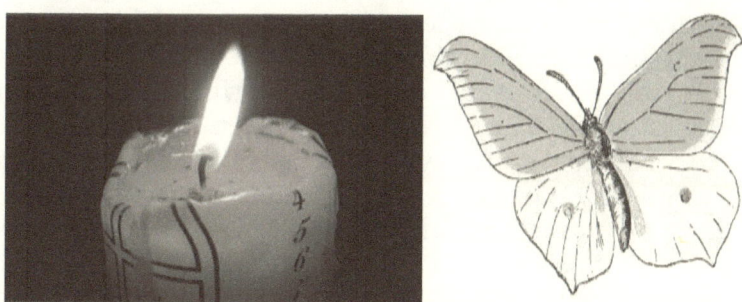

Politicians and public opinion polls are like a moth to a flame!

Day 313: Monday-September 14th, 2009/Day 282-Iraq began with more news from our friends in Iran. Ambassador Christopher Hill testified to Congress, "There is no doubt" that Iraq will maintain strong relations with Iran due to regional proximity **(MNFI, 12 SEP: *al-Hurra*)**. However, Iran should respect Iraq's sovereignty and stop arming extremists. Unless diplomatic efforts increase and enhance the relationship between the United States and Iran, it is highly unlikely that Iran will cease these tactics.

The US Senate has approved the President's $128 million request for military operations in Iraq and Afghanistan **(MNFI, 12 SEP 09: *AP*)**. While operations are winding down in Iraq, they are escalating in Afghanistan, so funding totals such as this are likely to continue into the foreseeable future.

The Commanding General for Multi-National Corps-Iraq, LTG Charles Jacoby, who I met last October at Ft. Lewis, Washington mentioned the job of Iraqi Security Forces. LTG Jacoby stated:

> We are witnessing the Iraqi Security Forces address challenges head-on. They are not backing down and are making steady progress towards taking full responsibility for the security of Iraq's people.
> **-MNFI, 12 SEP 09:** *Pentagon Press Conference*

While I believe there are still some issues the ISF needs to address in order to move forward, it does appear that they are serious about becoming better at protecting the populace.

Yesterday Shari had the family and several friends over to the homestead for one final hurrah for the summer. We have a beautiful pool, which Shari does a masterful job taking care of (see R&R photos). The weather has been beautiful recently and would be near 80 today, which is a bit warm this time of year in good old Wisconsin. My granddaughter Kayleigh also made an appearance and she loves to swim. I hope to be home for her 7th birthday, which is near the end of October. So a good time was had by all.

Finally today, the Green Bay Packers played Da Bears in the 2009 regular season opener. It was a typical 'black and blue' division tussle. While a bit sloppy on both sides on occasion, it was an exciting game and the 'green and gold' won 21-15. Needless to say, as our unit is based out of Darien, Illinois, just outside of Chicago, there were many disappointed Bear fans that took some ribbing from Packer backers. It was a little payback from when the Bears beat the Packers last December after we were mobilized, and they were hardly gracious winners! This is just another reason why this is still an excellent rivalry and the oldest in the NFL.

Way to go Pack

Day 314: Tuesday-September 15th, 2009/Day 283-Iraq began with the news that I would again be assuming the role of the G3. Apparently I did such a good job while Rich was gone when he had to fly back to the US to testify at a court martial hearing last month I am getting the job for the next 30 days! I guess the Oscar nomination for my acting role sealed the deal!

Actually, Rich will be moving over to become the Deputy Commander at the Gulf Region District once the GRD HQ inactivates and the members of the 416th TEC, myself included, have concluded our tour and gone home. So it has been decided that Rich should begin this transition immediately and he will be back and forth as his duties will overlap anyway. I told Rich I have no issue with this, or no choice for that matter, and do not mind the challenge. But as I mentioned previously, GRD has far too many meetings during the week and when I must attend all those meetings I am not getting any work accomplished. So in essence, with only about 30 days to go, my dress rehearsal from last month during Rich's absence, and the fact that things are winding down, it should not be that big of a deal.

Well, here we go again. Now I may possess a Ph.D. but that does not necessarily translate to understanding the military community with any great depth. However, with 34 years in the military in my 52 years of life, an analytical mind by nature, and good observation skills, I call 'em as I see 'em! There was a story in **Stars and Stripes** today that validated some of my previous ramblings in relation to the culture and organizational failures of the Army. As you have read all of these 'observations' you may have concluded that I am bitter toward the military. That is really not an accurate assessment as I love the military and have great respect for the people that serve our country. But I want our military to succeed and become stronger and it gets frustrating when you do not see that happening. I have also tried to validate my frustration with comments and stories from others so this book is not simply personal opinion on my part. But yesterday there was yet another article on PTSD and the military approach to this topic **(Stars & Stripes, 14 SEP 09, *Failing the stress tests?*, written by Megan McCloskey).** This article identified that the Army institutes a lot of different programs but really has no ability to measure the effectiveness of such programs! I have mentioned this 'band-aid on a gaping wound' or 'one-size-fits-all' mentality the Army often employs. The problem is that if they cannot measure success, how do they know their programs have been effective? The answer is – they don't! This type of mindset is much more reactive in nature than proactive. It was also cited that the Army does not even count how many soldiers are visiting mental health clinics on the front lines and are not tracking long-term mental health! So when a soldier leaves the military, the Army loses touch and has no way to account for these people that have served their country proudly. The statistics are already grim, but it may be deduced that they may be even worse if the Army accounted for the many soldiers that have left active/reserve duty after tours in Iraq and/or Afghanistan and suffered mental health issues that are unaccounted for. From my perspective, this story has just uncovered other glaring deficiencies in this large bureaucracy with few answers on the horizon. Hopefully because of this exposure and the constant scrutiny, the Army will continue to pursue better answers and more effective programs to assist those that have given so much. The real tragedy in all of this will be if they do not!

Day 315: Wednesday-September 16th, 2009/Day 284-Iraq was once again no BUA Wednesday. Next week Thursday would be GRD's second to the last update to General Odierno, and MAJ Sean Begley's debut as the voice of GRD.

The preparations for the drawdown and GRD HQ inactivation continue. There is much work being done by a lot of talented people at GRD, but one thing I must state is there is a clear lack of a unity of effort. Each section is doing their own thing and often there is little communication between sections as change occurs. Now I will state that because of the pace of change in a very dynamic environment this is a real challenge. But I will also state that since I have been here, I have noticed this communication vacuum and almost a protecting of information when it needs to be shared. GRD is filled with a lot of intelligent people and many leaders, but it really lacks that one guiding light to pull all efforts together despite all of the meetings we have during a week. Part of this may be attributed to an overwhelming workload. Some of it may be associated with people in unfamiliar positions and not completely understanding all of the nuances of the organizational structure. Much of it may be attributed to the fact that none of us have ever been involved in a headquarters inactivation. But I honestly believe that these organizational fissures existed since GRD has been here and the inactivation just makes some of the deficiencies more transparent. For example, I have highlighted that we have far too many meetings during the week; there is a real dichotomy of philosophy between civilians and the military; information sharing is a real challenge; many military leaders have never before worked in such an environment where they must deal with both civilian and military employees within the same organization; and many military personnel are in positions they were not trained for and are simply filling a gap. Then you have the micromanagers mixed in with the laizze-faire leaders and those caught in the middle of this leadership style sandwich! For example, we have a lot of military people who are doing the yeoman's share of the work. This should result in a readjustment of everyone's effort to reduce the workload for some and add to others not so busy at this stage. This is not and does not appear likely to occur. Because of all of the meetings that occur, many of which are redundant, there is not much work getting accomplished or direction provided. This translates to people having to work harder and longer than necessary. There is a lack of clear direction, which leads to wasted effort. Yesterday I spent hours working with the G1 and the HR section leaders developing a chart for a briefing. There was no clear guidance as to what was needed or desired, but there was some reference to an abstract concept by several people, which led to the effort. Then once the effort was accomplished, the data provided was not what was desired! Perhaps clearer guidance up front would have resulted in a better product. And while I would like to state that this was an anomaly within this organization, it was simply a microcosm of the cracks within the organizational foundation. Then, another officer spent several more hours and worked long into the evening to basically produce the same results! So the bottom-line for me as someone trained on this subject area is that there is a lot of room for organizational improvement.

Much of the talk recently has been about the war in Afghanistan and it seems most people have forgotten about Iraq. Well, the good news is that this is not unusual or unexpected, and while the situation here is still fragile, it grows stronger each day. The rhetoric from Tehran and Washington continues as today Secretary Gates commented that:

> The more that our Arab friends and allies can strengthen their security capabilities, the more they can strengthen their co-operation, both with each other and with us. I think this sends the signal to the Iranians that this path they're on is not going to advance Iranian security but, in fact, could weaken it.
> **-MNFI, 15 SEP 09:** *Middle East Online*

As we get closer to the national elections in January, this rhetoric will pick up considerably as Iran will continue to attempt to influence events and the future course of Iraq.

Day 316: Thursday-September 17th, 2009/Day 285-Iraq saw another visit from Vice President Biden. He is here to talk to the Iraqi Prime Minister to engage in bilateral discussions **(MNFI, 16 SEP 09:** *NY Times, AP, ABC, AFP, NPR, Reuters*). This is the Vice President's third visit to Iraq this year. This shows Iraq that we have not forgotten about them and are still here to assist them to become a stable country and a long-term global partner.

I watched a little of Admiral Mullen's reconfirmation hearing before the Senate yesterday and Afghanistan dominated the discussion **(Pentagon Channel, 16 SEP 09, 2100 GMT+3).** The strategic analysis completed by General McChyrstal, the Commander of coalition forces in Afghanistan, has still not been made public but it is creating great debate on Capitol Hill. Admiral Mullen was very non-committal on the topic of additional troops, as this is likely to be a political minefield, but he is laying the foundation for this request that will eventually come. Admiral Mullen has stated that it is likely more troops will be needed, but the exact number and the correct mixture is likely to fuel debate in the near future **(Stars & Stripes, 16 SEP 09,** *More troops urged* *for Afghanistan,* **written by Anne Gearan & JASG, 16 SEP 09,** *Military chief suggests need to enlarge US Afghan force***: New York Times).** Senators McCain and Lieberman believe we need to show commitment to Afghanistan and more troops will be necessary to achieve success, while Senator Levin and many others believe we should not send more US troops, but rather, train more Afghan security forces to take over this function for their own country. But as I have said quite often over the course of this book, as this Washington two-step goes on and the political posturing begins, troops are dying because they do not have the necessary resources. Remember what I said about our bureaucratic system and how slow it is to create change and make key decisions?

Finally, yesterday was a bit of a red-letter day here in Iraq. I started a 'braggers pool' a few weeks ago as to who could come closest to picking the exact day we would finally drop below 100 degrees as the high for the day here in Baghdad. This had to be validated by a BUA slide from the weather forecaster. The 16th of September was that day as we dropped to a cool 99 degrees! MAJ Sean Begley and Colonel Michelle Stewart both chose September 15th so they were declared the winners as no one had selected the 16th and the next day chosen was the 19th. Shari told me it was a whopping 44 degrees when I called yesterday morning and it would climb close to 70 for the high in Wisconsin Rapids. Fall has arrived there with falling leaves, cool nights, and beautiful days. In Baghdad, the evening lows are higher than the daily highs in Wisconsin and there are no leaves to fall! But it is finally cooling off a bit, which is nice, and it also signifies our approach to finally getting out of here. Hallelujah!

Day 317: Friday-September 18th, 2009/Day 286-Iraq saw some rocket attacks into the International Zone that injured some contractors working over by the new embassy compound and killed two civilians. Vice-President Biden was visiting with Prime Minister Maliki to discuss several topics concerning the way ahead for Iraq and long-term relations with the US **(JASG, 17 SEP 09,** *Biden confident in stable Iraq amid rocket attacks:* **AP).** I have no issue with the VP coming here and working the Iraq-US issue as he was also chairman of the Senate Armed Services Committee before he became VP and he has a solid background on the Middle East situation. But coincidentally, he also has a son serving over here. This is his third visit to Iraq this year! I would also assume he has visited with his son all three trips. I think that is great but it also seems unfair that that the thousands of other parents who have children over here are not afforded this same luxury. And let us hope these are not boondoggles disguised as political assistance visits just so he can see his son. As we have seen from the South Carolina governor recently, nothing is beyond possibility when it comes to our politicians!

It also appears as though the Iraqis are picking up on the political process very quickly. We are nearing national elections in a few short months. In the US, this is when you finally see our politicians work harder to get something accomplished in order to cash it in for votes. Najaf is a province to the southwest of Baghdad. There is a new governor in place there that is making some positive improvements. A law student in Najaf noted that "the new governor is working hard to accomplish projects. That is making him look good to the people, which also helps Prime Minister Maliki" **(MNFI, 17 SEP 09: *Wall Street Journal*)**. Is it coincidence that this is occurring with national elections looming? Perhaps!

The Prime Minister also visited southern Iraq provinces this week and pledged to implement water purification projects to solidify his Shi'a vote support base. In fact, the Prime Minister is supposedly urging leaders to intensify their efforts to get public works projects started to bolster his public image before elections **(MNFI, 17 SEP 09: *Wall Street Journal*)**. Is this just coincidence? Perhaps, but I think the Iraqis are really beginning to catch on to the political process over here!

Admiral Mullen is not pleased with the status of our efforts in relation to PTSD! I do believe I have stated this on several occasions and now the Chairman of the Joint Chiefs of Staff is stating the obvious **(Stars & Stripes, 17 SEP 09, *Mullen frustrated with PTSD outreach*, written by Leo Shane III)**. I hope further exposure of this inequity at this level within our government will urge more resources and research to be conducted so we can help our wounded warriors. This research can also assist ordinary citizens involved in car accidents or falls. Again, our government is quick to send our troops into harm's way, but awfully slow to take care of them when they return. Can you say bureaucracy!

Finally, today is the 62nd birthday of the US Air Force. We actually celebrated the event at GRD HQ yesterday with birthday cake, a short ceremony, and a wonderful rendition of the Air Force song. So while the Air Force may not be able to use their impressive array of jets and bombers in this type of war, they have been great at providing a lot of reconnaissance and surveillance assets, and they have many support personnel to assist in the efforts here and in Afghanistan. So happy 62nd US Air Force.

Happy Birthday and Aim High

▌▌ ▌▌ **Day 318: Saturday-September 19th, 2009/Day 287-Iraq** would be another busy day. Today we had our weekly transformation working group meeting where all sections report on progress within their area in regard to transferring functions and drawing down equipment as we get ready to inactive our headquarters in about a month. There is a lot of tension associated with this draw down as well. I am the 'keeper of the slides' for this briefing and there are about 80 of them! With 25 different sections reporting this is a real beast to keep organized each week. As I stated previously, it is often a case study in 'cat-herding' and something I would not want to do as a career.

We also had another doubleheader in regard to cigar clubs. An emergency meeting of the Cigar Aficionado Club at

GRD was called. This was due to the fact that a few of the members are with a contracting organization and they just received letters that they are being curtailed and sent home. So just last week the club president scheduled the meetings of the Cigar Aficionado Club opposite the Victory Cigar Club after I mentioned this little conflict to him a few weeks ago. So this was an immediate and sudden change in fortune for several club members.

Somalia, which is considered the number one failed state in the world, is heating up. Of course, you may recall all of the piracy that has occurred off of the coast of Somalia, and it is currently considered a safe haven for al Qaeda extremists. A few days ago the US military launched an attack on an insurgent camp in this lawless country, which is really like the wild west of America's yesteryear. I had heard that the US and other countries may get more aggressive in Somalia because it is becoming a breeding ground for extremists. The insurgents retaliated by killing innocent civilians **(Stars & Stripes, 19 SEP 09, *Witness: 11 killed in Somalia in revenge attack for US raid,* written by Mohamed Olad Hassan).** This area is likely to become a hotbed of activity in the near-term future which means the US is now involved in three of the top nine failed states in the world and two of the top four. This level of involvement may not be as we have seen in Iraq and Afghanistan, but it appears there is a new Marshal in town and he is ready to clean up the world's new wild west.

There is a new Marshal in town

❚❚❚ ❚❚ Day 319: Sunday-September 20th, 2009/Day 288-Iraq would find me placing the finishing touches on my second leadership book titled ***The EMS Leadership Challenge – A Call To Action.*** Rickey Pittman was my editor for this project and was very easy to work with. I hope to get that project packaged and sent to my publisher at BookLocker so it can hit the shelves by Thanksgiving or Christmas. Then I will be able to turn my full attention to this project when I return home.

News concerning Afghanistan from Washington is not good. The President is going to take his time making a decision concerning the future course there. As I have mentioned, he should take time to ponder the decision as it is extremely important, but taking too long indicates indecisiveness. It also translates into our troops being there without the adequate resources to be successful. So again Mr. President, as a leader, either provide the resources to our brave men and women fighting for this country, or get the hell out of there. Procrastination means more lives lost and piecemealing the necessary resources while staying involved is a weak political answer. The Afghan Ambassador to Washington also stated that the longer Obama ponders his decision he risks losing support at home and abroad **(http://www.bloomberg.com/apps/news?pid=20601087&sid=aOsI6x5z3b0)**. So a timely, well-thought out decision coming from Washington would be refreshing.

As you may have read this book you may have formed certain conclusions about me, about the war in Iraq, about a number of other issues I have identified. I hope this is the case as this book is intended as a vehicle to enlighten, inform, and educate, as well as stimulate thought and meaningful discussion. And yes, sometimes I come out of left field on issues, but I try and ground those statements in some fact by validating sources other than myself. One such issue concerns the care of our wounded warriors and how far we have yet to go on this issue. Admiral Mike Mullen, Chairman of the Joint Chiefs of Staff, is now speaking out on the topic and hopefully that will provide more exposure on the issue at hand and lead to meaningful change **(Stars & Stripes, 19 SEP 09, *Mullen: Community effort needed to heal war wounds*, written by Tom Philpott)**. There is a poignant example in this article of a mom who graciously allowed her son to provide his service to our country. Her son was severely wounded while serving and now needs constant care. His mother, Leslie Kammerdiener, is not a nurse, a psychologist, nor a physician. She is a mother who needs help taking care of her son. The VA, the US Government, and the Department of Defense are not stepping up to help her. This is only one of many heart-rendering stories of great sacrifice by our troops, and not so great sacrifice by those that sent them to war in the first place. To me, this is a crying shame and a shame on you federal government. Even in our economically challenged environment, as a parent I would hesitate or even dissuade my child from joining the military knowing what I do now, and how quick they are to send troops into harm's way, but they are not so quick when it comes to taking care of our troops upon their return home. Doesn't sound much like a grateful nation to me!

(-20) Day 320: Monday-September 21st, 2009/Day 289-Iraq found us one month out from leaving this place. As we continue to get closer I am doing my own personal drawdown to lighten my load going home. I have another footlocker ready to be mailed that is full of gifts, cigars, and clothing. I also organized my room yesterday and have two and half boxes packed and ready to be shipped. My coffee supply may come up just a hair short, but I have a little can of Maxwell House as a backup. I am drawing down my supply of Coffee Mate and Splenda as well, at both the office and my room. We also get to turn in our chemical protection gear before we head home. This is about a bag full of gear that we will turn in soon and will not have to lug back to Ft. McCoy. So I now will have plenty of room for my remaining gear and will go back with only three bags versus the five I arrived here with, plus the two large footlockers that were still coming! So my personal drawdown is going as planned. Hopefully the drawdown of US troops and the million plus pieces of equipment that must exit the country is going as smoothly, albeit a bit larger project!

There is a great deal of confusion at GRD HQ as people are turning every which way. Stress levels continue to be high, people are leaving but the workload is not reducing, at least not yet. Sections are not certain who is responsible for what product and confusion reigns supreme as we are also getting ready for a visit from the Chief of the Engineers, LTG Van Antwerp this week. All this means to me is more work and another busy week. We also have the first of three unit award ceremonies this week and LTG Van Antwerp will present some of the awards, which is likely to be primarily civilians this time around. The only award I want is a one-way plane ticket home! But until we reach that wonderful point in time I am certain that the mass exercise in confusion will continue at full pace.

The great state of confusion

CHAPTER 17
400 DAYS - Days 321-340
September 22 - October 21, 2009

Iraq is now a different country from the one I had seen first. However, we must realize that these gains are fragile and reversible.

--MNF-I Commander General Raymond Odierno/2008-present (also served as MNC-I Commander 2006-2008 and Assistant to Chairman of the Joint Chiefs of Staff/2004-2006 and Commander 4th Infantry Division 2001-2004)

Day 321: Tuesday-September 22nd, 2009/Day 290-Iraq
We begin our final month here in Iraq. The more John Maxwell books I read, the more I believe others need to read his books as well! While my work here is not all that interesting, just observing the dysfunctionalism and the human dynamics within the organization I work for really drive many leadership and organizational function points home. Everyone is now gearing up for LTG Van Antwerp's visit, which begins today. Once we get through the next week it will really be all downhill. But again, this is not a time for complacency.

I failed to mention to you the other day, Saturday evening to be exact, that Carlos and I attended the Cigar Aficionado Club meeting. That part I believe I mentioned. But as we got in my chariot and began to head toward the Victory Cigar Club at about 1900, this huge windstorm blew in. We could see some lightning in the distance and heard some thunder while at our first meeting, but this wind came on suddenly and was very unexpected. The wind blew fiercely for about a period of 15 minutes or so. It reminded me of a blizzard back home in Wisconsin, one that creates white out conditions so you can barely see the road in front of you. This time however it was sand as opposed to frozen ice crystals. This storm brought with it a little, but much needed precipitation. As we arrived at the CJ1/4/8 building and the site of Victory Cigar Club this sudden storm quickly subsided. It was not until the next day that I learned how unexpected the storm really was as a helicopter was forced down just to the west of Baghdad at a base which killed one soldier and injured eleven **(Stars & Stripes, 21 SEP 09,** *Black Hawk crashes at Balad Air Base, killing 1***, written by David Rising & JASG, 21 SEP 09)**. It was this short windstorm that caused the crash and claimed the life of another of our soldiers. So it is not just the extremists that are hazardous, it is now the weather!

Finally, we continue to move toward the national elections in Iraq. These elections will really set the future course for this country and determine if it will succeed or fail. As one citizen so aptly stated, "Our participation is necessary, it is the future of our country, if I and others do not vote, who will?" **(MNFI, 21 SEP 09:** *al-Rasheed*). Well said, sir.

Day 322: Wednesday-September 23rd, 2009/Day 291– Iraq I would send my fourth footlocker home. This one is stocked full of gifts and goodies. In fact, today would begin a string of about four days that I would send something home. I have three boxes loaded with containers from goodies Shari has sent over and a pile of books of I have read. I continue to drawdown here and only have a few boxes left that will go out within the next few weeks as we are now under a month to go.

Ramadan, which is an Islamic tradition that occurs each year during the ninth month of the Islamic calendar and features refraining from drinking, smoking, and eating from dawn to dusk. This is intended to teach Muslims patience, modesty, and spirituality. Ramadan culminates with an

Islamic holiday known as Eid al-Fitr. Muslims dress up and celebrate in grand style as they end their fasting ritual. This is one of the two most important Islamic celebrations during the year, with the other being the pilgrimage to Mecca **(JASG, 22 SEP 09: September significant dates and http://en.wikipedia.org/wiki/Ramadan).**

These types of celebrations and events also bring out the Islamic extremists who target innocent civilians or those who protect them. One security officer commented on the state of security during Eid al-Fitr:

> Thank God, the situation is stable in cooperation with police, army and our brothers operating the traffic. Today is the first day of Eid. People visit their relatives and visit places. When you see these security measures, you relax mentally.
>
> **-MNFI, 22 SEP 09:** *al-Hurra*

Shi'a leaders are also urging people to show unity as the national elections approach **(MNFI, 22 SEP 09:** *Reuters*). This means that coalitions are being formed and the next few months leading up to the elections are sure to be interesting and most likely filled with acts of violence.

Day 323: Thursday-September 24th, 2009/Day 292-Iraq was one filled with news from Afghanistan. Let the games begin! There were a variety of sources that cited a leaked copy of General McChrystal's strategy for the war. For me, it was not really any revelation and coincides as to what I have told you in regard to tactics required to win a counterinsurgent war. Not that I am an expert by any means. I am just well read on the topic and have observed the effects on the ground over here in Iraq. The bottom-line is to protect the people **(http://cnn.com/2009/POLITICS/09/21/afghanistan.mcc hrystal/index.html?iref=newssearch).** You can only do that with more troops initially, until enough Afghan security forces are trained to take this task over. Actually, much of what appeared in the general's report are techniques, tactics, and procedures that worked well and achieved success in Iraq.

General McChrystal's bottom-line is that he needs more troops or we cannot expect to win (http://www.washingtonpost.com/wp-dyn/content/article/2009/09/20/AR2009092002920.html; http://www.abc.net.au/news/stories/2009/09/21/2692414.html; and Stars & Stripes, 22 SEP 09, *McChrystal: Afghan war lost without troops boost*, written by Bob Woodward).** This puts the President in a bad situation, but this is what he gets paid the big bucks for **(Stars & Stripes, 22 SEP 09, *Obama at a crossroads on Afghanistan*, written by Leo Shane III)**. As I have stated before, most politicians will move as do public opinion polls, and right now public support for the war in Afghanistan is low. This translates into our political will waning and this plays right into the strategy of al-Qaeda. So the solution is very complex, yet quite simple. There are so many variables to this type of warfare that the right mixture of troops, trainers, advisors, and diplomats are necessary to achieve any sustainable and long-lasting success. This is also a multi-year effort. Historically speaking, most insurgencies last for decades, yet our government has never really explained this fact to the American people. So the formula to succeed in Afghanistan is complex and it is expensive. But the decision is quite simple. In my opinion, the President can provide the necessary resources to achieve this success, which includes more US troops initially, or he can pull the plug on the whole operation. The only way to stop the jets carrying flag-draped transfer cases back to the United States is to remove all US troops from Afghanistan. I am not even stating that this may not be the best methodology. But if we do so, al Qaeda will continue to plot and plan against the US and its allies. But to be honest, even if we remove or neutralize al Qaeda in Iraq and Afghanistan, they will simply pop up elsewhere. Again, this is truly a global war on terrorism and is likely to go on for decades yet to come. Will the US become involved in the many other failed states that can harbor al Qaeda extremists so they can train and plot without intervention? Will other countries finally step up their support as they too can be attacked? And the people in Iraq and Afghanistan are reluctant to believe the US will stay the course and know that we will eventually leave.

Then what happens? They have seen this video before and when we leave, the insurgents are still there and often wreak havoc upon those that assisted with the US effort. So why should they remain loyal to our efforts to assist them?

Procrastinating upon a decision on Afghanistan in light of the healthcare debate and our economic woes might be somewhat understandable. But unless the President has not been paying attention since he took over responsibility for this war, he should have some idea as to what he wants to do. And the longer he waits, the more US troops will die. So delaying a decision is not a solid strategy **(http://online.wsj.com/article/SB125350906414427191.html)**.

I give General McChrystal credit for stating what I have said in regard to the President's decision on the matter. Either provide the resources necessary to win the war in Afghanistan, or pull out immediately. There should be no half-ass attempt, or lame political maneuver that will achieve success as this leads us right back down the Vietnam path.

Because there are not enough troops, General McChrystal wants more troops in the more heavily populated areas to protect the people **(Stars & Stripes, 23 SEP 09, *McChrystal wants to focus on populated Afghan areas*, written by Greg Jaffe)**. So we shall see if the President is the consummate professional and leader, or simply another politician maneuvering for re-election who really does not care about the welfare of our troops except by providing a lot of lip service **(Stars & Stripes, 23 SEP 09, *Obama to re-evaluate war effort in Afghanistan*, written by Julian E. Barnes)**. But the time to decide is upon him. I sure he hope he doesn't pay attention to public opinion polls and makes a decision based upon the welfare of our troops and the best interest of our country. This may not be a popular decision, but it needs to be the best decision.

Day 324: Friday-September 25th, 2009/Day 293–Iraq was the day after the debut of MAJ Sean Begley as the 'voice of GRD'. Sean is a vice-principal of a high school in Indiana and also our Deputy G3. He has had the desire to do a briefing to General Odierno for a long time now and yesterday was the day. So SFC Maltes and I assisted in preparing him as much as we could. Sean has been to the SOC before to watch SFC Maltes go through all the motions prior to a briefing and he has heard her and myself on numerous occasions. So when it came time for him to present our information you could tell he was a bit nervous. But he did quite well, and when he returned to GRD HQ he admitted he was a bit nervous but enjoyed the experience. We told him he did so well that he can do our final BUA presentation on October 8th because by October 22nd we should be in Kuwait and getting ready to head home! It also shows that presenting information to a four-star general is not as easy as it looks. Good job, Sean.

The Begley Brothers at GRD HQ
(Sean is the one on the left)

We have also had a few rocket attacks onto VBC recently and some audacious attacks on helicopters in the broad daylight just outside of VBC in a location known as Abu Ghraib. You may remember this place as the site that exhibited the stupidity of some US troops as they humiliated and embarrassed Iraqi detainees back in 2004. General Odierno is not pleased with this recent upsurge in courage by the militants. But the kicker here is that it may not even be the work of Islamic extremists, but rather, organized criminals! It seems as though as Iraqi Security Forces have cracked down on the armed militias and the remaining insurgents, organized crime has taken a stronger foothold. The spokesman for the security forces, whose appearance reminds me a lot of a young Peter Sellers as Inspector Clouseau, stated: "After the success Iraqi Security Forces have achieved in tightening the noose on insurgent groups, we are seeing that some of them are turning to form well-organized criminal gangs" **(MNFI, 24 SEP 09: AP)**. So the beat goes on here, but it really was relatively peaceful during the Eid celebration at the conclusion of Ramadan, which leads me to believe that even though there are still pockets of violence around Iraq, there is hope for a brighter future. Perhaps our leaving for good will aid in this effort as our presence does seem to exacerbate the cycle of violence among some of the extremist groups. If the national elections go as many believe they will and the Iraqi people vote to remove foreign troops from their soil, then the timeline to depart will be accelerated by almost a year from the terms as laid out in the security agreement. That means the rest of the 124,000 plus US troops still here will be right behind me and my fellow 416th TEC compadres on the way home.

Inspector Clouseau?

Day 325: Saturday-September 26th, 2009/Day 294–Iraq was sure to be an interesting one. Our transformation working group meeting would be less structured than usual and there are many people who are very confused about this whole inactivation. There seems to be a real issue in communicating here as well. I have noticed that since I have been here at the HQ, as well as when I was over at the palace. But it is not only at GRD, it seems as though it is throughout the military community. Part of it may be because of operational security (OPSEC) concerns and sending things via email. Part of it may be a 'power' issue whereby people feel empowered by having certain information and do not want to give that power away, so consequently they do not communicate critical information. The rate of change is brisk here to be sure, but good organizations put procedures in place to ensure information flow is effective no matter that rate of change. But in a war zone, with so many people coming and departing over short periods of time, effective communication is very challenging, but certainly not impossible. But every time this communication faux pas occurs I think of Strother Martin's famous line in the 1967 movie **Cool Hand Luke**, "What we have here is a failure to communicate."

"What we have here is a failure to communicate."

There was an interesting poll that recently came out concerning the President and some of his decisions and/or policies **(JASG, 25 SEP 09, *In poll, public wary of Obama on war and health:* New York Times)**. The American people seem torn as to what actions he should take in Afghanistan. In regard to sending more troops to Afghanistan, the poll is split fairly evenly between sending more and decreasing troop levels. In regard to how long the American public is willing to allow troops to remain in Afghanistan, the range was between one and two years. However, interestingly enough, these same people felt that by having troops in Afghanistan the threat of terrorism has decreased, and if we withdraw the threat would increase which drew an overwhelming response. So if the President decides to pay attention to his image and appease the fickle public it would appear that he has a window of about one to two years to send more troops; the American people still feel that Afghanistan is a key in this global war on terrorism so simply pulling out is not a good option; and if he decides to pull troops out he needs another viable option to pursue. But not increasing troop levels and remaining in Afghanistan is probably the worst possible choice because it does little to assist our troops. So today, the decision is still his to make. Hopefully he does not take too long to do so.

Oh by the way, we still have a war going on here in Iraq in case you forgot! But the situation is stabilizing more with each passing day. But even as we continue to withdraw here there are many projects being completed and started. A recent Washington Times article lauded the efforts of US forces and their Iraqi partners in continuing to carry out infrastructure development projects throughout Iraq **(MNFI, 25 SEP 09: Washington Times)**. The article noted that US forces have become a part of the community and forged relationships with community leaders at the local levels. This is classic counterinsurgent strategy and it has clearly worked. This is the same type of strategy that General McChrystal wants implemented in Afghanistan. And to further validate the efficacy of these actions, a correspondent with HDNet World Report commented, "The logistical challenge of redeployment is huge while we are still trying to build infrastructure. Every week, if not daily, there is some sort of progress in Iraq **(MNFI, 25 SEP 09: Washington Times)**. Hopefully people realize that this is in part due to the fact that we have had enough troops in Iraq to complete such tasks, which is not the same luxury that currently exists within Afghanistan. Food for thought.

Day 326: Sunday-September 27th, 2009/Day 295 –Iraq saw more news from Afghanistan **(Stars & Stripes, 25 SEP 09, McChrystal welcomes debate on Afghan plan, editorial).** Apparently the general met with Admiral Mullen and General Petraeus in Germany to discuss his assessment of the Afghanistan situation. While the general may welcome debate on his assessment, keep in mind that he was handpicked for the assignment by the Secretary of Defense and is considered an expert in this type of warfare. Congress moved his promotion to a four star through its process very quickly and so here we are now. He has given his assessment as ordered and now the debate begins. It seems very hypocritical to me that the very people that wanted him in this position are now the same people questioning the strategy. He has also provided his troop request which the Pentagon is sitting on for about 10 days and then it will be another several days before the White House finishes analyzing the request **(Stars & Stripes, 27 SEP 09, *Afghan troop request in pipeline*, written by Julian E. Barnes and Mark Magnier)**. And all the while the debate rages on, mothers and fathers are losing their sons and daughters because they are not getting the support they need from Washington – again! If I seem bitter on this topic, good call on your part. We ask so much of our brave men and women in uniform, but we can't get people at the highest levels of our government to make sound, logical, non-political decisions in a timely manner. Keep all of this in mind when your son or daughter talks to you about joining the military. It is without doubt the greatest military the world has ever witnessed, but I can't state with any certainty that it is led by wise people. Do not confuse intelligence with wisdom! And that indeed, is a sad statement to make.

Day 327: Monday-September 28th, 2009/Day 296 –Iraq got off to an excellent start. Today I would again be taking the Army physical fitness test. The plan today would be to take an exit stage left at the appropriate time as opposed to continuing to the right as I had about 17 days ago! LTC Carlos Rodriguez came to assist me in this quest and would be at the critical corner along the running route to ensure I would take the correct path.

So I once again maxed out on my pushups and sit-ups and had 200 points heading into the running event. I also knew I needed to get on my pace early and then stay with it over the course of the two-mile loop. In order to accomplish this task, I also knew my left foot needed to hit the pavement

80 times per minute. This is an 8:34 mile pace and would put me at 17:08 for the two miles and 80 points. The key today would be to make that left and not keeping trucking to the right! So off I went and I got into my rhythm early and stayed with it throughout my journey. I ended with a 17:02 (6 seconds ahead of schedule) and garnered 81 points to finish with a 281! Not bad for an old guy.

The Packers also beat the St. Louie Rams yesterday; Brett Favre had a game winning touchdown pass to beat the 49'ers yesterday; Bucky Badger beat Michigan State on Saturday; and the Lincoln High School football team beat an old rival on Friday completing the weekend trifecta. So all-in-all, it was a good weekend and a good start to a Monday in Iraq.

There was some interesting news yesterday in relation to the current state of al Qaeda and other terrorist networks around the globe. Some terrorist experts believe that al Qaeda is actually in decline as some of their key leadership is being methodically picked off and their tactics of killing innocent civilians has discredited them across the Muslim world **(JASG, 27 SEP 09: *New York Times*)**. While there are still going to be acts of terror across the globe and cells popping up even in the US, their momentum has been severely curtailed and the pressure will no doubt stay on them. So perhaps using covert operations and taking out key leadership is not a bad option to sending more troops. The President must juggle the information he is being advised by his military commanders with what his national security advisors are telling him. So there may be more than one way to 'skin a terrorist', but non-action or delaying too long in making a decision should not be in his option package.

Left turn Clyde

Day 328: Tuesday-September 29th, 2009/Day 297 –Iraq was the day after a very long day. Of course, it all started with the physical fitness test early and then was followed by four meetings, three by the same group as they continue to try and figure out the way ahead. I should have told you that this whole process has had the proverbial monkey wrench thrown into it because those people and groups that are left behind after the GRD HQ inactivates in late October may have to move out of the building they just moved into! Last week I may have informed you that I attended the MNF-I Chief of Staff meeting for Colonel Goetz and Brigadier General Anderson, who is the MNF-I Chief of Staff, informed the group that GRD would be moving from our current location and that the building was going to be turned back over to the Iraqis. I found this interesting news that I had not heard previously, but then again, as I may also have already mentioned, effective communication is not a strongpoint here so I did not think much about it. However, when I informed Colonel Goetz of the news he was clearly surprised and immediately informed everyone within GRD HQ. Well, this sent all of the planners into a frenzy and now they are exploring options to not only right-size the organization but also to possibly move again. Now General Anderson did not provide any detail with his unexpected announcement but more planning is necessary.

So as we look to the east again, it appears that General McChrystal has submitted a request for additional troops **(Stars & Stripes, 28 SEP 09, *Gates urges patience on Afghanistan*, written by Kevin Baron)**. Now the posturing begins as politicians begin to more closely examine public opinion polls and their re-election platforms; military leaders examine how this will play out in Washington; and the President determines the efficacy of sending more troops to Afghanistan versus adjusting the overall strategy. While it is more important to get this thing right, it just amazes me that after eight years of war we still have no solid strategy for Afghanistan! Yet we expect Iraq to become a flourishing democracy in a few short years and we cannot even develop a coherent strategy in eight friggin' years! Is it just me or do you see something not quite right about this whole thing as well?

▌▌ ▌▌ **Day 329: Wednesday-September 30th, 2009/Day 298 –Iraq** saw my hometown, Wisconsin Rapids, Wisconsin mentioned in the ***Stars and Stripes*** news from the states section **(Stars & Stripes, 28 SEP 09, American Roundup)**. There was a picture of three young ladies and one gentleman, two from the Netherlands, one from Australia, and one from Austria who were visiting Wisconsin Rapids to learn more about the growing of cranberries. In fact, it is a little known fact that Wisconsin Rapids, located in the heart of central Wisconsin, is one of the largest cranberry producers in the world! Yes, you heard it here first.

We have talked a lot of Afghanistan recently but there is still news of progress here in Iraq. The current Prime Minister is in campaign mode as the national elections are only a few months away now. He stated that:

The coming election is the fundamental solution to all problems, and the way to bring about the change we all desire, to harness Iraq's resources to achieve Iraq's potential and the prosperity the Iraqi people deserve.
-MNFI, 29 SEP 09: *DPA*

Last week in Iraq there were approximately 600 weddings. A Ministry of State spokesman stated that "Iraqis are taking advantage of the holiday and the good security situation." He noted that the high marriage rates are an indication that Iraqi people feel safe **(MNFI, 29 SEP 09: *Reuters*)**.

So progress continues and General Odierno has flown back to Washington to inform Congress of such occurrences throughout Iraq. This may also be a time when he lays out what has worked here and what has not so it will assist Congressional and national security leaders to make some decisions concerning the future course in Afghanistan.

There is simply a lot of talk in Washington right now concerning Afghanistan as the political monster begins to rear its ugly head. An article on CNN.com highlighted that General McChrystal, an expert in this type of warfare, says that in order to win the hearts and minds of the Afghan people, which is ultra-important in winning this type of war, he needs more troops in order to protect them and root out the insurgents **(http://www.cnn.com/2009/POLITICS/09/28/afghanistan .obama/index.html)**. Admiral Mullen, the Chairman of the Joint Chiefs of Staff and General Petraeus, CENTCOM Commander and the architect of 'Surge I' which turned the situation around in Iraq, both concur with General McChrystal's assessment on the Afghanistan strategy he has submitted to the President.

John Kerry, chairman of the Senate Foreign Relations Committee, warns that the President should not rush into sending more troops until he decides upon the correct strategy. Former Secretary of Defense William Cohen also suggests the President must ask the difficult questions such as, why are we in Afghanistan? Why does it matter to the United States? What is our mission there? What are the costs associated with staying involved there? How many allies will stay the course with us? The President must answer these questions and then address the American public.

Current Secretary of Defense Robert Gates believes the President's deliberate approach to this issue is warranted. Gates stated that the situation in Afghanistan is more serious than was first realized (http://politicalticker.blogs.cnn.com/2009/09/27/gates-new-troops-to-afghanistan-wouldnt-flow-til-early-2010/). However, Gates went on to state that the Taliban and al Qaeda have defeated one superpower already (Soviet Union) and if we pull out it will be seen as a victory for them and will aid in their ability to recruit more terrorists and receive increased financing for their efforts, which would be a grave strategic error on our part. So it sounds as if Secretary Gates is trying to be politically correct with his comments, but is stating that we need to stay engaged in Afghanistan for awhile.

Then you have Vice-President Biden, who has a son serving in Iraq, pushing for less US presence in Afghanistan and targeting al Qaeda cells and leaders using special operations forces and unmanned aerial aircraft armed with missiles. Hubert Humphrey attempted to get Lyndon Johnson to pull out of Vietnam because he had not made a strong case to the American people as to why it was important to our national interest to be there in the first place (http://www.cnn.com/2009/POLITICS/09/28/zelizer.biden.afghanistan/index.html). We all know how this turned out, and because of this recommendation Humphrey fell from Johnson's grace and his inner circle of influence.

A recent Fox News Poll is also revealing in that while it is clear that the majority of Democrats oppose sending more troops and are against the war, Republicans go just the opposite way on both issues. However, when you analyze the poll more deeply, both Democrats and Republicans agree that because we are fighting terrorism in Afghanistan they feel safer as a result http://www.foxnews.com/story/0,2933,553203,00.html.

So what is it going to be? We have had eight years to figure this out and we still have no coherent strategy in Afghanistan while our brave troops continue to die at a record pace. While I do believe the President is taking the correct approach in reaching a conclusion, it seems as if our involvement there has been dismissed as a distraction rather than a priority. Now President Obama inherited the war from President Bush, but he has now been in office for nine months and still there is no strategy for an eight-year-old war where Americans are dying. Certainly healthcare and our economy are important issues to be sure, but bailing out auto industries and banks before attacking the Afghan war situation is questionable at best, even though there is an obvious linkage to the economic issue with cars and financial institutions. My point here is that there are many differing and diverse opinions in Washington. Getting the strategy right is incredibly important because that will in turn dictate the resources necessary to implement such strategy. Whether it is more troops on the ground or the building of more Predators, the strategy must first be sound. But again I must state that time is not on the side of our young men and women fighting so valiantly in Afghanistan right now as this political posturing takes place in Washington. We have the message and direction from our military leaders; we have implied intent from the Secretary of Defense; we have the opinion of the Vice-President; now we await the decision from the President. The problem is that not everyone has weighed in yet and this topic is sure to create a firestorm of debate in Washington. And I have outlined to you on many occasions how quickly our bureaucratic system works under such circumstances. So mothers and fathers, spouses and children, friends and acquaintances of those serving in harm's way, keep praying that wisdom somehow finds its way to Washington, and very soon, because the clock continues to move...tick...tick...tick!

Tick, tick, tick...

(-30) Day 330: Thursday-October 1st, 2009/Day 299–Iraq began our last month in Iraq. Our workload would begin to wind down, as within a few weeks we would be beginning our transitional training before heading home.

Last evening we watched General Odierno provide his Iraq assessment to the House Armed Services Committee **(Pentagon Channel, 1700 GMT+3).** It was like watching a much longer BUA and General Odierno told this congressional committee the same thing he has informed us about in relation to the status of Iraq. There was really nothing new for us to digest, but he did outline all of the progress that has occurred over here for those that are not so informed. Some of the questions asked were more difficult to answer than others, but I felt General O did a very commendable job providing answers to them. Kurdish-Arab relations, Iranian influence, political progress, evolvement of the Iraqi Security Forces, and even Task Force SAFE and the electrocutions that have occurred were all discussed. And while the congresswoman that pressed General O for an answer concerning a recent electrocution, she should be more focused upon the contractors that were hired by the United States government to perform such work. If the truth be known, the US military, through the creation of Task Force SAFE, is rectifying an existing problem that was created by contractors that are paid millions by the US government. These organizations are whom she should be addressing these difficult questions to. If she is unaware of this little factoid, then she has no business even being in Congress to be honest with you, or she was simply grandstanding and looking for votes from others that may not be informed of the real facts.

In fact, the situation here continues to stabilize to a point that General O may speed up the drawdown timetable slightly **(JASG, 30 SEP 09, *General says Iraq troop reductions may quicken*, New York Times and Stars & Stripes, 1 OCT 09, *Odierno: 4,000 more GIs leaving Iraq*, written by Leo Shane III).** If the national elections in January go smoothly, then the timetable to reduce US forces in Iraq may move more quickly. Again, because of the strain on military forces, any buildup in Afghanistan must be leveraged against the drawdown in Iraq. So even though General McChrystal wants 20,000 to 40,000 more troops, which equates to about 4-8 brigades or what are now evolving into Advise and Assist Brigades (AABs), there are really few places to get this number of troops except from Iraq. Considering most US combat troops in Iraq are in stand-by mode right now anyway, there probably is a push to move them out of Iraq and into the fight in Afghanistan. This may not occur until after January, and this would be consistent with what Secretary Gates has stated. So the delicate balancing act continues.

The President still wants to ensure he gets the right strategy before sending any more troops into harm's way **(Stars & Stripes, 30 SEP 09, *Strategy first*, written by Jennifer Loven).** I do not blame him for doing so, but it also sounds like a delaying action until we can determine what will occur within Iraq post-election period, and then we may begin to divert more troops toward Afghanistan. Nothing else in this whole quagmire of decision-making, or lack thereof, makes much sense. The only other explanation is that we have a lot of people in Washington who are completely clueless. Although this is not beyond the realm of possibility, it is highly unlikely. I believe there is a stalling tactic occurring disguised as a strategy review and a lot of political maneuvering. So what I think will ultimately occur is that the President will provide more troops to General McChrystal after the Iraqi elections in January. He will provide such numbers for a period of 18-24 months and use the newly developed metrics to analyze and benchmark progress. If there is no progress within this timeframe, then he will probably begin to drawdown troops in Afghanistan, but continue to use Special Forces and unmanned aerial drones to target al Qaeda and Taliban leadership. This will then become more of a containment action at that point than a war victory. This will also tell the insurgents that we are not going away and we will continue to make their day!

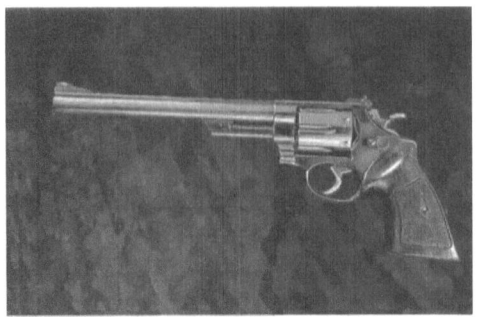

Go ahead, make my day

Day 331: Friday-October 2nd, 2009/Day 300-Iraq began our 300th day in Iraq. As I begin my last month here I will try and clean a few things up for you, although I will try and tie up all the 'hanging chads' in the final chapter.

To begin with, we had six rockets hit VBC last night killing one soldier and injuring three more. The rockets came in at about 2003 and I was returning to my barracks at what we now call Gulf Region District, formerly known as Gulf Region Central, and on October 22nd will re-dedicated as Camp Wolfe after Commander Duane Wolfe who was tragically killed in action on Memorial Day.

I was probably one to two miles from where the rockets hit, but that is still too close for comfort. The Phalanx system that we have around VBC is supposed to be able to shoot down such rockets. This weapons system is primarily used on naval vessels and is a large Gatling gun of sorts. Unfortunately, this system only intercepted one of the six rockets, while one also fell short of VBC. If the early warning tracking system sounded and alerted soldiers of the incoming attack, I did not hear it, but again I was in the car and the radio was on. Ironically, there was a test of this Phalanx system last night and many of our guys upon returning from chow were watching the one positioned near GRD HQ at the time of the attack. According to reports, the gun lit up the three-quarter moon night sky but they never thought we were under attack. Hopefully Shari did not hear anything of this attack, or my mother who watches the news religiously!

If you recall the fiasco we had getting our green footlockers over here last November, we were now getting ready to send them back to Ft. McCoy. Their journey back is much later than originally scheduled so I am glad I sent mine back by mail. But our First Sergeant was looking for volunteers to assist in laying out the footlockers for the customs inspector. Now I am all for helping out when possible to do so, but the time shifted from 1630 to 0430 in a matter of a few minutes. I thought to myself that 0430 is awfully early to trek over there to help, especially after they just requested assistance the night before. This told me that this whole thing was thrown together quickly, and what if the inspector did not show on time? So I decided I would not go in at 0430 but rather my regular time. If they needed assistance later then I would be happy to help. As it turned out, I made the right call on this one. This is not always the case, if you recall my right hand turn on the two-mile run! But not only did the customs inspector not show at 0530 when he was supposed to, he did not show until 1030! But I did assist in loading the footlockers back up into the metal CONEX container and there were many less footlockers going back than came over so it was an easy assignment.

Carlos and I also recently visited a place I never knew about here on VBC. It is called the Flintstones village and is another of Saddam's extravagant creations he built at the expense of his people. As you know, Saddam was a greedy tyrant who built several palaces around Iraq. Right across from the Flintstones village is the 'Victory Over America' palace he began to construct after the first Gulf War, even though he lost miserably. This palace has never been finished. But Saddam was a huge *Flintstones* fan and wanted a village built for his grandchildren modeled after the cartoon version. Now it is difficult to separate fact from fiction on some of this, but rumor has it that Saddam built this village for his grandkids as a distraction because he had the fathers of some of these children, his son-in-laws, executed for speaking out against his regime. Great father-in-law! I could not find any pictures of this village when it was initially built, as it is falling apart now after years of war and neglect. It has also become a bit of a graffiti park for coalition forces and others. But when you think of the Flintstones and the innocence associated with this cartoon that first hit the television screens back in the

1960s, it is hard to think of it as a place as evil as Saddam and his two psycho sons made it into. Yes, I was alive to watch the *Flintstones* when it was a primetime cartoon! Rumor also has it that Saddam's two sons, Uday and Qusai, defiled this ground where children used to play by raping and murdering women. So whether you believe this war was just and we should have become involved here or not, the world and the people of Iraq are better off without these three lunatics and criminals running loose to terrorize the innocent.

The infamous Flintstone village on VBC

Day 332: Saturday-October 3rd, 2009/Day 301–Iraq was hopefully my final haircut in Iraq. I stretched this one as long as I could and if I hit my three-week mark, this should put me at Ft. McCoy for my last one during this tour. I may even be able to stretch it all the way home and back to my regular barber Todd Twait back in good old Wisconsin Rapids. But this barber was the best one yet and he did an excellent job. He even trimmed my moustache and eyebrows, which no other barber has done over here. That warranted a nice tip for the young man and a nice way to end my barbering experience here in Iraq.

Iraq continues to evolve. Yesterday I had the privilege of watching three four star generals providing information to the American and British media concerning their respective areas of responsibility. First was General Odierno who was giving a press conference at the Pentagon to several reporters **(Pentagon briefing, October 1, 2009)**. Everything General O stated to the press corps we have heard on several occasions during the BUA over here. General O graduated from West Point in 1976 and was considered an average student. He was recruited there more as an athlete than an academic, but he has risen to top of his profession. He may not be the polished speaker that General Petraeus is, but he gets his point across just fine. A reporter asked him if the situation in Iraq is still 'fragile and reversible,' a phrase he and General Petraeus have used often. His response was that with each passing day Iraq becomes stronger and less fragile. An emerging trend in Iraq is the forming of political alliances to gain support from a broad spectrum of religions and ethnicities. One religious leader, Ammar Al-Hakim of the Islamic Supreme Council, said "Iraq is preparing for the next phase. This is democracy" **(MNFI, 2 OCT 09: *NY Times*)**. This is a very good sign indeed and another sign of lasting progress.

General McChrystal spoke to a group of reporters and scholars at the International Institute of Strategic Studies in London **(Pentagon Channel, IISS presentation, 1 OCT 09)**. General McChrystal was very politically correct with his responses and did not tip his hand on the recent discussions with President Obama on the situation in Afghanistan and his assessment of the situation there. General McChrystal was installed as the commander to provide his best assessment as to what it will take to achieve sustainable success in Afghanistan. He believes more troops is the avenue to go and is not a fan of scaling down military intervention as is Vice-President Biden **(JASG, 2 OCT 09, *McChrystal rejects scaling down Afghan military aims*: NY Times)**. General McChrystal was also a 1976 graduate of West Point and was not the party animal that General O was, but here he is with his fourth star, thus making the Class of '76 one of note.

Finally, I watched General Petraeus in an interview with Brian Williams on NBC who provided information on Iraq, Afghanistan, and Iran **(http://www.msnbc.msn.com/id/33127232)**. General Petraeus is the CENTCOM Commander overseeing all military operations in a very strategic value area, including both Iraq and Afghanistan. While he is pleased with the progress in Iraq, he is not ready to declare victory. His comments and assessment on Iraq were very measured and calculated, but he is pleased with the direction Iraq is moving. In regard to Afghanistan, he is less pleased with the current status of the situation, and he always includes Pakistan with any Afghanistan solution. He does not believe you can separate the two and simply focus upon Afghanistan. He also was very cautious with his comments concerning the recent meetings with the President and what the strategy will be in Afghanistan moving forward. Then, he talked of Iran and their quest for nuclear power and the influence they have within this region. While his command may be the smallest of the six combatant command regions around the globe, it is by far a hotbed of activity. General Petraeus, West Point class of '74, is an academic who finished in the top 5% of his class, owns a Ph.D., and is certainly the best guy for this job.

So this four-star media blitz on the 1st of October was very noteworthy. This may be setting the stage for any announcements coming from the White House. The Generals in place at CENTCOM in Iraq and in Afghanistan are in my humble opinion the best people for that area of responsibility. So we have the right leaders in the right place and this should not be underestimated or underappreciated. Now we shall see if this matches up on the political side of the ledger as we move forward with the very important decision concerning Afghanistan that will be forthcoming.

Day 333: Sunday-October 4th, 2009/Day 302–Iraq left only three more Sundays in Iraq. This week COL Goetz, LTC Rodriguez, and SFC Maltes would be leaving on the advanced party and heading to Ft. McCoy. I would only have a few more DFACs to rate on the second round and only a few more items to ship home. So time is winding down and my tasks at the Gulf Region Division are also getting much fewer. But the insurgents are letting us know that they still care as they lobbed in a few more rockets the other night, making it back-to-back nights that we had rocket attacks for the first time since I have been here. This time however there was no one injured and no damage done. But this is a disturbing trend we will need to keep an eye on.

Bucky Badger beat the Minnesota Golden Gophers yesterday and retained the Paul Bunyan axe. In fact, it was the sixth straight win for the Badgers over the Gophers. Way to go, Bucky!

Way to go, Bucky!

So what news of Iraq as I drawdown my experience here? Well, let us start with the bigger picture and work our way into more detail to support such statements. Iraqi Vice-President Adel Abd-Al-Mahdi stated this about the future of Iraq:

> The main goal of the next parliamentary elections is to build a prime government that would provide services, improve the standard of living for Iraqi people, to complete the withdrawal of foreign troops, end internal violence and build national unity.
>
> **-MNFI, 3 OCT 09:** *al-Iraq News*

Then we have the tense northern situation and disputed areas with al Qaeda in cities such as Mosul. One citizen there recently commented, "I haven't heard an explosion for a while. God willing, it's getting better" (MNFI, 3 OCT 09: Reuters). I have also noticed less violent activities in this area recently as well to support such claims.

Finally, in Iraqi Kurdistan there is a concerted effort to educate women on their rights **(MNFI, 3 OCT 09:** *PUKmedia***)**. In Iraq, this will be a major deal as women have been repressed here for centuries. But now women are being recruited for the armed forces, for the police, for governmental positions, and are making gains every day. There is always the infamous step backwards on occasion, but the gains women are making here is another sign of positive progress for Iraq and provides hope for sustained stability and a brighter future.

OK, just because this situation is getting so intriguing to watch, we must again talk of Afghanistan. The lines are being drawn in Washington as to which way this whole thing may go **(Stars & Stripes, 2 OCT 09, *Obama advisers split over Afghan strategy*, written by Philip Elliot).** It appears the military is leaning toward sending more troops to implement the strategy outlined by General McChrystal's assessment. This group includes key personnel such as the Chairman of the Joint Chiefs of Staff Admiral Mike Mullen and CENTCOM Commander General David Petraeus, although neither is saying anything publicly as of yet. White House staffers, who have little experience with activities on the ground, are leaning toward the more politically correct answer of less troops and more technology. Obviously not students of fourth generation warfare! The Vice-President is leaning in this direction as well. So this leaves Secretary of Defense Robert Gates, who is now leaning toward sending more troops **(http://www.cnn.com/2009/POLITICS/10/02/afghanistan.strategy/index.html)**. This also leaves Secretary of State Hillary Clinton, a Democrat, as the majority of Dems are lining up against such an increase. Secretary Clinton is staying very quiet on the topic thus far but is completing her own assessment of the situation. So this leaves President Obama. It will be his decision to make and may be a defining moment in his Presidency. This story is likely to get more interesting with each passing day and this coming week will see more discussions leading up to the ultimate decision. The President did meet with General McChrystal in Denmark to discuss the Afghan strategy **(Stars & Stripes, 3 OCT 09, *Obama, McChrystal discuss war strategy*, written by Julie Pace).** This discussion came after the President's pitch to the International Olympic Committee to select Chicago for the 2016 Summer Olympics. By the way, Rio was selected as the site for the 2016 summer games, the first ever in South America. Good for them. Let us hope this was not an omen of things yet to come in relation to the strategy for Afghanistan! What say you, Mr. President?

Day 334: Monday-October 5th, 2009/Day 303–Iraq I have started to read my last book here in Iraq. This one is a bit different than many of the others. It is called **Once A Marine** by Nick Popaditch. It is an account by a Marine who was seriously injured in Iraq in 2004 and follows his fight with the bureaucratic system to be take care of him after he gave this country so much in return. It is his personal story of sacrifice and struggle that many people would never otherwise have heard. I am only about a third of the way through the book, but at times it makes you angry and at other times it makes you cry when you read how our country has so carelessly not adequately taken care of the few that have sacrificed so much. I will provide more excerpts from this book in the days ahead, but it is a book every American should read and it is written in *Marinese*. This means it is very blunt, to the point, with a lot of cussing to emphasize key points. Many of the fourteen books I have read here have been written at a tactical level, which has a limited audience. I have enjoyed them, but I am not certain the majority of Americans would. This particular book is similar to what I hope to accomplish with mine, and that is the ability to reach all Americans no matter their strata in society. **Once A Marine** is written more at a tactical and guttural level, while my book runs the gamut from the strategic to the operational to the tactical level.

There are those days that really stick with you among those many that do not. Yesterday, Carlos Rodriguez and Mike Ryan went with me to the Paul R. Smith memorial, which is right around the bend from where GRD HQ is located. SFC Smith was a fellow engineer who was killed in action in April 2003, which was very early in this war. SFC Smith and his unit were assigned to block a highway between Baghdad and the airport. As they were doing so, they were attacked by about 100 insurgents. Smith and his men were assigned to guard several prisoners and quickly went about creating an impromptu POW holding area. An intense firefight erupted, and several Armored Personnel Carriers (APC) and a Bradley fighting vehicle from the 3rd Infantry Division were summoned to assist in the battle. One of the APCs (M113) was disabled by an enemy mortar. An area in this battle in the courtyard quickly turned into an expedient medical aid station. Because the enemy was surrounding the courtyard, Smith chose to fight to protect the wounded rather than evacuate them. So he climbed aboard the disabled APC and used the M50 machine gun mounted on the turret to repel the enemy advances. He immediately came under intense crossfire from a nearby tower and the trenches surrounding the courtyard. SFC Smith went through three boxes of ammunition fending off attack and saving the hundreds of wounded in the courtyard that day. SFC Smith bravely stood in the gap as had King Leonidas of Sparta centuries earlier to protect his countrymen. Tragically, SFC Smith took about 14 shots to his Kevlar body armor before a shot to the neck killed him and finally silenced the M50 that had assisted him that day. For this selfless act of bravery SFC Smith was awarded the first Medal of Honor of this war **(http://en.wikipedia.org/wiki/Paul_Ray_Smith)**. I was awed standing in this location as Carlos, Mike, and I relived this fateful day and were honored to stand on the same patch of turf that SFC Smith so valiantly protected over six and half years ago. Rest in peace, SFC Smith.

In honor of Medal of Honor recipient SFC Paul Ray Smith and the family he left behind

▌▐ ▌▐ **Day 335: Tuesday-October 6th, 2009/Day 304– Iraq** we received bad news from Afghanistan. To begin with, eight US troops were killed in a fierce firefight in northeast Afghanistan. This was the heaviest loss of life in this war in over 14 months **(Stars & Stripes, 5 OCT 09,** *Taliban storm US outposts, kill 8 troops***, written by Robert H. Reid and Rahim Faiez and http://english.aljazeera.net/asia/2009/10/200910455437690861.html and http://www.timesonline.co.uk/tol/news/world/Afghanistan/article6860616.ece)**. The situation there is unlikely to get better in the short-term.

Then, news came that SGT Ryan Adams from Rhinelander, Wisconsin was killed in action on 2 OCT 09. SGT Adams was from the 951st Sapper Company from the Rhinelander/Tomahawk area. This unit was called Charlie Company of the 724th Engineer Battalion and where I began my career as a commissioned officer back in 1989. While I did not know SGT Adams personally, I still know many of the guys from that unit that are either retired or still there, so this hits very close to home. SGT Adams is the 96th Wisconsinite to die in action (85 in Iraq and 11 in Afghanistan). My hometown of Wisconsin Rapids lost Corporal Matthew Grimm back in January 2007 in Mosul, Iraq. So many communities in Wisconsin have felt the sting of war and understand the price of freedom.

So we must again return to the Afghanistan debate as this decision begins to take shape. It appears that the President has a few options he can exercise **(www.foxnews.com/politics/2009/10/04/obama-faces-range-of-options-deciding-afghan-strategy/)**. Even though an option, simply just pulling out of the country as many Americans might prefer is not solid strategy. Scaling back the effort with less troops and more technology, while accelerating the training of Afghan Security Forces, is a viable option that will be debated vigorously over the next few weeks in Washington. Standing pat is not a solid option, and if we do, we are likely to see more incidents where we lose troops in large groups. Another, and perhaps the final option, is to ramp up and provide more troops as the top military leaders are requesting, while continuing to train the Afghan Security Forces and use Predators and Special Operations units to hunt down key al Qaeda and Taliban leaders. This amounts to the full court press. And while serious debate will abound from Washington, remember that time is not on the side of our troops by doing nothing as we have painfully seen over the last few days. That clock is still ticking.

Oh yes, the Packers visited Brett Favre and the Vikings on Monday night football (Tuesday morning in Baghdad). Brett got his payback this game, and quite frankly, I feel both sides (Packers and Brett) handled the situation last year very poorly. It was like a very messy divorce. So while I am happy for Brett and hope he continues to play as long as he is physically able to do so, I am sorry the Pack lost. But in less than a month, and shortly after my arrival home from this ordeal, the Vikes will come a calling to the not so frozen tundra of Lambeau for the rematch. There may be time to squeeze that entry into this book before I conclude it!

Good job Brett

Day 336: Wednesday-October 7th, 2009/Day 305–Iraq brought more discouraging news concerning the 19 electrocutions that have occurred here in Iraq since the war began **(Stars & Stripes, 4 OCT 09, *After losses, few answers*, written by Lisa M. Novak)**. It is shaping up that no one will be held accountable for these tragic deaths. While I have not spoken of this for a while, it is still an issue that continues and recently claimed the life of a young contractor. One of the groups responsible for ensuring the safety of our troops through the competent electrical wiring of structures is KBR. There have been two lawsuits filed thus far and an investigation by the Department of Defense Inspector General. But because of the time that has lapsed since some of these deaths, bases that once held such facilities are now closed or completely gone, thus leaving little to investigate. Now I do not believe there was any malicious intent by anyone related to this issue, but the fact remains that 19 people are dead due to negligence and incompetence yet no one will be charged or held responsible. Even the military, that should oversee any safety processes on their bases, are not blameless here. And our government, which enters into contracts with many such private organizations, must do a much better job of tightening up the language in such contracts and stop rewarding these organizations with follow-on contracts when they clearly have done such a poor job. So it is simply difficult to fathom the concept that no one or any group is responsible. Please! This whole thing just reeks as each side blames the other and everyone is using the Teflon defense hoping nothing will stick to them, which is apparently working well to this point. To me, there is plenty of blame to go around and this whole tragic situation is simply a microcosm of a much larger issue related to our involvement in both the Iraq and Afghanistan wars. Far too much waste and far too little oversight!

As I have highlighted the trials and tribulations of the Army's organizational structure and its inherent flaws throughout this book, my comments continue to get validated by other stories. For example, the ACUs (Army Combat Uniform) we wear I am not particularly fond of, nor are many of my compatriots **(Army Times, September 21, 2009, *New camo takes the field*, written by Matthew Cox and September 28, 2009, *Pattern of failure*, written by Matthew Cox).** There is far too much Velcro for any type of noise discipline, and the Velcro on the uniform does not stand up well after a few launderings. While it is a comfortable uniform to wear, it is not all that well-designed in my mind. However, I had thought that this particular digitized pattern and the uniform had been through extensive research in order to attain this one-uniform-for-all-environments. Apparently this was not the case and now the Army is analyzing the current uniform and its pattern to see if they can develop something better. In other words, they are now doing the legwork they should have done earlier! So it will costs millions more to conduct such a study and they will look to use the current uniform framework so they haven't wasted billions in their current investment. The US Marine Corps in my estimation has a much better uniform and they use two for different environments. The Army is now reaching the conclusion that perhaps they cannot create a one-uniform-for-all-environments! Again, these types of things do occur, but you would think with such an investment of taxpayer's money that there would be more oversight and accountability. Hum, thought we just went through this! The point is that a pattern is developing that I have been chronicling in that the Army organizational structure is broke and is in great need of an overhaul. But it is not just the Army, it is also our government that is slow to respond and is supposed to be providing such oversight. Clearly this is not occurring. And because it is not, we have billions of dollars wasted and 19 people electrocuted in just these two situations. Is anyone going to wake up and figure this out anytime soon?

Day 337: Thursday-October 8th, 2009/Day 306–Iraq brought more interesting news from the Afghanistan debate. The Secretaries of Defense and State, Robert Gates and Hillary Clinton respectively, recently stated that the US will be in Afghanistan for the long haul **(http://www.cnn.com/2009/POLITICS/10/05/clinton.gates and Stars & Stripes, 7 OCT 09,** *Gates says US not abandoning Afghanistan,* **written by Kevin Baron)**. So it would appear that simply withdrawing troops from Afghanistan is not one of the options the President can select from. But the debate rages on and makes great fodder for the press. There is supposedly division within the White House as to what strategy to pursue. James Jones, national security advisor to the President and a former Marine Corps general, is not convinced more troops are the way to go. While General Jones has had an illustrious military career, it is unclear how much he actually knows of counterinsurgent warfare. He could be one of those conventional-style warfare thinkers that cannot wrap his mind around fourth generation warfare, which we are engaged in currently. Senator John McCain has questioned General Jones' assertions and statements further adding fuel to the fire **(www.foxnews.com/politics/2009/10/05/mccain/)**.

The White House is also supposedly angry at General McChrystal over his public comments about the war **(www.telegraph.co.uk/news/worldnews/northamerica/usa /barackobama/6259582/)**. Well, was General McChrystal placed in his current position to provide an assessment of the situation in Afghanistan as a military professional, or is he to be a politician? Are we not paying him as a military commander? He has also forwarded his assessment to the White House long ago that has yet to be acted upon and sees firsthand what is occurring on the ground as a result of the lack of resources. And for an administration that keeps telling us it wants to be transparent, there sure seems to be a lot of contradictory evidence to this assertion. If we want our generals to be politicians then we should revert back to our heads of state being both, such as during the days of Napoleon and Alexander the Great. There is also some rumor that General Petraeus is being very politically correct and quiet on the topic, which some insiders see as posturing for future political aspirations **(JASG, 6 OCT 09, *Clear voice of Bush's Pentagon becomes harder to hear*: New York Times)**.

So clarity at this point there is not. It is likely the debate and strategic discussions will continue in Washington for some time. And so our under-resourced troops will simply just have to wait until such time that the President is ready to make a decision. Tick...tick...tick!

Day 338: Friday-October 9th, 2009/Day 307–Iraq I will refrain from talking of Iraq or Afghanistan. But yesterday COL Goetz, LTC Rodriguez, and SFC Maltes left for home. COL Goetz is heading back to Washington, DC where he works. LTC Rodriguez and SFC Maltes are heading to Ft. McCoy as our advanced party to ensure all is squared away and ready for arrival in a few weeks. It is an exciting time to be sure.

COL Goetz was my rater on my Officer Evaluation Report (OER) and Major General Eyre is my senior rater. I received a sterling performance evaluation from both as I signed the report yesterday. This was the third goal of my three coming on this mission if you might recall. Hopefully by writing this I do not jinx myself, but goal number one was to leave here alive. Goal #2 was to leave here with everything I came with, including limbs, eyesight, etc. And goal #3 was to do the best job possible whatever I was assigned to do. My OER validates goal #3 and so far so good on the first two!

Because I am the JOAT (Jack-Of-All-Trades) and our drive to inactivation is basically complete, I am now assisting Carlos Rodriguez with the redeployment efforts. While there many 'hanging chads' out there to tighten up, everyone is working hard to get us home. Once again, the problem is lack of communication as the left hand is not certain what the right hand is doing. So all I had to do was be the information conduit, open the lines of communication, and find that everything is almost completed for our departure.

Yesterday marked the final BUA that our happy little group would be doing. In two weeks someone other than myself, SFC Maltes, and MAJ Sean Begley would have to brief General Odierno. This was Sean's second briefing and once again he discovered that this task is not as easy as it may appear. He was a bit nervous again and stumbled slightly out of the starting gate, but quickly recovered and did a fine job the rest of the update. But another benchmark in our redeployment activities has been achieved as we continue to get ready to leave Iraq.

I also had to spring into action as a paramedic yesterday. After my return from my cardio workout and lunch Sean informed me there was a young lady down the hall that had simply passed out. I thought she had already been evacuated to the hospital, but he said she had not and thought I might want to take a quick look at her. So I did. She was lying down on a couch when I arrived, was very lightheaded, and not feeling well. She was being attended to by some first responders that weren't really helping her all that much. So I took her history for the medics that had been summoned to transport her to the hospital. They had not even taken any vitals on her when I arrived. While we ruled out any type of head injury, I also quickly ruled out the possibility of a stroke or any cardiac problems. Clearly however, she would become more lightheaded as she tried to sit up and then also became nauseated. Then the medics showed up and I was not impressed with their patient care skills at all. I suggested they start an IV on her as she may be slightly dehydrated and while one of the young Specialists got the right equipment out, they did not start one before they left! They did start some high flow oxygen on her with a non-rebreather mask, but they forgot to inflate the reservoir bag before placing the device over her mouth and nose! I also used to be a medic in the military long ago and these guys were very young and obviously inexperienced in dealing with a medical condition. They might be great at treating traumatic injuries, which is what they see a lot more of over here, but they were not so proficient at caring for this medical condition. They finally decided to simply evacuate her and transported her to a clinic and not the hospital! I suspect because she need would more conclusive diagnostics accomplished she would consequently be transported to the hospital eventually. All of this keystone medic stuff makes me appreciate my guys back at the Wisconsin Rapids Fire Department even more and what we have been able to accomplish at the paramedic level of EMS there. My guys are very good at what they do and what I just observed I would have chewed them out over. As I am also an EMS instructor, I was very unimpressed with the care this poor young woman received and thankfully it was not more serious.

The Major League Baseball playoffs are underway. Hope springs eternal in all cities that have teams in this year's playoffs, which unfortunately does not include my team, the Milwaukee Brewers. The Brew Crew had a tough year and not much quality pitching as they finished 80-82, a year after making their amazing playoff run. Hopefully they can keep their young nucleus of players such as Ryan Braun and Prince Fielder. In fact, Fielder had a monster year and is very likely to finish in the top three in the National League Most Valuable

Player voting. But for now, we will just have to wait until next year like the majority of the baseball teams in the majors.

We'll get 'em next year Brew Crew

Day 339: Saturday-October 10th, 2009/Day 308–Iraq we must return to the Afghanistan debate. Well, it appears as though we will get to the real answer to the strategy that will be employed there through the process of elimination. I guess the reduction of US forces in Afghanistan is off of the table of possible options **(JASG, 7 OCT 09, *Obama rules out large reduction in Afghan force:* New York Times)**. So is Vice-President's Biden's counterterrorism only option that called for the use of technology such as Predator drones and Special Operations units. While this may be a part of a larger option package, it will not be the sole option. So there have been a series of high-level meetings in Washington this week and they have been deliberate and methodical, which is the right way to go about making a decision this important. The President is hearing all options and every side of the debate before he makes a decision, which again, is very wise. But the White House press secretary is saying the President is weeks away from making a decision! In my opinion, this is not as wise, as our troops and the situation in Afghanistan do not have time on their side. So we certainly want the President to make a wise decision and not a hasty one, but it should not take forever either.

Much of this debate has made its way into the media and many people have weighed in. This to me is not counterproductive, but healthy. Secretary Gates believes any advice from top military advisors and civilian officials should be kept private **(Stars & Stripes, 6 OCT 09, *Gates: Keep counsel to Obama private*, written by Jeff Schogol)**. While I have a great deal of respect for Secretary Gates, and certainly believe commanders and civilians alike should follow their respective chains-of-command, secrecy is not open government either. Mothers and fathers have a right to know what is going on as they provide the most precious resource we have in this counterinsurgency fight – their sons and daughters. Let us hope that the historic announcement we heard yesterday that President Obama had been awarded the Nobel Peace Prize does not weigh in on any decision he may make. It is a great award to be sure, but it seems a bit odd that his nomination was submitted only two weeks after he became President! It appears the award is for his potential and not his body of work. I believe this President is capable of great things, but as the old adage goes, 'the proof is in the pudding'. While I believe he has great potential, it is still too early to determine if his decisions in regard to the economy, healthcare, and foreign policy have been good ones. And I will reserve opinion on all of this until I see what he will do with Afghanistan. While Washington takes its sweet time and continues to be distracted by other issues we still have a lot of troops in harm's way. I just do not see a sense of urgency here and this is what angers me the most. This kind of leadership does not impress me in the least. But again, I will reserve a final grade for our Nobel Peace Prize winning President until such a decision on Afghanistan is made and then I will comment again on this particular topic.

(-40) Day 340: Sunday-October 11^h, 2009/Day 309–Iraq was my dad's 76th birthday. My dad was in the US Navy during the Korean War and is not real pleased with our politicians and their inability to make sound decisions. He also wants me to get out of the Army as he knows impending deployment to Afghanistan is likely down the road. This entire affair has been difficult for Shari and my parents and I really do not want to put them through this again. While I could suck it up and do it despite all of my disenchantment with the military's organizational structure and the poor decision-making coming out of Washington, I will need to make this very important decision about my military future within a year after my return. But this decision is not mine alone to make and there are other very important people I will need to consider. But today is #76 for Donald Rowell and so Happy Birthday Dad.

This war has been unique in that our soldiers and marines have had to quickly adapt and become diplowarriors. They have had to use just the right mixture of military power sprinkled with a lot of diplomacy **(Stars & Stripes, 6 OCT 09, US troops adapt from battlefields to politics, written by Heath Druzin).** Those conventional-style thinkers that could not adapt to this new environment have struggled. And keep in mind that our troops have had to adapt during a war not in between. They had not been trained adequately for such warfare and while many adjustments have been made to strategic and operational plans during the eight years in Afghanistan and six years in Iraq, the military faces a larger dilemma by retooling training programs for fourth generation warfare, while maintaining the curriculum to ensure our troops are still proficient in third generation warfare in which we excel.

In my opinion, this war has become more of a police action with pounding the best, developing key relationships, being out among the populace, kicking down doors, conducting room to room searches, arresting key suspects, maintaining peace, etc. Most of their proficiency has been via on the job training and not some formal school. So our troops ability to adapt in mid-fight is also very historic and noteworthy and one that seems to be getting overlooked. While our military has certainly made adjustments in the midst of past wars, I do recall such a dramatic shift moving from third generation tactics (firepower, firepower, firepower and a little maneuver and technology) and fourth generation warfare which is more of the police action I just highlighted for you. These are two completely different and diverse types of warfare. But much like our friend the **chameleon** (sorry John

Maxwell), our troops have been able to adapt to the fight and are achieving great success.

It is now our politicians and the American people that seem unable to adapt as their expectations center around a short, decisive war using our overwhelming technological and ordnance superiority. They too, like some of our military commanders, have not been able to mentality adapt to the type of fight that we are actually engaged in. Perhaps it is a mental block due to past historical examples. Perhaps it is the inability to mentally adapt to a fight we were not prepared for. Perhaps it is our media not telling the entire story to the American public. Perhaps it is our politicians not communicating well with their constituents. Perhaps it is a combination of these variable and more. But our troops have adapted and they deserve our support. Ignorance of reality is no excuse for inaction or a lack of support. There seems to be enough of that in Washington!

Happy Birthday, Dad!

CHAPTER 18
400 DAYS – Days 341-360
October 12 – October 31, 2009

We have real enemies in the world. These enemies must be found. They must be pursued and they must be defeated.
--President Barack Obama
2009-present

Day 341: Monday-October 12th, 2009/Day 310-Iraq is Columbus Day. We celebrated the holiday here yesterday as many people took the day off. Again, it was already my half day and once again, as I have done on every holiday over here, I went in to work after lunch. But it was very slow and there is little to do at this stage so I left a little earlier than usual to enjoy some football and a meal in my room.

Happy Columbus Day

We also began our final full week in Iraq. Yesterday I attended an Operation Proper Exit ceremony at Al Faw Palace. I may not have mentioned this previously, but this is the second such event since I have been here. This is a great program where soldiers who are injured in Iraq come back so they may have some closure to their situation and to assist in their mental well-being. The support for this program is tremendous and you can tell it means a lot to the returning warriors. And while I am aware if the program really hits it mark by being therapeutic for the soldiers, if it is, then it is that much better and is one of those programs that should be sustained. This was one of those lump-in-the-throat moments as we all took a moment to thank these kids for giving so much. Whatever gratitude and support we can provide they more than deserve.

As we wind down our mission in Iraq you have two key areas that will ultimately determine success here. One is the security of the people. LTG Jacoby, the Commanding General of Multi-National Corps-Iraq, stated "we are very close to success in Iraq. So now it's a question of having the will and the determination to finish the task" **(MNFI, 10 OCT 09; KOMO TV)**.

The other piece of this success puzzle deals with diplomacy. The US Ambassador Patricia Haslach, the Deputy Chief of Mission in Iraq stated, "It's going to take a commitment from the government of Iraq to make the necessary changes...that means the elected officials and the ministries have to make the hard decisions" **(MNFI, 10 OCT 09: Reuters)**.

So again, the final chapter on Iraq will not be written until 5-10 years after we have left. If they are successful in protecting their people from terrorists and outside influence, and the government can provide essential services and civil capacity to the people, then we may be able to claim Iraq as a success story. And no, I do not plan on coming back to write Volume II to apprise you of this!

Well, while it is still hovering around triple digits here in Iraq, we are expecting snow in central Wisconsin! While I have not been a fan of cold weather for some time now, this year I look forward to it. It also signals the end of this trying year. But it also appears that the cranberry season may produce a record yield. I have already told you that Wisconsin is the biggest cranberry producer in the country. What I did not tell you is that my hometown of Wisconsin Rapids just added a 125,000 square foot expansion to its existing structure, thus making it the largest cranberry processing facility in the world **(http://www.wisconsinrapidstribune.com/apps/pbcs.dll/article?AID=20091010/CWS03/)**.

Way to go Wisconsin Rapids

Day 342: Tuesday-October 13th, 2009/Day 311-Iraq began with an article from an author whose books ***Fiasco*** and ***The Gamble*** I have enjoyed while over here. This Thomas Ricks article focused on something very unique in Afghanistan, and probably another story that will not get much American media attention. This story deals with how the US Marine Corps is using its female warriors **(MNFI Early Bird News, 9 OCT 09)**. It appears as though the female marines are able to get into places that their male counterparts cannot. More importantly, they can reach an important segment of the Afghan people that has really gone neglected for most of this war. One elderly Afghan gentlemen stated, "Your men come here to fight, but we know the women are here to help." These teams of women are termed FETs (Female Engagement Teams), not to be confused with the same Army acronym that stands for Forward Engineer Team! But it appears that this unique and innovative concept is realizing some success and for that the US Marine Corps deserves kudos for the effort. Just more validation of my theory that the USMC is a more flexible, adaptable, and creative organization than the larger bureaucratic machine known as the US Army. But before I completely throw the US Army under the bus and back over them again, Mr. Ricks does go on to state that one of the main barriers to this program and more extensive use of such teams is not the Afghan men, but rather, US Marine and Army officers. Go figure!

Well, today it finally happened. I knew it was only a matter of time but I thought with only about a week to go I was home free. Not so fast! As you may recall, I have a vehicle to drive around in. While many vehicles have been turned in and people from where I live over at what we call Gulf Region District, formerly Gulf Region Central, and soon to be re-designated as Camp Wolfe, these people have been forced to ride a bus back and forth. I have also told the roads here are incredibly poor and very rough with jagged gravel roads in certain areas. Every Sunday I have conducted a thorough preventative maintenance check on my sedan. All turn signals, oil levels, brake lights, headlights, etc. have been in perfect working order. I have had the chariot in twice for its scheduled service and was going to do my final service stop on Monday and then turn in the vehicle on Tuesday (the day before we leave) if I could get away with it for that long. No one has inquired about my sedan thus far and I was not about to initiate the conversation. So I went to work out as usual today with no issues. Then at about 1630 several of us from the G3 section were going to go over to Gulf Region District and partake in some cake, coffee, and munchies to celebrate the Navy birthday. Well, everyone beat me out the door and must have gotten another ride over as when I hit the parking lot everyone had departed. I started the old chariot and before I left the parking lot I knew something was amiss. So I stopped the car, got out, and sure enough, the left side front tire had gone flat as a pancake. My first mishap of any sort! So as not to panic, I calmly opened the trunk and thought to myself, "I can simply change the tire quickly and still make the celebration." In the trunk I found a fully inflated spare tire and the car jack. However there was one glitch. No friggin' tire iron! So Plan C it was. I moved the car carefully back to its parking stall and then hoofed it down the road to AMECO. This is only a short distance from GRD HQ and where I take my car for service. I talked with a very nice elderly gentlemen who said, "No problems" and informed me they would tow the car over, change the tire, and do the normal service on the car. With that excellent news, he gave me a lift back over to the car so I could get my notebook out and give him the key. I would pick up the repaired chariot tomorrow evening after work.

So tonight I would be a part of the bus crowd, which would leave GRD HQ at about 1815. So I did a little more work, caught my ride, and got back to my room at about 1830. With the thought that perhaps the party was still going on, I walked over to the building where it was being held and found out it had just ended! The guys were packing up all the food. I could smell chicken wings and noticed a pair of tongs with frosting on them but no cake in sight. Good thing I had a piece at lunch to celebrate the Navy's 234th birthday. So I grabbed a piece of American cheese sitting on a platter and went back to my room for the evening. Oh well, it was a Navy birthday I am sure to remember.

Happy #234 US Navy

Day 343: Wednesday-October 14th, 2009/Day 312-Iraq began with news from Afghanistan. One of the books I have had the pleasure of reading over here is ***The Accidental Guerilla*** by counterinsurgent expert David Kilcullen. Mr. Kilcullen has been to Afghanistan on several occasions and an advisor to General McChrystal. His opinion is that at least 25,000 troops are needed to achieve success in Afghanistan **(http://www.cnn.com/2009/POLITICS/10/09/amanpour. kilcullen/index.html).** He believes there are three critical problems that must be addressed in Afghanistan. The first is to stabilize the legitimacy of the government. This will take diplomatic effort, not military power. The second is the elimination of safe havens for terrorists across the border in Pakistan. This will take serious diplomatic efforts and cooperation with the Pakistani government. The third are additional military resources, or in other words, enough troops to clear and hold territory. This is not all that different than Iraq's situation so one may surmise that if we succeeded with such strategy in Iraq, why would you not expect it to work again in Afghanistan? While we have heard how different both countries are and rightfully so, there are also many similarities. Mr. Kilcullen believes the biggest problem facing Afghanistan is the civilian leadership. Once again, it boils down to leadership and if diplomatic efforts cannot assist in strengthening this area, then all the military might in the world will not do much good in the long run. Perhaps this is the dilemma our President is contemplating as he is making his decision.

And in Iraq, security of the populace is improving. One Iraqi Security Force General said the reason for the improvement is that civilians are providing valuable intelligence about potential violent extremists **(MNFI, 13 OCT 09: *al-Iraqiya*)**. The General went on to state:

> The average of three terrorist cells are dissolved on a weekly basis. This is a great achievement by Iraqi Security Forces; however, the main source of intelligence is our Iraqi citizens who are providing our security forces with tips. The citizens are considered the eyes of the security forces.
> **-MNFI, 13 OCT 09: *al-Iraqiya***

Security of the people is one large piece of the puzzle in a counterinsurgent war and success is being achieved in Iraq as a result. We hope to reach this same state in Afghanistan, but it will take time, patience, and funding. If our government is not willing to commit to this reality, then we should remove our troops from harm's way immediately and move toward another strategy with the understanding that we will not achieve the same level of success in Afghanistan that we have been able to do in Iraq. Your call, Mr. President, and by the way, the clock continues to tick...tick...tick.

Day 344: Thursday-October 15th, 2009/Day 313-Iraq was more like it. As we continue to prepare for our departure, we are trying to anticipate all of the little 'ankle biters' that might get in our way. So far, so good and we begin our redeployment briefings tomorrow.

I also got my car back yesterday. It was serviced, the tire was repaired, and it was clean! So my reliable little Nissan Maxima would finally be turned in on Tuesday. Busing it was not all that bad however, but it did stop me from getting my cardio workout in with no wheels, but I would get back on track today.

I also had a pleasant surprise from one of the friends of the lady who was ill last week. She thanked me for helping that day and gave me a quick update that her friend had indeed been sent back home to the states, but that she was also expected to be fine. That was very good news indeed.

Iraq is getting ready for their big elections, and judging from what occurred in Iran and Afghanistan recently, General O wants to ensure we do everything we can to assist them with conducting fair elections. There has been a debate on closed or open lists for the candidates. Many people want their choice of candidates so they prefer an open list. Some of the politicians prefer a closed list to limit the number of candidates. While an open list would be more of an administrative nightmare, it certainly would be more fair. Iraqi Prime Minister Maliki recently called for an open list and has really been posturing to be re-elected **(MNFI, 14 OCT 09: *al-Sumaria, Khabaar*)**. As I have said before, he has turned into a pretty savvy politician.

Normalcy continues to creep into Iraq as Baghdad residents recently praised security gains for local businesses now thriving and people feeling safer **(MNFI, 14 OCT 09: *Washington Post*)**. It has been relatively quiet recently for the most part, which allows progress and stability to take root. But as I have mentioned, we will really not be able to write the final chapter on Iraq for another 5-10 years and well after we have left the country for good.

The Afghanistan debate continues, and now we hear the reality that even if more troops are committed to this country it may not save it from falling into a terrorist haven **(www.foxnews.com/politics/2009/10/13/mcchrystal)**.

This news of course will make the President's decision even more difficult as placing more troops in harm's way is no guarantee of long-term success. And after watching an **ABC News Frontline** presentation on the status of Afghanistan, the situation is very complex, but so has been the case here in Iraq. Pakistan is a key to any success in Afghanistan. You cannot separate Afghanistan from Pakistan when developing strategy, much like the situation between Iraq and Iran. But unlike Iran, we have a better relationship with the Pakistani government than we do with Iran's. However, the al Qaeda and Taliban leadership move into pockets in Pakistan where US troops cannot go after them. The Pakistani government is trying to assist, but due to corruption and a lack of resources, progress has been extremely slow. Throw in the poppy situation, which is a major source of income for farmers and the country, but it also finances the Taliban and al Qaeda. It is huge mess to be sure, but one that can be solved given the proper resources, the political resolve necessary, and public support. Otherwise, we are simply we wasting our time and needlessly placing troops in harm's way. So this decision could very well be one of those rare defining moments of President Obama's presidency and his ultimate place in history. If you are truly an American and not simply following along party lines, you should hope he makes a wise and thoughtful decision.

Day 345: Friday-October 16th, 2009/Day 314-Iraq as single-digit midgets most everything at this point is our last while in Iraq. In other words, this will be the final Friday in Iraq. This will be our final football weekend in Iraq. Tomorrow is our final Victory Cigar Club meeting. You get the idea. So it is getting exciting and I cannot wait to get the heck out of here.

We began our redeployment training today. Today was simply what is termed a post deployment health assessment, or PDHA for short. This process was relatively short and painless. We were all to fill out a short questionnaire online prior to attending. It was a basic health assessment form with several questions concerning both physical and mental well-being. Upon arrival at the Troop Medical Clinic (TMC), we watched a brief 10-minute narrative PowerPoint presentation, of course, and then talked with a healthcare provider about our online questionnaire. The whole process took less than half an hour. That is the good news. The bad news is that I hope this is not how the Army screens for signs of post-traumatic stress disorder or suicidal tendencies. If there is no more than this before we are released from active duty, then it is very clear that the military does not have a good handle on how to more adequately identify potential problems. Simply asking a few questions in a setting established to push people through like cattle is not a good answer to this major problem as far as I am concerned. But I will reserve final commentary until I am completely through the redeployment process. I certainly hope there is more than this!

I read an interesting piece from the *New York Times* recently about the eerie comparison between Obama and Afghanistan and John Kennedy and Vietnam **(http://www.nytimes.com/2009/09/27/opinion/27rich.html)**. Now I will admit that as a military officer I am a bit biased toward military options, but not exclusively. And after the Bay of Pigs fiasco, John Kennedy still listened to his military commanders, but not solely. In fact, he ignored military advice in relation to Vietnam and instead opted for the military to conduct advisory operations only. It was not until after his assassination that this policy was reversed in favor of combat operations. Joe Biden is today's George Ball who was the Undersecretary of State for the Kennedy administration. He advised Kennedy against a more proactive military option, while the Secretary of Defense (Robert McNamara) and Secretary of State (Dean Rusk) supported the Joint Chiefs of Staff call for combat troops. Our current President is also a student of history and will make this lonely decision based upon the right set of circumstances, much as did Kennedy. At least we hope so. But there is also the possibility that even if we send more troops, we will never truly achieve any sustainable success without a stable Afghan government. This may be a more difficult task than ridding the country of the Taliban and al Qaeda! So whatever decision is ultimately made will truly be a crap shoot, and sacrificing more US lives for a losing proposition is certainly not wise. So if the President carefully considers all options, envisions an end-state, and keeps his decision from being a purely political one, then we should support whatever decision he ultimately decides upon. Of course in our society, this is utterly impossible. Good luck Mr. President.

Day 346: Saturday-October 17th, 2009/Day 315-Iraq I will talk of some of the advances in medicine that are coming out of this war in relation to head injuries **(Stars & Stripes, 14 OCT 09, *Battlefield of the brain*, written by Melissa Healy).** Due to the IEDs that have led to head trauma in the form of traumatic brain injuries and concussions, a great deal of funding has been allocated to research in this area. Whatever research is accomplished will ultimately benefit many people not involved in a war that are victims of sports injuries, car accidents, falls, etc. Teaming up with their civilian counterparts and the National Football League, physicians are researching new helmets that can absorb more impact and cushion the head, and new imaging devices to better and more quickly diagnose head trauma. So this research is very encouraging and will hopefully lead to some breakthroughs that will benefit all head injured patients.

On the home front, Shari is already preparing for Halloween and Christmas. She and her friend Dawn recently took a trip to Minneapolis to shop at the Mall of America and she has a great start on Christmas. All is well at home and I am so looking forward to returning.

My EMS leadership book is almost complete and will hopefully be out before Christmas. Unlike my first experience with a publisher, working with Angela Hoy of BookLocker, Inc. has been a great experience. I have had much more creative input into the book itself and the cover design. While I do not expect to become rich due to my writing, I do hope to enlighten and educate people. Plus, I just enjoy writing.

Speaking of leadership, MAJ Frank Witsberger was recently doing some 'dumpster diving' to determine if any sensitive information was being discarded from people within our organization. Frank fished out a copy of an American Heritage magazine because it had an article on Vince

Lombardi, and Frank knows I am a big Packer fan. It was a good article chronicling the career of Vince Lombardi who is obviously an icon in Wisconsin and the NFL **(American Heritage, Fall 2009)**. I grew up as a kid during the Lombardi years and became spoiled at Packer success. I am still a big fan of the Packers although not so pleased with the state of professional sports in our country and all of the money paid to guys to play a game while the kids protecting our country and our freedoms, or out protecting our communities such as police officers and firefighters, are paid far less for their contribution to society. But despite this societal inequity I still enjoy watching football.

Lombardi was a man with many leadership attributes. The debate always comes up if Lombardi could coach as effectively in today's football world as he did in his day? It makes for a great and spirited discussion, but a leader is a leader no matter the period they exist in time. To me, a great leader is adaptable (chameleon-like leadership) and flexible enough to adjust to any environment and be successful. Obviously we will never know the answer to the Lombardi debate, or whether George Patton could have been successful in an all-volunteer military. This is why it is simply a debate and nothing of scientific value. The same may be stated of leaders today whether they could have been successful back in the more authoritarian days of the 40s, 50s, and 60s. But insight into this question for me would be if the person in question exhibited adaptability capability, then the chances would be favorable that they could transcend time and be a successful no matter the era. If the individual in question was so rigid he or she could not adapt to his/her environment, then chances are good that they would not be as successful in other environments. Keep in mind that leaders today can be successful in one organization, take a job in another, and not be anywhere near as successful. Simply taking one's style and transporting to another organization is no guarantee of success. But there is no doubt that Vince Lombardi's persona has certainly transcended time, and his quotes and leadership style are still talked about today. The Super Bowl trophy bears his name. He was most certainly a great football coach and of that there is no debate, and he helped to solidify the legend that is the Green Bay Packers.

Vince Lombardi – The Legend

Day 347: Sunday-October 18th, 2009/Day 316-Iraq was my last Sunday in Iraq. What to do? Well, last night Mike Ryan and I attended our last Victory Cigar meeting. We had Cuban cigars, which were very good, but to be honest, as I do not consider myself a connoisseur, I would be hard pressed to tell the difference between a Cuban and let's say a Dominican cigar. Nevertheless, it was still a good cigar.

Before Mike and I went to the cigar club meeting the security company for the Gulf Region Division, Aegis, had a cookout that Mike and I attended. This was nicely done as well and one of the leaders of Aegis, who is British, gave a very nice speech about the relationship between Aegis and the Gulf Region Division, and between the United States and the United Kingdom. He also said something I found very interesting and insightful in that success in Iraq for many of us who have been here may not be measured in victory parades upon our return home, or claims of strategic success in defeating all Islamic extremists that threaten our way of life. But when we leave, if we can truly state that we have left Iraq a better country through our efforts of rebuilding than we found it in 2003 when we first arrived, then this should be viewed as a success.

Also yesterday we had the remainder of our redeployment briefings. The first few were done by a Chaplain and this was perhaps the best briefing I have had since I have been here. Chaplain Middlebrook is a fire and brimstone kind of preacher for the 10th Combat Support Hospital (CSH). He did an excellent job of explaining our role in reintegrating back into our lives that we left almost a year ago. He also said something that struck home that I had not previously considered. He said that we have been absent for a while now and our families have filled the void we have left. We are now going to reappear and pop back into their lives. It would be foolish to believe we will pop back in where we left off a year ago. We have changed, and they have as well. And he said we must keep in mind that it is we that should find the strength to find out where we fight back into their lives, not vice versa. Excellent advice indeed and words I will take to heart.

We then watched an excellent video on what I would call marriage counseling of sorts. The video was created by a minister I had not previously heard of. He was very entertaining to listen to and provided great insight into how men and women think much differently. I will try and gain a copy of this video, which is about an hour long and take it home so Shari and I can view it together. As I have been single most of life, marriage can be work and it is something that needs to be worked on like anything else. I know I never should stop learning about Shari or take her for granted. I know this deployment has been equally if not more difficult for her and I think this type of enlightenment can strengthen a marriage. I always felt marriage counseling was for people who were having problems within their relationship. As a novice at this thing called marriage, how wrong I was. This type of forum can be used to make relationships stronger and provide greater insight into one another as there can be no doubt that men and women are wired differently. So this was an excellent day, but filled with some sad news that brought us back to the reality in which we and our fellow uniformed members find ourselves. A former Colonel from our unit, who just retired last year, lost a son fighting in Afghanistan. This is very tragic, as are all of our losses here and there. Whatever decision the President is about to make, I hope he makes it soon.

As this is my last Sunday here in Iraq, I guess I can provide you with my final DFAC rankings for Round 2 on this day. Here is the comparison:

ROUND 1:
Sather
Stryker
Coalition
DeFleury
Sports Oasis
Dagger Inn
Iron Horse (MND-B)
Desert Diamond

ROUND 2:
Sather (5.5)
Stryker (7.5)
Dagger Inn (9.0)
Sports Oasis (9.5)
DeFleury (10.0)
Coalition (10.5)
Pegasus (11.5)(MND-B)
Desert Diamond (-)

My second ranking was a bit more scientific in nature as I examined four critical components of dining experience. I examined the areas of freshness, selection, atmosphere, and other intangibles. The scoring was on a scale of 1-5 with 1 being excellent and 5 being poor. So a lower overall score is better. As you can see by the ranking that Sather was a clear-cut winner and Stryker remained a solid number 2. Then the rankings got a bit jumbled for one reason or another from the first ranking order. Desert Diamond remained last, but as I mentioned a few weeks ago, this DFAC is currently under renovation and you eat out of a tent right now. While it may have been unlikely that it moved up in the rankings, because there was no way to reevaluate the DFAC it remained last in the list of eight. My ranking are further validated by a connoisseur of fine dining, LTC Tony Jocius. His first three were exactly the same as mine, but then we had some slight differences of opinion. All in all, this DFAC ranking was a nice distraction from the monotony of live here on VBC. I also do something similar with Hooters restaurants when I travel as I have visited 394 different locations thus far and my tour is far from over. Another thing I love about Shari is that she is often the first one to suggest a trip to Hooters because they do have excellent wings, cold beer, a nice atmosphere, and sports on all the time. What is not to like to like. And yes, I do have a top five locations for Hooters as well, but that may be for another book!

Day 348: Monday-October 19th, 2009/Day 317-Iraq was my last full day in Iraq. Yesterday was very slow paced as I took my half-day off as usual to work on this book, a few articles for the home crew, and continued to pack. Then I went into GRD HQ, checked my email, and then about five us went on a tour of a few places over at Camp Slayer.

There must have been a group of about 70-80 people that went on this guided tour, which included the Victory Over America palace, a Republican Guard facility, and the Ba'ath Party House. The Victory Over America palace was never completed but stands as a testament to Saddam's greed. This would be another location that he barely visited built in his honor at the cost of his people. A real piece of work! This place would need a great deal of work to repair as there is a lot of rubble throughout the massive structure. The view from the top of the building was impressive as it overlooked the Flintstone's village, which I mentioned previously, communications hill (this is the one on Camp Slayer), and the 'Perfume Palace,' which is rumored to be an old brothel for higher-ranking Republican Guard members.

Next stop was a mural of Saddam at a Republican Guard facility on the other side of Flintstone village. While many people took pictures, some of the Iraqi on the tour with us threw things at the mural or gave it the old middle finger as even though he is gone his twisted legacy will live on for generations.

The last stop was the Ba'ath Party House where some business was conducted in this resort area. In fact, this was the facility in which Dan Rather interviewed Saddam before the war began. It is also a place where a couple of Tomahawk missiles struck early in the war and basically turned into an unusable shell of a structure. There was a room that was particularly eerie that used to be a conference room with theater-style seating. A Tomahawk missile hit this location with about 250 people inside of this room. Only 50 bodies were ever pulled from the room, which meant that about 200 were vaporized by the huge explosion and never recovered. Another Iraqi on the tour said Saddam can rot in hell and this building should be torn down. All-in-all not a bad three-hour tour, and no getting stranded on a desert island (Gilligan fans)!

After a few more last minute tasks at GRD HQ, we finished the day with a cookout at Gulf Region District. So we had a nice relaxing day, and as the Green Bay Packers beat the Detroit Lions 26-0, we should be back in Wisconsin to watch their next game!

Perfume Palace and Victory over America Palace

Day 349: Tuesday-October 20th, 2009/Day 318-Iraq was our last day in Iraq. Finally! Yesterday we had our awards ceremony, which I could have done without. While I think awards are great, many such awards given out over here have really diluted the entire effect as far as I am concerned.

Yesterday was pretty uneventful as we continued to pack and prepare for our departure. We did have our awards ceremony and after 34 years in the military they still remain pretty much the same. Some people get the award they deserve, others get awards they do not deserve, and others do not get the recognition they should. Will things ever change? I know I implemented an awards program back at my fire department and while some guys complain about it, I can honestly say we have kept politics and friendships out of the process and the people that have won awards have truly deserved them based on merit and performance.

I received the second highest award called a Defense Meritorious Service medal. This is the highest award I have ever received and will sit atop my ribbon holder. This was all very nice but actually more than I deserved in my mind. I really was not provided the opportunity in the positions in which I was placed to do anything truly meaningful. I could have done much more to assist the Army and the people of Iraq, but because of the Army's antiquated system of placing personnel into mismatched positions, it really limits commanders as to where they can place personnel. Yet another structural inequity! And while I got more than I should have, a few higher-ranking people or active duty personnel got the now infamous bronze medal for performance. This medal has been given out like candy here and really diluted the meaning of the award in my eyes. It should never be confused with the bronze medal with a V device, which designates valor in battle. The two awards are polar opposites in my eyes and mean a lot more to the active duty people than to us reservists. Others, especially civilians, appeared to receive awards due to the position they held, whether they were worthy of one or not. Such is life in the military, and when you hear the people who created the award packets tell you there will be some real bitching after the ceremony, you know politics was involved in the some of the selections as there always is. So the award received is not really based solely upon merit or performance, but this process has not changed in my 34 years in the military and likely never will. This unfortunately is just the way it is. I was proud to get the Iraqi campaign medal for my time here and really would have preferred the Joint Service award, which is a lower ranking award, but as I worked with many other outstanding people from other services and other countries, it seemed more appropriate. Oh well!

One interesting final piece of news from Iraq that you may recall is the referendum to boot us out of Iraq early that was originally scheduled to occur this past summer was pushed to the national elections early next year. Now it appears as if the entire referendum may be dropped! It seems the Iraqis are okay with the current security plan which calls for all US troops to be out of Iraq by the end of 2011. They see troops and equipment leaving Iraq and there does not seem to be any big push to conduct this referendum at this point **(Stars & Stripes, 17 OCT 09,** *Iraq lawmakers drop plans for referendum to speed US withdrawal,* **written by Liz Sly & MNFI, 18 OCT 09:** *LA Times*). To me this is simply another sign of progress here in Iraq.

REDEPLOYMENT COMMENCES:

Day 350: Wednesday-October 21st, 2009/ Redeployment

Well, today was day 350 and we began our journey home. I will reserve my final thoughts about Iraq until the final chapter, but as we leave I wish the Iraqi people well and hope they take full advantage of the opportunity that lies before them. I hope they do not become a mini-me, a mini-US that is. The Iraqis are not like us and should not be expected to be. They can hopefully develop a good government that will take care of the people that Saddam neglected for so many decades. Hopefully they can find peace, but it may be a long time before this ever occurs. But if they can keep al Qaeda from redeveloping cells here in Iraq then I believe we and the Iraqi people have been successful.

Well, of course, we again begin the 'hurry up and wait' cycle all over again as we begin our long journey home. Our day yesterday began a little rough with some minor transportation issues to the airport, but as we included plenty of 'fudge-time' into the planning, it was no major deal. We easily made it to BIAP with plenty of time to spare and then waited for about three hours before our flight took off. While there, we took one more opportunity to eat at Sather, my number one ranked DFAC on VBC.

We took off and picked up some of our unit brethren to our south in Tallil. From there it was on to Ali Al Salem air base in Kuwait where we then proceeded to move to Camp Virginia, about a 20-minute drive. When we arrived at Camp Virginia, we had a quick briefing orientating us to the base and then we went to eat in their dining facility, which was not bad. From there it was to our tents, which were a bit more Spartan-like living conditions than we have enjoyed, but none-the-less it was one step closer to home.

Heading home at last

Day 351: Thursday-October 22nd, 2009/ Redeployment

While at Camp Virginia we basically just decompressed. We ate, worked out, relaxed, slept, and really did little else as we all reflected back upon the events of 2009. Our brethren from the north had still not arrived, but Camp Virginia is a good place to just chill even though it was around 100 degrees! They have a McDonald's, a pizza place, a Starbucks, a Green Beans (another coffee shop), a Taco Bell, a Baskin-Robbins, a small but decent post exchange, and a very nice USO. In fact, the USO had a couple of different movie rooms, an Internet café, a reading area, and a couple of video game kiosks. Some of the guys engaged in a game called Guitar Hero 5 and Sean Begley turned out to be the star. Perhaps America's Got Talent is in his future!

I bought a few more souvenirs, hit the Starbucks and Green Beans, and then hung out at the USO for a few hours. It was just a nice way to unwind and relax. Several members from my unit commiserated with one another on the trials and tribulations of this deployment. Interestingly enough, I was hard pressed to find too many individuals that claimed they had a productive and meaningful deployment. Go figure!

So two nights sleeping on cots in a tent was enough to peak our fun meters, although it is not that much different than camping. But believe it or not, I look forward to the softer bunks at good old Fort McCoy, which then puts me an hour from home.

I called Shari last night, and while not telling her our trip plans in detail, she knows I am on the move and heading home. She also dropped a little bomb on me last night. No, nothing about our relationship or anything of that nature, but I guess we have a new addition to our family! I know what you are thinking. It has been about four months since I was home on R&R and now the old guy is going to be a papa again! Admit it, this is what you were thinking was it not? Well, Shari and I enjoy being grandparents and have 'been there and done that' in regard to the parent thing. Being a parent is a lifetime commitment after all! The new addition of which I speak is yet another dog in the Waite family! For those of you with scorecards, this makes four pooches! I told Shari she should just work in a kennel instead of bringing home another dog! I know she like critters, much like Ellie Mae Clampett from *Beverly Hillbillies* fame, but this is ridiculous! I love animals as well, but this new addition is Siberian Husky and is already making himself at home I guess. His name is Toby. When I was much younger, I always dreamed of getting a Husky as I think they are beautiful animals, but with three already we may need to get a kennel license soon! Buster and Bailey, our two Dachshunds are getting older and play with one another, while Sadie Mae, our cross breed between a pit and a chow looks tough but is a big wimp. This was Jeremy's dog that somehow we, or I should say Shari, adopted. Sadie is young and energetic and likes to play with Buster and Bailey, but they are more finicky and will play only when the spirit moves them. Toby will seem to round out the kennel as he and Sadie Mae are now running mates. I guess if you are going to have three, especially after having two for a prolonged period of time, you might as well have four so everyone has a playmate! So with four dogs, three grandkids and two more on the way, and the Waite Estate, Shari has had her hands full.

Toby

Well, we left Camp Virginia after less than 48 hours there and headed back to Ali Al Salem air base. Once there it was once again hurry up and wait. We sat through a few briefings, weighed in with our gear, met up with the rest of our unit who had traveled from northern Iraq, and then proceeded through the US Navy's custom ordeal. This process needs some work let me tell you. First of all, we had about 370 soldiers who would be heading back to Fort McCoy for demobilization. Each soldier had between three to four bags. Each soldier had to lug each bag through a long customs line and we worked up a serious sweat and emptied every last piece of equipment in each bag. Then of course we had to repack it all. We went through two X-ray machines and two wand areas. All-in-all it was a very redundant and like herding cattle, but we made it through this ridiculous process and then proceeded to wait another three hours before we would head to the airport.

Day 352: Friday-October 23rd, 2009/Redeployment
We, the 416th Theater Engineer Command, along with a transportation unit from Iowa finally left Kuwait at about 0500 and our first leg home took us back to Leipzig, Germany. There were about 370 of us that crammed onto this large metal bird with all of our gear. Once again, I was amazed at how the psychics of air transportation actually works as this big metal bird took all of us and our tons of luggage and launched us toward home. This first leg in this long journey took about six hours, with a two-hour layover in Germany. Then it was back onto the plane and eastward as we went back in time much like the old show the *Time Tunnel* and today promised to be a long one. Remember, heading east we would pick up a total of eight hours by the time we landed back in Wisconsin!

Our next stop was approximately eight hours later at a little place called Portsmouth, New Hampshire. As we landed there a large cheer went up in the passenger compartment as we were now officially back on US soil. Here we met a group that you also may never have heard of before called the 'Pease Greeters.' This group consists of retired military or just simply people who care about our troops. As we exited the plane we were greeted by customs agents who took our customs card. As we entered the terminal the hallway was lined with hundreds of the **'Pease Greeters'**. There were hugs, handshakes, wishes of 'welcome home troops', hot coffee, Dunkin Donuts, ice cream sundaes, snacks, and many other assorted sundries. The colors in New Hampshire were also spectacular as the multi-colored trees suggested that fall had arrived here, although the peak of the colors may have already passed. The leaves are already dropping in Wisconsin as we have already had snow a few times, but no accumulation to speak of.

After a lot of visiting with our wonderful greeters from New Hampshire, we also had a group picture taken that they will place on their walls inside the terminal along with the many others. We also had a nice little ceremony conducted by the Pease Greeters that they do for every unit either going to war or coming back. This group has been recognized nationally for their outstanding efforts, and this little excerpt in this large book is only a small token of appreciation for what all of these wonderful Americans do for our troops. It meant a lot to me and to my fellow soldiers, as I am certain it does for all other troops. Well done, Pease Greeters.

While there was little rest to be had today and we were packed into the large jet like sardines, we were treated and fed well by the flight crews, and we watched four movies along the journey. The first was **Night at the Museum II** with Ben Stiller, which was good; **Ghosts of Girlfriends Past** with Matthew McConahey and Jennifer Garner which was alright; **Marley and Me** with Owen Wilson and Jennifer Aniston which was good but sad; and then **Duplicity** with Clive Owen and Julia Roberts. So anytime you watch four movies on a flight, you know it was a long one! But since we were heading home it was all good.

Then we had a little delay as the aircraft and a ground interface unit could not communicate to one another so we had an additional two hour delay in heading to Volk Field in Wisconsin and then on to Ft. McCoy from there. At Volk Field we were greeted by some other members of our unit and some very cool, rainy weather. In fact, the long walk from the aircraft to the school bus that would haul us back to Ft. McCoy was a long, cold walk and a much different contrast to the 100° weather we had just left behind in Kuwait. But it was still good to be home.

Once we arrived at Ft. McCoy at about 2100 hours we turned in our weapons, gathered our gear on this cold, rainy Wisconsin evening, and then it was off to our barracks for the evening. While this process was less than ideal and could have been a bit more organized to expedite luggage retrieval, it was not that bad. It just seemed like it after a 15-16 hour flight and very little sleep. I finally hit the rack about 2300 and slept very well knowing I was back in the good old US of A and this long year was almost behind me.

Day 353: Saturday-October 24th, 2009/Redeployment

Today was my stepson Jeremy's 25th birthday. I do not think many people really understand how many important events one misses when being deployed. Bet no one in Congress, or Washington, DC for that matter, missed many important family functions! But I digress.

We began our day at 0630 with a redeployment briefing and then went to the many stations we needed to complete in order to out-process. These stations included a medical screening, an auditory test, a dental exam, and then a review of health insurance and our ID cards. With about 400 soldiers, the process went well but there was a lot of waiting in line. I got through most stations this day, but still would need to complete my final few tomorrow. During the waiting today, I had the opportunity to speak with members of the Iowa National Guard transportation unit that came back with us, watch the **Simpsons** movie while waiting in line for my dental exam, and read a bit of my new book titled **Dragon Days.** This book, written by John Poole, covers the background of fourth generation warfare and how it came to be and reasons why the US has so much trouble adapting to this type of fight. It also spends a considerable amount of time discussing China and how they are positioning themselves very nicely with covert and discreet acts to sabotage our interests so they do not have to face us in conventional warfare. Thus far, about 100 pages in, the book is very good.

We completed our stations at about 1700. I finished going through an audiology exam for my last station of the day, but not real sure why this was necessary after all of my testing last year. There was nothing in my records at Ft. McCoy of this extensive exam, nor did the people at this station have access to it online. This appears to be another failing in the system. With all of the technology available you would believe that with the insertion of a social security number that your immunization records, dental file, and other medical information would be at the fingertips of those rendering such care. In fact, this is how it is supposed to work but it simply does not. So it is a case of more time wasted and more money spent, but it appears that this wasteful spending is acceptable to our elected officials and the general public; otherwise, it would be changed and forced to work the way it is supposed to.

Shortly after our stations were completed, we all gathered at place on base called McCoys. Yes, creativity may have be lacking in creating this name but you never forget on what base you are located! We were going to once again suspend General Order #1B stating that no alcohol is permitted while mobilized! Here we go again. This makes no sense to me either, but then, so much has not during this deployment that this is just another in a long line of items. Consider the fact that we are back from war. We have proudly served our country and done all that was asked of us and now they tell you no alcohol until you arrive at your home of record. Please! The crazy train is still moving down the tracks at a high rate of speed! So if I understand this cuckoo's nest policy correctly, it is alright to go and serve your country and even die for it if necessary, but you cannot have any alcohol while you are doing so! Again, I realize there are good reasons for such a rule in a war zone and every one carrying weapons, but in case the Einstein's in Washington do not realize this yet, Ft. McCoy is not in a war zone last time I checked! Where has all the common sense all gone?

But I digress again, as we again found a way to get an exception to policy letter signed so we could partake in some beverages. Our First Sergeant arranged for us to select from either a rib eye cut or a New York strip. I chose the New York strip as this is my favorite cut of steak. We grilled our own slab of beef on three massive grills on the patio and I had two delicious Leinenkugel Honey Weiss beers. That was plenty for me after not drinking for months, but it sure tasted good.

Many of us from the 416th TEC commiserated over dinner about our Iraq experience and discussed where many of our fellow members would now be heading. The majority of our AGR personnel are being sent to other locations, or what is called a permanent change of station (PCS) because they have already been at the 416th for an extended period. Some of the reservists would also need to find new homes because the transformation process that occurred while we were gone also took away their previous positions. In my case, I would be returning to the 416th and most likely move back to the G3 section, although we were to find out more information later today. So it is was great to be home, but many of my fellow brethren would be looking at new locations and all that this entails such as moving, new schools for the kids, etc. All-in-all, however, it was a nice relaxing evening and I returned to the barracks early to hit the rack as the barley and hops and the jet lag seemed to hit all at once. Time to count deer now that we are back in Wisconsin!

Happy 25th Jeremy

Day 354: Sunday-October 25th, 2009/Redeployment

We were back at our stations at 0700 to continue our out-processing. I got my blood drawn, an influenza vaccination, and checked my final financial situation and was done with all stations by 0930. Then we had until about 1330 before we turned in some more equipment and had a VA brief to wind up the day. During our lull in activity I walked over to the PX, grabbed a cup of Joe, and called home. I got my brother Brad at home and talked Packer football with him for a while, but had no luck getting Shari. Before our evening meal however I was able to get in touch with her and told her that tomorrow this would all be behind us. For that, we were both thankful.

Day 355: Monday-October 26th, 2009/Redeployment

Was the last wakeup in this long experience. Today Shari would pick me up and I would begin to figure out my future and how I could best fit back into her life. The day many of us had patiently waited for was finally here. There is nothing about this experience that I will miss. I will miss a few individuals, but truly hope this book makes some money for our wounded warriors, which will make this lost year a bit more bearable and meaningful. So this will be the last entry in this documentary, an odyssey that really began March 1st, 2008 with the news that I would be heading to Iraq. I will try and wrap up all hanging chads for you in the final chapter as I attempt to capture the feelings and observations of the past 354 days. While the period from March 1st, 2008 to this day in time is much longer than 400 days, this number was actually to begin upon my official mobilization last November 6th. Counting my terminal leave, which is the final 16 days of leave I accrued while in Iraq, my orders will take me to November 13th when I again become a civilian. Even at this date puts the official total less than 400, as my original orders stated. But when you consider the preparation to even get to November 6th, which included a two-week information operations school; a week in Darien with USACE personnel to discuss the finer points of contracts and construction projects; three weeks at the Regional Training Center at Ft. McCoy; 16 more days at the MRX at Fort Lewis, Washington; and the additional five weekends we had in preparation of this little adventure, this was far more than 400 days that I would spend away from my family. However, as my orders state 400 days, which generated the name of this book, that is the number that my universe centered around, so the book title shall remain the same.

Happy #7 Kayleigh

In three days my granddaughter Kayleigh will turn 7. She is little Miss Personality and I look forward to spending some time with her when I return. Because she is the only one of the family whose picture has yet to appear in the book, I must put it in even though it is a few days early. Otherwise, there would be a storm a brewin'.

Then at 1230 CST, as scheduled, Shari arrived at Ft. McCoy to pick me up and we were again reunited. We kissed and hugged, and I whispered in her ear that it was finally over. And so this not so magical journey had finally come to an end and my three goals successfully achieved: 1. to come back alive; 2. to come back with all parts I left with; 3. and do the best possible job while in Iraq, no matter what it was. Mission accomplished.

Mission accomplished

CHAPTER 19
400 DAYS
EPILOGUE

As of November 1, 2009 there were 4346 US troops killed during Operation Iraqi Freedom.

The Mission

When you review American military history, you find that we are not a military of occupation, but rather a military of liberation. Other than conflicts on our own turf, we typically liberate and leave. Looking back upon the first Gulf War, we moved through Iraq so quickly that by the time we reached the outskirts of Baghdad it was already time to head home! Then in early 2003 we headed back to Iraq, but there would be a much different execution of this war and an ending that has yet to be written.

Once again, the American military war machine, the greatest the world has ever witnessed, quickly cut through Iraq and reached Baghdad post haste. Despite Baghdad Bob's claim that the Iraq Army was kicking American ass and winning the war decisively, once we stormed into the capital of Iraq we were unsure what to do next. The first few years of the war we made many mistakes. Not enough troops on the ground, the horror of the Abu Ghraib detention center, lack of political adaptability, and the inability to understand the Iraqi culture led to many failures as the United States attempted to rebuild an entire nation using a script that had never before been written!

But as the years went on, the climate began to change. With the Sunni Awakening, Muqtada Al-Sadr seeking change through politics instead of violence, and 'the surge' orchestrated by two American generals that history will look back upon as two of our greatest, David Petraeus and Raymond Odierno, the tide slowly began to change. During the tumultuous period known as 'the surge', coalition casualties escalated. But the strategy that was followed was sound and the US military began to better understand and respect the Iraqi culture. We also began to do a better job of protecting innocent Iraqi civilians and improve the quality of life for them. As a result of these actions, sustained stability and incremental progress began to occur across this cradle of civilization. It is also important to note that the Iraqi government began to take ownership of its country and this is the point that the United States began to plan for drawing down its footprint within Iraq.

The insurgency of today is termed fourth generation warfare. This type of warfare is the only form of war the United States has ever lost (ex: Vietnam, Somalia). The United States is unparalleled in third generation warfare which is a more conventional style using all of our technological and firepower superiority. Third generation warfare is more symmetrical in nature and one our military is more comfortable and adept at fighting. But fourth generation warfare is very asymmetric and fought in streets and villages where the enemy hides in plain sight. Counterinsurgent warfare is dirty, it is difficult, and success comes in very small increments. Most insurgencies last at least a decade and many experts today feel this is the type of warfare we will face for years to come. So having a strategic plan, meshing all instruments of national power together, putting the right people in the right place at the right time, having effective leadership at all levels, communicating openly and aggressively, and understanding that young men and women will die throughout such a protracted battle must all be factored in to the equation and the final end-state.

The insurgent hopes to attrit the public support and political will of the counterinsurgent. If this occurs, then the insurgent has succeeded. These are the reasons the counterinsurgent must continue to press the fight, hunt down key leaders and eliminate them or bring them to justice, not allow them to rest, limit their freedom of movement, cut off their financing, surgically dismantle their organization, and get more countries to understand that this is a global war that affects everyone, not just the United States. Remember, these terrorists are also killing innocent civilians, including fellow Muslims. Such acts of terror have alienated the terrorist from the local populace, which is also a key part of their recruiting base. And for the future of our children and grandchildren this is a fight we cannot afford to lose.

There can be no doubt that this war has been costly both financially and in human terms, especially for the families that have lost a loved one in the war. Looking back at recent history would further suggest that war strategy is hardly an exact science. While it may still be debatable whether we should have gone to war with Iraq, it should be more transparent that we indeed needed to take the fight to those that changed American consciousness on September 11th, 2001 and killed over 3000 innocent civilians, including 343 firefighters.

Many people still question the validity of the Vietnam War. What was the national strategy at that point in our history? What did the United States stand to gain? Did over 58,000 Americans die for basically nothing? Did our administration actually believe winning in Vietnam would stop the spread of communism in Southeast Asia? Has Vietnam, still considered a communist state, hindered American goals since that war? What about the Korean War? The same questions apply. The first Gulf War was a pre-cursor to the second, but protecting Kuwait and forcing the Iraqis out of that country seemed like a worthy cause. At that point in American history moving Iraqi forces out of Kuwait did nothing to correct the root cause of the problem. Instead, it simply delayed the inevitable.

The cost of freedom has always been high. Americans often forget that being able to talk about politicians without fear of retaliation, move about the country without having to worry about improvised explosive devices (IEDs), watch their favorite sport whenever they desire, take a vacation at Disney World, etc., all comes at a price. We enjoy these privileges because of the blood that has been shed on foreign soil by thousands of American troops. Hitler, in his twisted world, wanted to expand his empire, and had he not made some poor military decisions such as fighting a two-front war, it is not beyond the realm of possibility that the United States may have become a Nazi target at some point. While the Cold War brought a stalemate between two superpowers, do not believe for a minute that the Russians did not consider spreading Communism throughout the world, including the United States!

As the terrorists who perpetrated 9/11 brought the fight to our shores, we had no choice but to go on the offensive and begin to systematically root them out and destroy their infrastructure. Al Qaeda seeks to be the instrument that perpetuates the uprising of the radical Muslim movement. However, they do not seek to be in control of such a movement. Rather, they simply desire to be the beacon for other Islamic extremists to follow. So this centralized ideology and decentralized operational capacity makes it very difficult to cut off the head of this monster. Even if Osama bin Laden were to be eliminated, his ideal would still live on. It is virtually impossible to kill an ideal and unrealistic to believe we can eliminate all insurgents. These insurgents are also a transnational organization, which means they are a global network and not just indigenous to the Middle East. This is why this is truly a global war on terrorism and not just an American problem.

Similar to the whack-a-mole arcade game where you hit one on the head and another one pops up elsewhere, this analogy is a microcosm of fourth generation warfare. If we suppress the radical movement in Iraq, Afghanistan, and Pakistan, it will simply resurface elsewhere. This is why it is imperative that other countries become more engaged in this effort and continue to pressure the insurgents so they cannot rest and they cannot plot more major attacks unimpeded.

The American public must also better comprehend the type of fight we are waging and why it is not only important today, but will be for generations to come. It must be clearly understood that the insurgents we fight today are much different than their predecessors. This global war will be an ongoing struggle for decades to come and this is what makes this newest form of insurgency so dangerous. Unless the counterinsurgent (i.e., us) is prepared to stay the course in a multi-year struggle, exhibit great resolve, and be able to endure the flag draped transfer cases that will undoubtedly continue to return to U.S. soil, the scale will tip toward the terrorists. This is a battle that will continue for decades and the counterinsurgent must press the fight. We must continue to push them, hunt them down, and give them no rest, no quarter. In this quest we must not fail.

It is also important to note that terminology is often very misleading. The term 'winning' should not be translated by the American public as a situation where US forces defeat and eliminate all insurgents in Iraq and Afghanistan, and consequently, both governments stabilize and prosper as democracies. This may have been the original and very tainted perspective by many high-level US officials, but I am not sure the term 'winning' is the most appropriate word to use. Incremental progress and sustained stability are perhaps better and more realistic measurements of success in this newest type of warfare. If coalition forces can degrade the insurgency to a point that each country's respective security forces can handle them effectively; each country can protect its borders by severely limiting the number of foreign fighters and weapons crossing into and out of its sovereign territory; each country's respective military is capable of self protection from invading enemies; and each government can stabilize and work to provide its respective populace with essential services and employment opportunities; then success may be claimed after a sustained period.

It should be clearly articulated that past perceptions about what constitutes winning a war, especially one of an irregular nature, are very different than what occurred in World War I, World War II, or the first Gulf War. So Americans, including many in the military, must better frame what success in Iraq and Afghanistan realistically looks like.

One percentage of the American population is fighting very hard for the other 99% to ensure we do not have another 9/11 on American soil. But this fight is dirty and it is hard. The insurgent is very patient and hopes to attrit the coalition of its will to continue the fight. They look to reduce public support and create apathy. They are more than willing to outwait the counterinsurgents level of support. So a concerted world effort, continued pressure, coordinated intelligence, and assisting other nations to secure their territory so there are no large safe havens for terrorist cells to train and plan unimpeded, will help to neutralize the insurgent. And even with all of this being accomplished, there is no guarantee of success. With a military already stretched beyond its limits, billions of dollars already sunk into this effort, and undoubtedly more flag-draped transfer cases returning to the US, these variables will make the clock tick much faster in Washington where patience will grow thin, and many politicians will 'roll with the poll' and drift with public opinion to secure more votes.

In conventional style warfare, the American military currently has no equal. Throughout my 34-year military career all emphasis has been upon conventional style warfare. Our military has been involved in this style of warfare for some time. Even during the Vietnam era, when guerilla tactics were used by small cells of insurgents, the American military was more concerned with the Soviet Union running rough shod over Europe and refused to adapt or change tactics. As was witnessed in the first Gulf War, such as it was, the American military machine cut through Iraq like a hot knife through warm butter. However, once we arrived at the outskirts of Baghdad, the war was over. Of course, the basic premise of this war according to those in power at the time was to push the Iraqis out of Kuwait - nothing more, nothing less. But it did set the stage for the second Iraq War post 9/11. In this case, the American military machine moved quickly through Iraq. The problem was that once the conventional style fighting was over there was no plan or preparation to become peacekeepers and nation builders! We train for war, not for peace. So as the Americans shifted gears and tried to impose their will on the same people who were initially grateful for their liberation, they were now becoming frustrated with our continued presence. But setting up a government, a competent security force, reestablishing an infrastructure for the country after decades of autocratic rule and neglect was no easy task to be sure.

The Vietnam War was a protracted one and military commanders were really never allowed to unleash our superior firepower. They were never resourced properly by the American government, but the military really never changed tactics or shifted gears either. This is one more reason why this war was so unpopular and the vets from this war were treated with such disrespect. So if Vietnam was such a matter of vital national strategic interest, then why did our politicians not treat it as such? These questions have not and may never be answered.

Such has not been the case in the Iraq War. While commanders on the ground would have appreciated more troops initially and subsequent to the surge, our politicians have basically allowed the commanders to dictate the fight. As such the commanders on the ground observed what strategy was working and what was not. From these observations and lessons a new strategy was created and effectively implemented.

As I mentioned on Day 200, the halfway point of this adventure, it is interesting to observe that there are many countries around the globe that have serious issues. Many of these troubled areas of the world will require some form of intervention, whether on the diplomatic, economic, or military front. And in the future the United Nations will need to take the lead on many of these challenges, with the more fortunate countries assisting in such efforts. But it is also apparent that as the world's lone superpower at this moment in history, the United States will continue to lead the charge in many areas around the world. And if military intervention is necessary, we are very likely to face an insurgent enemy once again. The good news is that the US has learned many lessons in Iraq and Afghanistan that can be applied in similar type of engagements. In fact, US forces are becoming very adept at counterinsurgency. The bad news is that these types of conflicts are long, drawn-out affairs. As such, the US will need a clear and coherent strategy that is articulated and understood by not only leaders at the highest levels of government and within the military, but also each soldier, marine, airmen, sailor, and coastguardsmen on the ground.

Perhaps equally important, is the public opinion of the people of the United States in relation to how much they are willing to support in terms of lives and dollars. Once public support ends for these types of conflicts, so too will United States intervention. And it is interesting to note that of the many states that are failing, the US is currently engaged in three of the top ten according to the failed states index at **www.fundforpeace.org**. The question will then become: With so many troubled areas out there, which failed states will we select to intervene in and for what reason? And it may also be extrapolated further into the immediate future that the US is likely to be engaged in such conflicts for decades to come. The alternative is to revert back to isolationism, which is highly unlikely. But the result of trying to do too much may imperil the future of the US, just as occurred with the Roman and Persian empires. They became weaker as they overextended their capabilities.

The other issues that the military will be faced with are training challenges and sustaining the same size military with a dual role. The military of tomorrow must not only be able to engage in and win conventional-style, force-on-force warfare, they must also be able to quickly and adeptly transition (chameleon-like) to a counterinsurgent force and achieve sustainable success in these protracted irregular conflicts. This is not an easy thing to accomplish. So unless the UN decides to really step up to the plate in the future, the US will really need to pick its battles very carefully because we cannot be everywhere and we cannot be everything to everyone. Now if we can get the political machine in Washington to grasp this concept, then we may be heading in the right direction!

But I will leave you with a few thoughts that I hope I have adequately detailed for you in relation to this war. The first would be the American propensity for attaching hard timelines on everything we do. Much like I communicate to my officers on the Wisconsin Rapids Fire Department, most people can see issues that are black and white, but it takes an adept leader to see the many shades of grey. Political agendas are often framed too black and white for their constituents, while challenges such as war, healthcare, and the economy are incredibly complex and contain multiple shades of grey. Many experts believe that if we leave Iraq too soon, it will collapse back into chaos and sectarian violence. But what is *too soon*? When is the right time to leave? These are questions that adept governmental and military leaders must answer based upon conditions on the ground at that point in time. Perhaps midnight on December 31st, 2011 is not that time. Perhaps noon on April 24, 2012 is a better time! My point is that the decision should be conditions-based and not premised upon a hard timeline that politicians always seek. No parent wants to lose a son or daughter by remaining in a war too long. But many Americans have been conditioned in an instant gratification society and have lost much of their ability to be patient and persistent. The culture in Iraq is much different than ours and our inability to understand this culture was one reason why we floundered there for several years, which resulted in many lives needlessly lost and billions of dollars wasted. So even though we have been in Iraq six plus years, we really only began to get a good clue on a coherent and effective strategy in 2006. So if you factor this into the success equation, in addition to the unrealistic timeline that we are expecting Iraq to become a stable and functioning democracy capable of its own security in less than five years when it took us much longer to achieve sustainable success in our own country, it all seems a bit perplexing to me! Conversely, as I present the other side of the argument, we are trying to change a culture that has known war and sectarian division for centuries. If we are trying to create Iraq in our own image, then I would submit that this line of thought is seriously flawed. Given the fact that our government and military cannot overcome many of our own cultural inequities, are we seriously expecting this to occur in a country much older than our own? Is anyone seeing the delusion in all of this? But even with this being stated, I believe there is hope for Iraq if we are wise and they are committed to the idea of a diverse society without widespread sectarian violence. Ultimately, it will be up to the Iraqi people to determine which path they shall travel. The Iraqis have a very unique opportunity to forge a new future, but in reality it simply may not occur. This war should not be looked back upon as one that America lost—no matter the eventual outcome. We have provided the Iraqi people an opportunity for a brighter future, which is something many other countries will never get. It will be up to the Iraqi people to decide what to do with this opportunity.

The State of the Army

Some other observations I have made over here are in relation to the rank of O-6 (Colonel). Many of these officers, primarily Army-types, truly believe the sun rises and sets upon them. Now you may sit back and say, wow, this Waite dude needs to chill. He seems very disgruntled about the whole situation over there. Well, perhaps there is some truth to your thought process, but I must state in my defense that as someone who has been educated in the disciplines of leadership and organizational behavior, taught the subject in both the military and the fire service, been a practitioner of leadership for over a quarter century, and conducted extensive research on the topic, I feel pretty well qualified to make certain observations in relation to this subject. Famed author Thomas Ricks **(*Fiasco & The Gamble*)** wrote an article that I reported on back in April concerning why we should close West Point and all other military schools, including war colleges. This article led to a very spirited debate on **'power row,'** but Mr. Ricks made some very valid points. Now as a reservist, but also someone who has served on active duty, I can find more agreement with some of his points than do most active duty officers. His argument partially validates my theory that reserve officers think more diversely and are more multi-dimensional in their thought process because of their civilian experience. In a war, where having good communication skills, and the ability to effectively develop relationships and work with people, most reserve officers are much more adept at this than their active duty brethren because they practice these skills much more frequently in their civilian occupations. Conversely, by and large, most active duty officers are more adept at fighting the conventional style of war and all that that entails. But in case no one is paying attention, we are not and most likely will not be for many years to come, fighting conventional style wars. And military institutions that have produced many of the active duty officers here in Iraq and Afghanistan further validates Mr. Ricks point that it is the system they were trained within that needs to evolve. Only then will we produce more enlightened and diverse-of-mind thinkers within the active component.

I would tell you that as a civilian employer I would not hire many of the Colonels (O-6) I met in Iraq because many lack the ability to respect others, they complain about issues that are intended to assist at-risk soldiers, they have poor oral communication skills, they have very poor personnel management skills, and they clearly do not understand the finer nuances of the art of leadership. With this stated I have also met some outstanding Colonels here as well.

It may be surmised by some military professionals that many of the not-so-outstanding Colonels I encountered were staff officers. The fact of the matter is that some officers are excellent at staff work because they can manage situations and issues very adeptly. They are organized and what I call 'weed divers.' They pay close attention to detail and really dig into the weeds in everything they do. However, these types of officers typically do not make good commanders because true leaders are always looking ahead to the bigger picture and cannot afford to get buried in the weeds. They have staff officers to do this for them! The problem with the military system is that in order for these types of officers to get promoted they must hold leadership positions. Placing a 'weed diver' with poor personnel management skills in front of soldiers can lead to disastrous effects. I have personally witnessed this on more than one occasion. This system has yet to be fixed in my 34 years in the military community and is unlikely to change anytime soon. Placing someone without any leadership ability in command often results in poor morale within the unit, poor performance, and bad retention rates. But these officers are simply products of their environment, so I submit that this is yet another process within a larger system that is broke and in great need of repair. And you need not take my word for it. Look back at some of the comments made by civilians at the Pentagon concerning the Army and its culture, or read Thomas X.

Hammes and pay close attention to what he states about the military's personnel system.

My Experience

Looking back, my time in Iraq was certainly life-altering. I am still upset that this little adventure took me away from my wife and family for a year, which is time I will never get back. I do not think that the position I functioned within maximized my skill sets, but many other soldiers I have talked to have stated the same fact. But it was a privilege to have served my country, to have done my part to maintain the freedoms we enjoy, and to assist the Iraqi people to lead a better quality of life.

I found most Iraqi people very friendly and kind. They have been quite appreciative of our efforts here and I hope their future is bright as we begin preparations to end this war and leave the country in their hands.

It has also been very interesting to have served with the other service members here. In a joint service environment, I have had the privilege to work not with only soldiers, but also marines, sailors, and airmen. I have also worked with many civilians over here. Then we have the Brits and Aussies who have been great to work with and they have been great allies for the US. The Gurkas (Nepalese) that have guarded the Gulf Region Division areas of operations have also been incredibly friendly and capable, and are perhaps my favorite group to have worked with. The Ugandans have guarded many of the other facilities and have been excellent to work with as well.

I must admit that I will not miss Al Faw Palace or my job there and look very much forward to getting home and resuming my life, although I also know that it is forever altered by this experience. But I say this in a good sense, as I believe this deployment has enriched my life and offered me a perspective of the world I had never before observed. I wish the Iraqi people great success and sincerely hope they have a better future ahead of them. But my time here is at an end and the time has come for me to continue my life back in Wisconsin.

I may have mentioned a few times in this book that it often felt like the Bill Murray movie – **Groundhog Day** in Iraq. Much like the movie, every day you got up felt much like the one before! While there were very unique days to be sure, and minor nuances in each day as well, many ran together and you simply lost track of the day of the week quite often. Such can be the monotony of war, especially in a staff position. And while the days were often long (yes I know they are all just 24 hours!) the weeks did seem to pass by quickly. I would get in to the palace at about 0630, check emails, get some work accomplished, monitor the BUA, go to the post-BUA huddle, then get a few more things done and before you knew it, it was already midday. Then it was off to work out and grab some lunch and then back to the palace. After some more work, perhaps some professional development, it was time for the evening update and the day was done. Then back to the barracks at about 2000, perhaps a late snack, watch a little television, and read about ten pages or so from whatever book I was reading at that time. This was the typical battle rhythm. Then the next morning, you would get up and start the day all over again. Just like the movie!

When history looks back upon this war, it will reveal many lessons. The question then becomes, will people in future generations actually review this history in relation to future conflicts? This obviously was not the case in relation to this war. All the information was available in order to be successful, but politicians and military commanders alike failed to heed those historical lessons and therefore, as philosopher George Santayana predicted, as a result of this failure to learn we relived history. But then, unlike anything ever observed before, we finally began to look at history in mid-war! Led aptly by General David Petraeus, who undoubtedly will go down not only as one our generation's greatest military commanders, but of all generations, the tide began to turn when historical examples were applied. General Petraeus had the answer when he was first here as the Commander of the 101st Airborne. The problem was no one else was paying attention in Washington or in Baghdad! General Petraeus is certainly a case study in consistency. He had the answer from the beginning, and when given the opportunity and the authority, he applied these principles to the Iraq War and we grabbed success from the jaws of defeat! Conversely, General Odierno was like a bull in a china shop when he first arrived in Iraq as Commander of the 4th Infantry Division. His heavy-handed tactics were actually fueling the insurgency as opposed to quelling it. But he was certainly not alone in this tactic. In fact, he was in the majority and that is why we struggled and floundered during the early portion of the war. But as MNF-I Commander, General O gets it now and is expertly applying these counterinsurgent principles. He is a study in evolvement and adaptation. I highly doubt that George Patton, as great as he was, and a student of history by the way, could have made such as transition during a campaign.

Loose Ends

Let us go back through this book and clear up a few items, which I promised I would comment on in this chapter.

1. **To begin with, what about my pre-deployment training and did it assist me with my duties in Iraq?**

The answer is a resounding *No*. Now I will quantify this statement with the fact that many of the lessons learned in this war were eventually incorporated into pre-deployment training. The problem is that this training is not being tailored to fit the needs of the individual or the unit deploying. As I have mentioned frequently throughout this book, the one-size-fits-all philosophy that the Army is known for does not work well in today's world and is quite antiquated. Customizing the training based upon the unit's mission would be far more meaningful. Knowing how to fire an M-50 is important, although I never had to do it in Iraq and was well aware of that before heading there, but I can still see some value in this training. Breaking it down and putting it back together was a complete waste of time. First of all, if you asked me to do it now, I could not. And secondly, our unit has no M-50 heavy machine guns! So, I mention this is because this wasteful training not only took me away from my family longer than necessary, it also cost the taxpayers a lot of money. When you talk of wasteful spending, there are plenty of areas to examine I can assure you.

2. Was I prepared to conduct my duties assigned in Iraq?

This answer is a resounding *No*. From the time I was notified I would be mobilized, my job changed three times! I even went to a two-week school at Ft. McCoy and another two and half week readiness exercise at Ft. Lewis in preparation for my duties. I will not tell you that the information I gleaned from these experiences was not beneficial in some fashion, but as I did not function in these positions I was preparing for it again became a matter of wasting time and money. In fact, most of my job was on-the-job training and creating my own path as I went forward. While I thrive in such an environment, I really needed no preparation for this and I could have spent more time with my family and less of the taxpayer's money as a result. So the bottom-line here is that trying to figure out who to place in what position on a Joint Manning Document (JMD) is often an exercise in futility. But this is the Army way and another example of the round peg in the square-hole philosophy. Matching skill sets to actual jobs in the theater of war is a much better methodology. A more flexible and adaptable version of a Joint Manning Document would have placed me in several jobs that I could have filled that would have been meaningful and assisted the war effort more effectively. Now, with all of this being stated, someone had to do the job that I had. I tried to do the best job I could given the lack of preparation. But I think this whole situation is simply a microcosm of the flaws in the entire Army structure. There is a systemic flaw in the way they do business, and while other services are figuring things out, the Army continues to resist change. For example, I have already outlined the antiquated format the Army uses for professionally developing its officers. The Navy and Air Force use systems and methodologies used in the civilian sector to reach more students who seek to further their professional development. The Army, for whatever reason, chooses to share this knowledge with only a select few. This close-to-the-vest mentality does not serve the Army or its Officer Corps well.

Another example is the war effort in Iraq. When faced with a counterinsurgent war after streaking to Baghdad using conventional warfare methodology we quickly ground to a halt. You had a lot of very bright people either making really poor decisions, or not making any at all. Most officers reverted back to what they were comfortable with, fighting a conventional style war, which was entirely the wrong strategy to employ. Now you can say this was a failure of the Army to adapt, but some commanders, including General Petraeus as Commander of the 101st Airborne Division, seemed to have things figured out early in the war. So why didn't the others figure this out? Well, the answer to this question is quite complex, but the short answer is because Americans have an inexplicable knack for not learning from past experience, we did not understand the culture we were in, there was no overarching strategic plan to guide military commanders, there was no meshing of the national instruments of power, there was a serious clash of personalities in Baghdad as the military and civilian leadership were often at odds, there was a clear lack of understanding the situation on the ground in Washington, there was a multitude of ineffective leadership, we had the wrong personalities in positions of power, we had political games being played in relation to the Presidential re-election, we had an insufficient command structure in place at the beginning of the war, and we failed to recognize the type of conflict we were engaged in. However, through all of these challenges the military and political strategy evolved and adapted. While sustained success and incremental progress continue in Iraq, and the correct strategy is being employed for a similar outcome in Afghanistan, it has come at an incredibly high cost. Forget the billions of dollars that have been wasted due to the lack of accountability and oversight by the US government and the out-of-control private contractors used in both theaters of war. The greatest cost has come to our young men and women in uniform, especially those who have made the ultimate sacrifice. If we ask our troops to go to a foreign land and perform their duty to the best of their ability, then they and their families should expect no less from the highest levels of government and the military. Unfortunately, for the first three to four years of the war in Iraq, these high levels of government let our troops down. The bigger question at hand is this: Have we learned enough through all of this to ensure this never happens again? As I have looked back upon history and how we have responded in similar crises decades removed, I will state that I am not very confident that we will remember these lessons. Many Americans have already forgotten what happened eight years ago on a bright September morning in New York! The key to success in any conflict, whether symmetrical or asymmetrical in nature, is planning. Counterinsurgent experts will tell you that they key to success is to do a complete and thorough analysis before beginning any military action. This was not done in Iraq. Others who understand the art of leadership will tell you that *a failure to plan is a plan to fail.*

The Sacrifice

The next thing I want to ensure I leave you with is the sacrifice that the families of deployed troops make. As the saying goes, when a soldier is deployed so too is his/her family. My wife has made enormous sacrifices, as have the spouses of thousands of deployed troops. It has not been easy and Shari and I have made mistakes along this journey. But I love my wife for everything she has done. This sacrifice becomes even more profound the more times someone is deployed, not any easier. Many of these multiple deployments end in divorce, financial problems, Dear John letters, etc. It would be interesting to know how many of the suicides of our service members were a direct result related to one of these issues. Then compound the fact that soldiers are under great stress doing their duties in a foreign land and are several thousand miles away from home. This often simply becomes too much of a burden to bear and ends with tragic consequences. So much of these variables all fit together like the pieces of a large puzzle. A protracted war with no clear national strategy or military end-state, two wars going on simultaneously, an Army-Marine Corps centric war, a force too small to handle two wars and multiple deployments, a suicide rate that now surpasses that of the general public, a struggling economy, ineffective leadership at multiple levels, a bureaucratic organization too slow to react swiftly, etc. I would venture to guess that the divorce rates and domestic abuse rates are both up within the Army and Marine Corps as well. So this war has claimed many victims, not all of which have been in Iraq. Then you can extrapolate any results to the Afghanistan War and you have a real problem. Americans that have not directly or indirectly been affected by this war do not understand such consequences and the high cost of war. And more troubling is the fact that many of the politicians who voted to send our uniformed military members to war without a coherent strategy or end state get to sit at home every night, do not miss a son's little league game or a daughter's birthday, can sleep in their own bed, see their wife and deal with everyday issues. I hope they sleep well at night with this knowledge! And as the families of deployed troops continue to sacrifice, I just want you to know of their burden as well as they receive little attention and get no medals. But they also deserve great credit. Please let that fact never be forgotten. And while I truly believe our political system is in the WTF league of professional BSers, at least they do not resort to violence when they do not get their way. At least not yet anyway!

Final Thoughts

Finally, I am certain many people will ask about my experience in Iraq and was it worth a year of life. I will state that I will never get a refund on that year. It will be one less year I can spend with my soul mate in this lifetime. It will be one less year I have with my aging parents. For that, I am sad. I cannot say my deployment was a completely wasted year (reminds me of an Iron Maiden song) because I finished my

EMS leadership book, I wrote this book, I completed my first course in Naval War College, I wrote several other articles and blogs to keep my hometown informed, and I reached greater depths in my professional development by reading fifteen books related to the Iraq War. So I am more professionally developed on the military side of my dual career than I have ever been. I am also in the best physical shape I have been in since I was about 18 years old and just ready to join the United States Marine Corps. So I would not state that it was wasted year, but if I would have had the option to stay home and not go to Iraq, this is the option I would have selected.

I tried to do the best job I could, given the position I had. I still believe that I could have done so much more over here to assist the Iraqi people, but unfortunately was not given that opportunity. Once again, it is a case of the inflexible square peg trying to fit into that immovable round hole that the Army uses to slot personnel.

After this deployment, observing the challenges within the military establishment, the lack of leadership at the highest levels of government, and knowing that when is all said and done, we as soldiers are just pawns in a political game to be used on a whim, makes me question my involvement in a military I have served proudly for over 34 years. Perhaps it is simply the depression of war, the death and suffering all around a war zone, or perhaps it is merely a deeper understanding of it all. And maybe it is a little of both of these variables.

As this book is being finalized and readied for publishing, I am at a crossroads in my life. I return home and search for normalcy that many military members often can never recapture. But unlike those brave kids that had to clear buildings day in and day out in hostile territory, watch their comrades in arms being killed, and not knowing if this day was to be their last, my job was not nearly as stressful. So normalcy should be easier to find. But I am also close to retirement age in relation to my job in the fire department. Shari would like me to quit both the fire department and the military and I would hate to risk another deployment and jeopardize our relationship as a result, but I do know I have some time to mull my military options over before needing to make such an important decision. The same could be said of my job as fire chief. I love the job and the people I work with, but perhaps it is time to move on. Even if I retired from the fire service and the military, I would then seek to put my Ph.D. to work for me. This may mean leaving an area I love and aging parents that I am very close to and love very much. But this experience has changed me as my father, an old Navy veteran of the Korean War, told Shari it would before I deployed. I believe it has made me a better person, who has searched for and come closer to my spiritual self, has a greater depth of knowledge of the war in which we are engaged in, appreciate my wife even more than I did before, understand the strengths and deficiencies of our military machine, and have a much better global view of the world as a result of my experience. While I do not miss Iraq, I do wish the Iraqi people well and hope they succeed in their quest for a brighter future, and I will say prayers for all my comrades I leave behind. So as the sun sets on this chapter of my life, I hope and believe that other doors of opportunity will open. I might be able to continue to assist the fire service by writing, or working for the State of Wisconsin or the International Association of Fire Chiefs. I might be able to assist the military by teaching higher-level leadership courses at West Point or Annapolis. I do not believe I will get rich by simply writing although I sure enjoy it, but I feel the need to assist the fire service and the military in whatever capacity I can. As I try to figure all of this out, there are still millions of Iraqis that are searching for a brighter future and to them I conclude this book my simply saying:

**Ma'a salama (goodbye)
&
Bettawfeeq (good luck)**

End of tour in Iraq

POST WORD

Bucky meets Iraq

This portion of the book is intended to bring some important points home concerning my thoughts throughout deployment.

The first thing I should tell you is that despite trying wholeheartedly to not be affected by my deployment, I failed miserably. It took about two weeks or so after I returned home to really get back into the swing of things at home. As I would not return to the fire department until after the first of the year, I had a solid two months to get re-acclimated to life on the home front. This included some painting, some yard work, putting up Christmas decorations, hosting Thanksgiving and cooking two 20-pound turkeys, and talking fire department business with my second in command. So it was a gradual return to normalcy but it still took some time. I am certain this return to a state of normalcy varies for each individual and is dependent upon the number of deployments, variables in his/her life at a particular moment, type of mission he/she performed in Iraq, and the amount of support and stability awaiting upon his/her return. Most people adjust, but others are not so fortunate and may turn to alcohol, drugs, suicide, domestic abuse, or perhaps murder.

The month since my return has been filled with stories from the Fort Hood shooting which simply validates my claims that there is a systemic problem within the Army. While I was saddened by this tragic event, knowing what I now know, I was not shocked or surprised.

I have also noticed that there is little news concerning Iraq in the American media. Such a shame that all of the good news and the successes are not being broadcasted to Americans as this war winds down. To me, this is a failure of our media to follow a story to its conclusion.

The President has also just reported that he is sending another 30,000 troops to Afghanistan. While this was not unexpected, much of the strategy that will be used is very similar in nature in relation to what was successful in Iraq. One must then ask, why did it take the President so long to make a decision then? I do not know the answer to that question, but the decision seems to be the correct one. If we cannot realize some success as we did in Iraq in the 18 months these troops are deployed then we should look at revising the strategy or pulling out completely. But that story and historical documentation is for another book.

So the first of the year (2010) I will return to the Wisconsin Rapids Fire Department as fire chief. I look forward to getting back into the swing of the job I love so much and leading the fine firefighter/paramedics of our department. But even as I do so I will turn an eye forward and look to find another job (my second career) as an Associate Professor at some university to teach leadership and share my vast experience in this area. While I love the fire service, and have not ruled out the possibility of staying in this field in some capacity, or perhaps even as chief of another organization, I really would love to get into teaching future leaders in this country. I have much to share and much to teach of higher levels of leadership that I do not believe is being taught in many educational institutions in this country. But my job in Iraq is now complete and I also ponder my future in the military. This decision will come well after this book has been published, but I have not really thought about Iraq since I have returned to the US and do not care to go back there or Afghanistan, so it may be time to seriously contemplate retirement from the military after 34 years of service to our country. This point comes to all who live long enough to get there and my time is approaching quickly.

I can also tell you that at this point in time in regard to screening returning service members for PTSD the VA is doing a much better job than the Department of Defense. I am a bit disappointed in the lack of serious screening by the military in this area and this fact simply validates my concerns further. However, the VA is beginning to get much better in this area, but the flaw comes in the fact that it is the veteran him/herself that must seek this assistance out, as the military does not automatically refer troops to the VA, nor does the VA initiate such contact. This is very unfortunate for many of our returning troops and another flaw in a very large, bureaucratic system.

While much of this book I pointed out the flaws of the Army, it is still the greatest Army the world has ever seen. Its strength is not in its firepower, its speed of maneuver, nor its technology. And despite its many flaws in systems and processes, the Army's strength lies in its soldiers. Despite these inequities, they continue to perform remarkably in difficult situations and environments. If we can just figure out viable solutions to some of the larger systemic issues then we can provide the support these warriors deserve from the highest levels of the government and within the military. It is out of deep respect and admiration for our soldiers that I point out such flaws within this book.

I am certain I will be asked to speak of my experience in Iraq to several groups during the course of the next few years and that will be fine with me. It provides me with another opportunity to plug this book and talk of the sacrifice that our troops and their families have and will make. If I can make some money for the Wounded Warrior program along this journey then that will really make this past year more worthwhile.

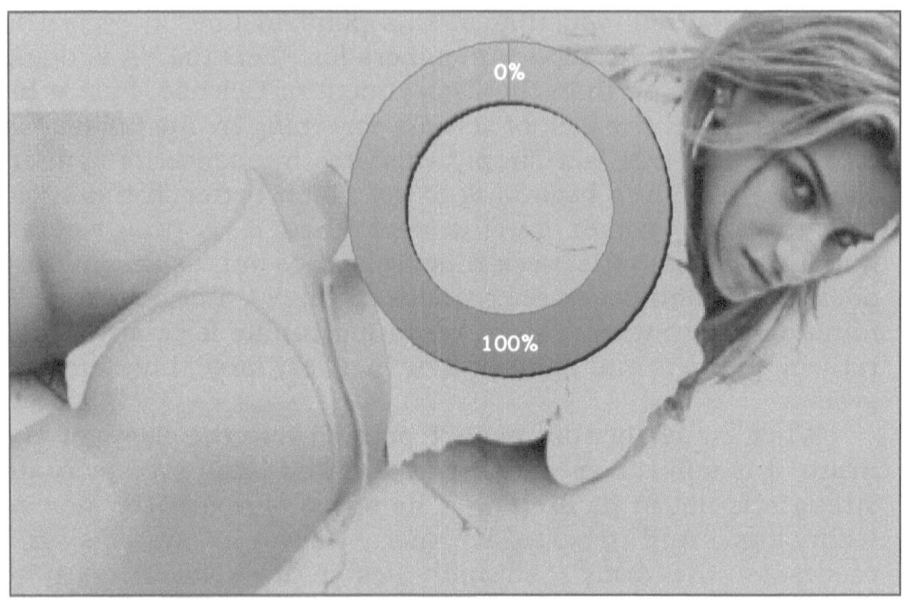

The End
(Yes, this is the other timekeeper we used in Iraq)

INDEX

416th TEC - 2, 38, 73, 107, 176, 189, 192, 217, 218, 225, 229, 232, 236, 242, 243, 263, 299, 316, 368, 372
adhocracy - 85, 210
Aegis (Security) - 285, 359
Ahmadinejad, Mahmoud - 111, 129, 143
al Qaeda - 114, 115, 129, 141, 155, 245, 258, 290, 306, 313, 320, 324, 327, 333, 338, 355, 365, 379
Al Sadr, Muqtada - 118, 129
Barnich, Terry - 100
Barzani, Massoud - 211, 220
Begley, Sean MAJ USA - 38, 235, 300, 303, 314, 315, 342, 366, 422
Biden, Joseph VP - 192, 220, 272, 302, 303, 324, 331, 344, 356
bin Laden, Osama - 231, 290, 379
Bremer, Paul - 171
Bush, George President - 115, 144, 216, 325
Clinton, Hillary - 12, 29, 30, 31, 32, 59, 128, 334, 340
Colbert, Stephen - 111, 120, 123, 127, 130, 149, 166, 280, 413

Crocker, Ryan - 22, 30, 191, 283, 417
DCP (Deployable Command Post) - 426
Desorcy, Tim - 184
Eyre, Michael MG USA - 95, 105, 106, 227, 229, 232, 250, 342
Friedman, Thomas - 126, 144
Fuller, Van Col USAF - 135
Fultz, Ted COL USA - 232, 250
Galula, David - 166, 172, 240
Gates, Robert - 4, 18, 39, 59, 61, 64, 80, 83, 90, 126, 140, 141, 196, 199, 202, 207, 211, 220, 246, 276, 280, 285, 289, 302, 322, 324, 327, 334, 340, 345, 414
Girone, Michael LTC USA - 75, 106, 107, 191, 198, 408
Goetz, Joseph COL USA - 250, 255, 274, 321, 332, 341, 342
Graham, Lindsey - 75, 84, 231
Gregris, Mark MAJ USA - 189, 194
Gulf Region Division - 45, 223, 264, 271, 289, 332, 359, 387, 410, 419

Hammes, Thomas X. - 166, 172, 241, 264, 277, 387
Hansel, Harley MSG USA - 189
Holbrooke, Richard - 256
Holczer, Kurt SFC USA - 217, 218
Hoy, Angela - 357
Hussein, Saddam - 97, 200
Jacoby, Charles LTG USA - 297, 349
Jocius, Tony LTC USA - 2, 191, 361
Jolie, Angelina - 201
Joling, Jason - 184
Jones, Drew (Stepdaughter) - 44, 181, 182, 183, 238, 244, 416
KBR (Kellogg, Brown, and Root) - 49, 81, 85, 88, 89, 131, 148, 212, 231, 232, 292, 339
Kerkman, Dave - 184
Kertis, Dick (Father in-law) - 37, 87, 90, 183, 237, 411, 418
Kertis, Kathy (Mother in-law) - 37, 87, 90, 183, 216, 411
Kilcullen, David - 166, 172, 202, 240, 352
Kocher, Jeremy (Stepson) - 44, 180, 183, 367, 370, 373, 424
Kocher, Josh (Stepson) - 18, 37, 84, 87, 88, 119, 180, 238, 411
Kocher, Matt (Stepson) - 18, 180, 183, 184, 238, 416
Kubisiak, Randy - 47

Lagerquist, Grace (Grandmother) - 137, 206, 259, 419
LaPorte, Ryan LTC USA - 104
ledocracy - 41, 86, 210, 220, 264, 435
Maliki, Nouri al - 27, 30, 31, 33, 35, 42, 47, 79, 94, 99, 101, 145, 150, 162, 193, 196, 201, 202, 205, 220, 225, 236, 244, 251, 270, 282, 303, 304, 354, 412
Maltes, Gilda SFC USA - 20, 21, 64, 99, 132, 164, 199, 235, 258, 314, 332, 341, 342
McCain, John - 231, 257, 275, 303, 341
McChrystal, Stanley GEN USA - 67, 75, 76, 89, 112, 149, 222, 230, 231, 233, 246, 256, 272, 274, 276, 287, 312, 314, 318, 322, 323, 327, 331, 334, 341, 352, 354, 420
McKiernan, David GEN USA - 62, 67, 75, 112, 230, 246
McWilliams, Scott SFC USA - 292
Mullen, Michael ADM USN - 62, 64, 125, 142, 256, 276, 302, 304, 318, 323
Nagl, John - 153, 166, 172, 240
Obama, Barack President - 49, 59, 66, 75, 78, 111, 112, 114, 201, 202, 246,

274, 275, 325, 331, 334, 345, 355
Odierno, Raymond GEN USA - 2, 7, 8, 9, 11, 17, 21, 22, 29, 31, 32, 33, 34, 36, 38, 39, 41, 42, 43, 47, 54, 59, 67, 70, 96, 97, 112, 124, 127, 142, 145, 149, 160, 198, 205, 206, 207, 211, 212, 223, 235, 236, 237, 247, 274, 281, 300, 314, 315, 323, 326, 330, 342, 389, 421
Peters, Ralph - 75, 96, 97
Petraeus, David GEN USA - 22, 31, 34, 35, 36, 67, 75, 76, 96, 97, 111, 136, 318, 323, 331, 341, 388, 391
Pittman, Rickey - 203, 306
Prater, Mitch CSM USA - 285
Pratt, Rich LTC USA - 223, 250, 255, 262, 269
Pritchard, Bob 'Mad Dog' LTC USMC - 21, 104, 109, 170
PTSD (Post Traumatic Stress Disorder) - 61, 103, 299, 304, 400
Ramadan - 254, 276, 311, 316
Ramsey, Rick LTC USA - 194, 411, 424
Rice, Condoleeza - 46
Ricks, Thomas - 12, 13, 34, 131, 350, 385

Rodriguez, Carlos LTC USA - 250, 253, 279, 319, 332, 335, 341, 342, 420
Route Irish - 108
Rumsfeld, Donald - 4, 173, 231
Ryan, Mike LTC USA - 250, 253, 279, 335, 359, 420
Samaris, Chuck LTC USA - 6, 8, 135, 193, 223, 250
Sanchez, Ricardo LTG USA - 97, 121, 171, 172, 173, 246, 407
Sendlebach, Donna SFC USA - 194
shamal - 208, 417
SIGIR (Special Investigator General for Iraqi Reconstruction) - 35
Song, Sean MAJ USA - 21, 40, 41, 104
Staab, Randy MAJ USMC - 20, 21, 37, 104
suicide (prevention) - 61, 68, 103, 120, 121, 131, 155, 203, 204, 209, 212, 222, 225, 252, 283, 286, 392, 393, 398
Taliban - 67, 128, 231, 245, 267, 287, 293, 324, 327, 337, 338, 355, 356
Talibani, Jalal - 238
TF SAFE - 85
United States Army Corps of Engineers (USACE) - 100
Waite, Allyson (Sister) - 46, 59, 180

Waite, Brad (Brother) - 157, 180, 373, 415

Waite, Don (Father) - 57, 156, 157, 183, 188, 216, 259, 281, 345, 346, 347, 415, 423

Waite, Jane (Mother) - 57, 58, 216, 259, 411

Waite, Kayleigh (Granddaughter) - 180, 206, 298, 374, 424

Waite, Lexie (Daughter) - 82, 180, 370

Waite, Shari (Wife) - 5, 6, 7, 9, 18, 25, 26, 28, 29, 32, 41, 46, 49, 52, 59, 63, 73, 77, 78, 85, 119, 127, 135, 138, 140, 141, 142, 148, 160, 164, 179, 180, 181, 182, 183, 184, 185, 186, 187, 197, 207, 217, 219, 223, 226, 237, 238, 244, 248, 259, 262, 285, 298, 303, 311, 328, 346, 360, 361, 367, 373, 375, 392, 394, 395, 411, 414, 416

Weaver, Ellen (Aunt) - 183, 206, 417

Wisconsin Rapids Fire Department - 104, 200, 343, 383, 399

Witsberger, Frank MAJ USA - 357

Wittenberg, Jennifer - 418, 424

Wolfe, Duane LTC USA - 100, 328, 351

Young, Scott - 184

REFERENCES

Note: All information contained within this book is unclassified and obtained from open source media, including MNFI, MNCI, CENTCOM, and JASG unclassified media highlights.

Articles:

(alphabetical order)

Barber, Phil. (2009). *Lombardi rules.* American Heritage. Vol 59, No. 3, pp. 36-41.

Metz, S. and Wipfl, R. (2007). *COIN of the Realm: US Counterinsurgency Strategy.* Strategic Studies Institute.

Multi-National Forces-Iraq. (2008). Joint Campaign Plan. *Economic Line of Operation* (unclassified). MNF-I Web page.

Ricks, Thomas. (19 April 09). *Why we should get rid of West Point*, The *Washington Post.*

Books:

(alphabetical order)

Couch, Dick. (2008). *Sheriff of Ramadi.* Annapolis, MD: Naval Institute Press.

Galula, David. (1964). *Counterinsurgent Warfare.* Westport, CT: Praeger Security International.

Hammes, Thomas X. (2007). *The Sling and the Stone.* Minneapolis: Zenith Press.

Kilcullen, David. (2009). *The Accidental Guerilla.* Oxford: Oxford University Press.

Mansoor, Peter R. (2008). *Baghdad at Sunrise.* New Haven, CT: Yale University Press.

Maxwell, John C. (2005). *The 360 Degree Leader.* Nashville: Thomas Nelson Inc.

Nagl, John. (2002). *Learning To Eat Soup With A Knife.* Chicago: Chicago University Press.

Popaditch, Nick. (2008). *Once A Marine.* New York: Weider History Group.

Ricks, Thomas. (2006). *Fiasco.* New York: Penguin Press.

Ricks, Thomas. (2009). *The Gamble.* New York: Penguin Press.

Robinson, Linda. (2008). *Tell Me How This Ends.* New York: Public Affairs.

Sanchez, Ricardo. (2008). *Wiser In Battle.* New York: HarperCollins Publishers.

West, Bing. (2008). *The Strongest Tribe.* New York: Random House Publishers.

Yon, Michael. (2008). *Moment of Truth in Iraq.* Minneapolis, MN: Richard Vigilante Books.

OTHER:

(alphabetical order)

Peters, Ralph. (2009). *Burying Military Reputations.* Sent on 17 MAY 09 by LTC Michael Girone on NIPR.

Waters, Tom (14 November 08). USACE representative at Ft. McCoy, unclassified briefing, 1220 CST.

Manuals & Texts:

(chronological order)

U.S. Department of the Army. (1999). *Military leadership (FM 22-100).* HQ TRADOC, Ft. Monroe, VA: Author.

U.S. Department of the Army. (2003). *Mission Command: Command and Control of Army Forces (FM 6-0).* HQ Department of the Army, Washington, D.C.: Author.

U.S. Department of the Army. (2006). *Army Leadership (FM 6-22).* HQ Department of the Army, Washington, D.C.: Author.

U.S. Department of the Army. (2006). *Counterinsurgency (FM 3-24).* HQ Department of the Army and Marine Corps Combat Development Command, Washington, D.C.: Author.

U.S. Department of the Army. (2008). *Stability Operations (FM 3-07).* HQ Department of the Army, Washington, D.C.: Author.

400 Days Picture File

Chapter 10

General George Casey
http://en.wikipedia.org/wiki/File:Casey_Blue_ASU.jpg
Quote:
http://www.mnf-iraq.com/index.php?option=com_content&task=view&id=24572&Itemid=128

Arr matie
http://en.wikipedia.org/wiki/File:Piratey,_vector_version.svg

The truth is out there
http://en.wikipedia.org/wiki/File:Glieseupdated.jpg

LTC Waite at Al Faw during MNF-I Engineer Summit
Picture taken by Kendal Smith, GRC PAO
April 19th, 2009

Epic battle with a desert warrior
http://en.wikipedia.org/wiki/File:%D0%9C%D1%8B%D1%88%D1%8C_2.jpg

Power Row
Picture taken by unknown soldier
April 22nd, 2009

Happy Birthday Army Reserve
http://en.wikipedia.org/wiki/File:United_States_AR_seal.svg

Katelin Kerkman's class project
Picture taken by Mitchell R. Waite
April 24th, 2009

British flag
http://en.wikipedia.org/wiki/File:Flag_of_the_United_Kingdom.svg

Taxi Driver
http://en.wikipedia.org/wiki/Taxi_Driver

Iraqi boy with toys distributed by Gulf Region Division personnel
www.grd.usace.army.mil/news/photos/fullphoto.asp?Link=giveaway2.jpg&PhotoID=378

Chapter 11

Secretary of State Condoleezza Rice
http://en.wikipedia.org/wiki/File:Condi_rice.jpg
Quote:
http://thinkexist.com/quotes/condoleezza_rice/2.html

Happy Birthday Sis & Kubi
Sister – taken by Mitchell R. Waite on April 7, 2006
&
Kubi – taken by Mitchell R. Waite December 21, 2007

The Good Idea Bus
http://en.wikipedia.org/wiki/File:2008_blue_bird_vision_tinted.jpg
http://en.wikipedia.org/wiki/File:Gluehlampe_01_KMJ.png

Map of Baghdad area
http://www.globalsecurity.org/military/world/iraq/images/baghdad-map-areas2.jpg

Mother's Day bouquet and Mom
Picture:
http://en.wikipedia.org/wiki/File:Flower_bouquet20091225.JPG
&
Mom – taken by Mitchell R. Waite June 14th, 2004

Australian flag
http://en.wikipedia.org/wiki/File:Flag_of_Australia.svg

Sixth Sense?
http://commons.wikimedia.org/wiki/File:Saturn99Stupidity.jpg

The Pie of Time
Pie of time sent to me by LTC Rick Ramsey on December 14th, 2008

Happy Birthday Josh
Picture taken by Shari Lynn Waite
January 9th, 2007

The Long Walk to the Dentist
http://commons.wikimedia.org/wiki/File:Stasi_Hallway.jpg

Happy 50th Anniversary Dick & Kathy Kertis
Picture taken by Shari L.. Waite
November 14th, 2009

Mirror, mirror on the wall?
http://commons.wikimedia.org/wiki/File:Mirror_baby.jpg

Taking Chance
http://en.wikipedia.org/wiki/File:Phelpschance.jpg

Chapter 12

Iraqi Prime Minister Nouri al-Maliki
http://en.wikipedia.org/wiki/File:Nouri_al-Maliki_with_Bush,_June_2006,_cropped.jpg
Quote:
http://www.spiegel.de/international/world/0,1518,druck-566852,00.html

Overhead protection for De Fleury DFAC on VBC
Picture taken by Mitchell R. Waite
June 4, 2009

Lipstick on a pig
http://commons.wikimedia.org/wiki/File:Pig_USDA01c0116.jpg
http://en.wikipedia.org/wiki/File:DiorLippenstift.jpg

D-Day June 6th, 1944
http://en.wikipedia.org/wiki/File:1944_NormandyLST.jpg

Bond – James Bond
http://en.wikipedia.org/wiki/File:Aston.db5.coupe.300pix.jpg

The Little Dutch Boy
http://en.wikipedia.org/wiki/File:Hans_Brinker_Madurodam.jpg

Colbert arrives in Iraq
http://www.army.mil/klw/submissions/print/catR/FORSCOM/Bradford/Bradford%20-%20Articles.pdf
Photo by Lee Craker, Multi-National Corps—Iraq Public Affairs

Stephen Colbert and GEN Odierno
http://en.wikipedia.org/wiki/File:Colbert_haircut_army.mil-40677-2009-06-11-130625.jpg

The Blob
http://commons.wikimedia.org/wiki/File:Blob_Gif.gif

Jungle cat ready to pounce
http://en.wikipedia.org/wiki/File:Jaguar-schwarzer-panther-zoologie.de-nk0005.JPG

And that's the way it is
http://commons.wikimedia.org/wiki/File:Walter_Cronkite_on_television_1976.jpg

The Bucket List
http://en.wikipedia.org/wiki/File:Wooden_bucket.jpg

Styrofoam-gate
http://en.wikipedia.org/wiki/File:Carnegie_Deli_Strawberry_Cheesecake.jpg
http://commons.wikimedia.org/wiki/File:Hefty_square_styrofoam_food_container_open.JPG

Chapter 13

Secretary of Defense Robert Gates
http://commons.wikimedia.org/wiki/File:Robert_Gates,_official_DoD_photo_portrait,_2006.jpg

Quote:
http://www.mnf-iraq.com/index.php?option=com_content&task=view&id=24572&Itemid=128

Happy Birthday Honey & the Army
http://commons.wikimedia.org/wiki/File:US_Flag_Day_poster
_1917.jpg
Shari – Taken by Mitchell R. Waite on November 11th, 2007
http://www.westpointmwr.com/army_birthday.jpg

Fair elections?
http://en.wikipedia.org/wiki/File:Scale_of_justice_2.svg

Bad driving
http://en.wikipedia.org/wiki/File:Demoltion_Derby_Action_Sh
artlesville.JPG

Tip of the Hat
http://commons.wikimedia.org/wiki/File:Coolidge_after_signi
ng_indian_treaty.jpg

TV in Iraq
http://commons.wikimedia.org/wiki/File:Television.svg

Just a trim please
http://www.grd.usace.army.mil/news/photos/fullphoto.asp?
Link=LCDSC00311.jpg&PhotoID=337 &
http://en.wikipedia.org/wiki/File:G_a_custer.jpg

Kickin' it up a notch
http://commons.wikimedia.org/wiki/File:Emeril_Lagasse_boo
k_signing.jpg

Have a cold one for me Dad
http://en.wikipedia.org/wiki/File:PintJug.jpg
Father: Picture taken by Mitchell R. Waite on
Father's Day 2005

Happy 50th Bro
http://commons.wikimedia.org/wiki/File:Varanus_komodoensis6.jpg
Brad: Picture taken by Mitchell R. Waite on December 25th, 2006
http://commons.wikimedia.org/wiki/File:UserDuncHarris.jpg

Fleet foot?
http://commons.wikimedia.org/wiki/File:Talaria.svg

I am here to pump you up
http://commons.wikimedia.org/wiki/File:Crunch.gif
http://en.wikipedia.org/wiki/File:Push-up.png

The Fuzzy Everest Blast Off Cardio Routine
http://en.wikipedia.org/wiki/File:Fuzzy_dice.jpg
http://en.wikipedia.org/wiki/File:Everest_kalapatthar_crop.jpg
http://en.wikipedia.org/wiki/File:STS120LaunchHiRes.jpg

The Waite Control Program
http://commons.wikimedia.org/wiki/File:Cheeseburger.jpg
http://en.wikipedia.org/wiki/File:Salad_platter.jpg
http://en.wikipedia.org/wiki/File:NCI_Visuals_Food_Taco.jpg
http://www.cafemed1.com/images/gyro.jpg
http://en.wikipedia.org/wiki/File:Gyros.jpg
http://commons.wikimedia.org/wiki/File:Pepperoni_pizza.jpg
http://en.wikipedia.org/wiki/File:DQ_Crispy_Chicken_sandwich.JPG
http://en.wikipedia.org/wiki/File:Chocolate_chip_cookies.jpg

The Green and Gold Cardio Routine
http://en.wikipedia.org/wiki/File:GreenBayPackers_100.svg
http://en.wikipedia.org/wiki/File:Super_Bowl_29_Vince_Lombardi_trophy_at_49ers_Family_Day_2009.JPG

The Patient is in Trouble
http://commons.wikimedia.org/wiki/File:Clown_in_surgery.jpg

Tie a yellow ribbon 'round the ol' Oak tree
Pictures taken by Mitchell R. Waite on July 3, 2009

The Waite Estate
Pictures taken by Mitchell R. Waite on July 3, 2009

Drew's 'Sweet 16th'
Pictures taken by Mitchell R. Waite on July 3, 2009

The 'man-cave'
Pictures taken by Mitchell R. Waite on July 3, 2009

Happy 26th Matt
Picture taken by Shari Waite on December 13, 2009

The boys of the WFRD
Picture taken by Diane Sloat on July 14, 2009

My sweetie
Picture taken by Jennifer Joling on July 4, 2009

GRD soldier delivering toys to children
www.grd.usace.army.mil/news/releases/Hi-Res/Donation1-sm.jpg

Chapter 14

Ambassador Ryan Crocker
http://commons.wikimedia.org/wiki/File:Ryan_C_Crocker.jpg
Quote:
http://iraqfoundation.org/news&announcements/announcementsannouncement//PPM43_070910_crocker_testimony.pdf

Angelina Jolie the humanitarian
http://en.wikipedia.org/wiki/File:Angelina_jolie_lugar.jpg

Happy Birthday Ellen
Picture taken by Mitchell R. Waite on November 17, 2003

A shamal at VBC in July
Pictures taken on July 29, 2009 by Mitchell R. Waite

The Ostrich Syndrome
http://commons.wikimedia.org/wiki/File:Struthio-camelus-australis-grazing.jpg

Food shortage?
http://commons.wikimedia.org/wiki/File:Breakfast_of_Champions.jpg
http://en.wikipedia.org/wiki/File:Bannana2500px.JPG
http://en.wikipedia.org/wiki/File:Chocolate_pudding.jpg
http://www.freeclipartpictures.com/clipart/food43.htm
http://bakingbites.com/wp-content/uploads/2009/04/puddingcups.JPG

Another epic battle in the desert
http://en.wikipedia.org/wiki/File:Rat_diabetic.jpg

Go Pack Go!
http://commons.wikimedia.org/wiki/File:Lambeau-field.jpg

Happy Birthday Coast Guard
http://en.wikipedia.org/wiki/File:USCG_S_W.svg

My next place of employment ☺
http://en.wikipedia.org/wiki/File:2009-02-22_Hooters_in_Morrisville.jpg
http://commons.wikimedia.org/wiki/File:Hooters_Calendar_Girl_Melissa_Poe.jpg

Death from above
http://www.historycommons.org/events-images/a118_predator_firing_hellfire_2050081722-16359.jpg

GRD soldier training Iraqi maintenance workers
www.grd.usace.army.mil/news/releases/Hi-Res/Generator%202-sm.jpg

Chapter 15

General David Petraeus
http://en.wikipedia.org/wiki/File:David_H._Petraeus_press_briefing_2007.jpg
Quote:
http://council.smallwarsjournal.com/showthread.php?p=10347

Happy Birthday Dick
Picture taken by Jennifer Wittenberg on November 14, 2009

Keep that hourglass draining my pretty!
http://en.wikipedia.org/wiki/File:Wooden_hourglass_3.jpg

The G3 shall hold the line
http://en.wikipedia.org/wiki/File:Leonidas_Thermopylae2.jpg

Happy Birthday Gram
Picture taken by Mitchell Waite on August 28, 1988

The Golden Hour or best care available?
www.med.navy.mil/sites/nmrc/Pages/ccc_rm.htm
http://www.militaryspot.com/gallery/showphoto.php?photo=518

Cat herding at GRD!
http://en.wikipedia.org/wiki/File:ARTrussellCfullsize.jpg
http://en.wikipedia.org/wiki/File:Collage_of_Six_Cats-01.jpg

PowerPoint Warrior
http://en.wikipedia.org/wiki/File:Slideshow.jpg
http://en.wikipedia.org/wiki/File:Casing.jpg

Book 'em Danno
http://en.wikipedia.org/wiki/File:2010_mavericks_competition.jpg

I would like to thank....
http://commons.wikimedia.org/wiki/File:Academy_Awards_1988.jpg

GRD employee handing out toys donated by the Gulf Region Division
www.grd.usace.army.mil/news/photos/fullphoto.asp?Link=hand1.jpg&PhotoID=92

Chapter 16

Chairman of the Senate Foreign Relations Committee Senator Joe Biden
http://commons.wikimedia.org/wiki/File:Joe_Biden_official_portrait_crop.jpg
Quote:
http://joebidenquotes.info/quoteson/iraq

A Slow-to-catch-on cocktail
http://en.wikipedia.org/wiki/File:Cocktail1.jpg

McChrystal's crystal ball?
http://en.wikipedia.org/wiki/File:Quartz_crystal.jpg

Go Bucky!
http://en.wikipedia.org/wiki/File:AmericanBadger.JPG

JOAT (Jack-Of-All-Trades)
http://en.wikipedia.org/wiki/File:Jack_playing_cards.jpg

The fall of the Roman Empire
http://en.wikipedia.org/wiki/File:Colosseum_in_Rome,_Italy_-_April_2007.jpg

Mike Ryan and myself at Cigar Aficionado Club meeting
Picture taken by Carlos Rodriguez on September 6, 2009

Ready for the PGA?
Picture taken by Carlos Rodriguez on September 6, 2009

The sad tale of Humpty Dumpty
http://commons.wikimedia.org/wiki/File:HumptyDumpty.jpg

Sometimes you feel like a nut, sometimes you don't
http://commons.wikimedia.org/wiki/File:Almond.JPG

Politicians and opinion polls are like a moth to a flame!
http://commons.wikimedia.org/wiki/File:Gonepteryx_cleopatra_(Millot).jpg
http://en.wikipedia.org/wiki/File:Candle-calendar.jpg

Go Pack Go
Picture sent on Facebook by Kevin Siehr on August 24, 2009

Happy Birthday Air Force Sep 18
http://en.wikipedia.org/wiki/File:USAF_logo.png

There is a new Marshal in town
http://en.wikipedia.org/wiki/File:US_Marshal_Badge.png

Senior moment for Brett
http://www.packerpalace.com/blog/images/bf-senior-moments1.jpg

Chapter 17

General Ray Odierno
http://commons.wikimedia.org/wiki/File:Flickr_-_The_U.S._Army_-_U.S._Army_Gen._Raymond_T._Odierno.jpg
Quote:
http://www.spacewar.com/reports/Odierno_warns_on_Iraq_security_as_he_takes_US_command_999.html

Inspector Clouseau?
http://photos.upi.com/topics-Iraqi-military-spokesman-and-US-Brigadier-General-speak-on-the-Iraq-War-in-Baghdad/b105f38bfcba64aa0db41764de6db2cf/Q.jpg

What we've got here is a failure to communicate
http://commons.wikimedia.org/wiki/File:Communications-tower-w-antennae.jpg

Left turn Clyde
http://en.wikipedia.org/wiki/File:Orangutan.jpg

Tick...tick...tick
http://en.wikipedia.org/wiki/File:RailwayStationClock.jpg

Go ahead, make my day
http://en.wikipedia.org/wiki/File:S%26W_Model_29-2.jpg

The infamous Flintstone village on VBC
Picture taken on October 2nd, 2009 by Mitchell R. Waite

Way to go Bucky
http://en.wikipedia.org/wiki/File:BuckinghamUBadger.jpg

Medal of Honor recipient SFC Paul Ray Smith
Picture taken on October 5th, 2009 by Mitchell R. Waite

Good job Brett
http://commons.wikimedia.org/wiki/File:Brett-Favre-Jets-vs-Rams-Nov-9-08.jpg
http://commons.wikimedia.org/wiki/File:David_Martin82_Brett_Favre4-Edit2.jpg
http://commons.wikimedia.org/wiki/File:BFAVREVIKE.jpg

The Begley boys
Picture taken on August 25th, 2009 by Mitchell R. Waite

The New Brew Crew
http://commons.wikimedia.org/wiki/File:DSC03365_Ryan_Braun.jpg
http://en.wikipedia.org/wiki/File:Prince_Fielder_(929557698).jpg

Happy Birthday Dad
Picture taken on April 7, 2006 by Mitchell R. Waite

Chapter 18

President Barack Obama
http://en.wikipedia.org/wiki/File:Official_portrait_of_Barack_Obama.jpg
Quote:
https://www.brainyquote.com/quotes/authors/b/barack_obama.html

Happy Columbus Day
http://commons.wikimedia.org/wiki/File:Columbus_Taking_Possession.jpg

Wisconsin Rapids is Cranberry Central
http://commons.wikimedia.org/wiki/File:CDC_cranberry1.jpg

Happy 234th Birthday Navy
http://commons.wikimedia.org/wiki/File:Billthegoat.gif

Vince Lombardi
http://commons.wikimedia.org/wiki/File:2009-0620-WI011-GB-Lambeau.jpg

Perfume Palace and Victory over America Palace
Picture taken on October 19th, 2009 by Mitchell R. Waite

Jeremy's 25th
Picture taken on November 10th, 2007 by Jennifer Wittenberg

Kayleigh's 7th
Picture taken on July 12th, 2009 by Mitchell R. Waite

Sunset in Baghdad
Picture taken on October 12th, 2009 by Mitchell R. Waite

Chapter 19

Honoring fallen hero
http://thinkprogress.org/wp-content/uploads/2008/04/pic1.gif

Sunset in Iraq
www.grd.usace.army.mil/news/photos/fullphoto.asp?Link=sunset.jpg&PhotoID=129

Post Word

Bucky in Iraq
Picture taken on August 8, 2009 by Mitchell R. Waite

The End
Morale donut sent to me by LTC Rick Ramsey on December 15, 2008 via email

Military Acronyms

A-D

AAB	Advise & Assist Brigade
ACH	Advanced Combat Helmet
ACU	Army Combat Uniform
AFN	Armed Forces Network
AGR	Active Guard & Reserve
AM	After midnight (Morning)
APFT	Army Physical Fitness Test
AQI	Al-Qaeda of Iraq
BCT	Brigade Combat Team
BG	Brigadier General (one-star)
BIAP	Baghdad International Airport
BOG	Boots on Ground
BUA	Battlefield Update Assessment
C7	Corps Engineer
CENTCOM	Central Command
CF	Coalition Forces
CG	Commanding General
CHOPS	Chief of Operations
CHU	Containerized Housing Unit
CIF	Central Issue Facility
CJ 1/4/8	Personnel, Logistics, and Finance at MNF-I level
CJ2	Intelligence at MNF-I level
CJ3	Operations at MNF-I level
COIN	Counterinsurgency
COL	Colonel
CPT	Captain
CSH	Combat Support Hospital
CSM	Command Sergeant Major
CST	Central Standard Time
CUB	Commanders Update Briefing
DCP	Deployable Command Post
DFAC	Dining Facility
DIME	Diplomacy, Information, Military, Economic
DOS	Department of State

E-H

EFP	Explosively Formed Penetrator
EOBC	Engineer Officer Basic Course

EOD	Explosive Ordnance Disposal
EOF	Escalation of Force
EST	Eastern Standard Time
FOB	Forward Operating Base
FRAGO	Fragmentary Order
GAO	Government Accountability Office
GoI	Government of Iraq
GMT	Greenwich Mean Time
GRC	Gulf Region Central
GRD	Gulf Region Division
GRN	Gulf Region North
GRS	Gulf Region South
GWOT	Global War on Terrorism
HPA	High Profile Attacks
IDF	Indirect Fire
IED	Improvised Explosive Device
ISF	Iraqi Security Forces
ITAO	Iraqi Transition Assistance Office
G1	Personnel at GRD
G2	Intelligence at GRD
G3	Operations at GRD
G4	Logistics at GRD
G5	Plans at GRD
G6	Information Management at GRD
GEN	General (four-star)
H1N1	Swine Flu virus
HPA	High Profile Attacks
HQ	Headquarters

I-L

IA	Iraqi Army
IDF	Indirect Fire (rockets, mortars)
IHEC	Iraq Independent High Electoral Commission
IO	Information Operations
ISF	Iraqi Security Forces
IZ	International Zone
JAM	Jaysh Al Mahdi
JASG	Joint Area Support Group
JMD	Joint Manning Document

JOAT	Jack-Of-All-Trades (my own creation)
JOC	Joint Operations Center
KIA	Killed in Action
KRG	Kurdish Regional Government
LN	Local Nationals
LNO	Liaison Officer
LT	Lieutenant
LTC	Lieutenant Colonel
LTG	Lieutenant General (three-star)
LZ	Landing Zone

M-P

MAJ	Major
MG	Major General (two-star)
MIA	Missing in Action
MNC-I	Multi-National Corps-Iraq
MNF-I	Multi-National Force-Iraq
MNSTC-I	Multi-National Security Transition Command-Iraq
MRAP	Mine Resistant Ambush Protected
MRX	Mission Readiness Exercise
MSG	Master Sergeant
MWR	Morale, Welfare, and Recreation
NCO	Non-commissioned Officer
NEC	New Embassy Compound
NWC	Naval War College
OCS	Officer Candidate School
ODP	Officer Development Program
OER	Officer Evaluation Report
OIC	Officer In Charge
OPSEC	Operational Security
PAO	Public Affairs Office
PRT	Provincial Reconstruction Teams
PSD	Personal Security Detachment
PT	Physical fitness Training
PTSD	Post Traumatic Stress Disorder
PX	Post Exchange

Q-T

QRF	Quick Reaction Force
RFI	Request for Information
ROTC	Reserve Officers Training Corps

RPG	Rocket Propelled Grenade
R&R	Rest & Relaxation
RSOI	Reception, Staging, Onward movement, and Integration
RTC	Regional Training Center
S3	Operations
S4	Logistics
SET	Security Escort Team
SFC	Sergeant First Class
SIGIR	Special Investigator General for Iraqi Reconstruction
SGM	Sergeant Major
SME	Subject Matter Expert
SOC	Strategic Operations Center
SOF	Special Operations Forces
SoI	Sons of Iraq
SOT	Staff Orientation Training
SPO	Security, Projects, and Operations
SRC	Soldier Readiness Center
SVIED	Suicide Vest Improvised Explosive Device
TAC	Teach, Advise, Counsel
TEC	Theater Engineer Command
TF	Task Force
TIOC	Tactical Information Officers Course
TMC	Troop Medical Clinic
TTP	Tactics, Techniques, and Procedures

U-Z

UAV	Unmanned Aerial Vehicle
UK	United Kingdom
USA	United States Army
USACE	United States Army Corps of Engineers
USAF	United States Air Force
USCG	United States Coast Guard
USG	United States Government
USMC	United States Marine Corps
USN	United States Navy
VBC	Victory Base Complex
VBIED	Vehicle-borne Improvised Explosive Device

WIA	Wounded In Action
WMA	Wisconsin Military Academy
WTF	Ask your kids about this one!

1-10

1SG	First Sergeant

Symbols
Appearing periodically throughout the book

The Iraq Campaign Medal ribbon will appear throughout the book and precede the day in which a US service member was killed as recorded on the Website:
www.globalsecurity.org/military/ops/iraq_casualties.html
and/or http://www.icasualties.org/Iraq/index.aspx

http://en.wikipedia.org/wiki/File:Iraq_Campaign_ribbon.svg

The honor flag will appear throughout the book and celebrate the lives of specific individuals who have made the ultimate sacrifice in the Iraq War.

http://www.mfr.usmc.mil/4thmardiv/HQBN/MPCo/bflag.jpg

The HOT-meter will appear throughout the book, primarily during the warmer months of the year, to document increases in temperature. It will be recorded using the forecasted weather and the actual reading of a digital thermometer at the Ugandan guard shack located at the end of the entry bridge leading to Al Faw palace.

http://en.wikipedia.org/wiki/File:The_sun1.jpg
http://en.wikipedia.org/wiki/File:Raumthermometer_Fahrenheit%2BCelsius.jpg

The Groundhog Day reference will appear throughout the book as many of the long days in the string of 400 ran together and seemed very similar to one another.

http://en.wikipedia.org/wiki/File:Groundhogday2005.jpg

The 'Power Row Salute' will appear throughout the book to cite the departure of a valued member. Power Row consisted of the GRD LNO; the MNC-I LNO; the MNSTC-I LNO; and the CENTCOM LNO.

http://www.militaryspot.com/gallery/showphoto.php?photo=721

Throughout this book I will allude to many topics that appear void of logic and defy explanation. As I discuss such issues, look for Mr. Einstein to tip you off.

http://en.wikipedia.org/wiki/File:Einstein1921_by_F_Schmutzer_4.jpg

Throughout this book I will allude to the premium placed upon flexibility and adaptability. Often such expectations are unrealistic and unfair, but as the old adage goes, 'I never promised you a rose garden.' As I highlight such issues, look for Gumby and the ever-adaptable chameleon to tip you off.

http://upload.wikimedia.org/wikipedia/commons/7/76/Gumby_and_Pokey_-_Bendable_Figures.jpg
and
http://en.wikipedia.org/wiki/File:Cam%C3%A9l%C3%A9on_Madagascar_02.jpg

During my deployment there were numerous factoids that came up that I believe few people are aware of. So, much like Cheer's Cliff, I feel it my duty to inform you of certain information that perhaps you may not know. If you see the Cheers sign, then you know there is an interesting factoid somewhere near.

http://en.wikipedia.org/wiki/File:Cheers_Boston_2005.jpg

As my tour drew to a conclusion in Iraq, I used the Big Wave of Hawaii Five-O fame to highlight key dates toward this end.

http://en.wikipedia.org/wiki/File:2010_mavericks_competition.jpg

In the world of politics, the military, war, and global events, there are many key concepts that are intimately linked together. Some of these relationships are more obvious than others. I will do my best to link some of these key ideas together for you throughout the course of this book by using the concept of the bouncing ball. Please keep in mind that some of these linkages may be real and some may be perceived, but they are interesting to discuss nonetheless. It will be left up to you to decide if these relationships have any merit.

http://en.wikipedia.org/wiki/File:Basketball.png

www.ingramcontent.com/pod-product-compliance
Lightning Source LLC
Chambersburg PA
CBHW020720180526
45163CB00001B/49